Have you been to our website?

For code downloads, print and e-book bundles, extensive samples from all books, special deals, and our blog, please visit us at:

www.rheinwerk-computing.com

Rheinwerk Computing

The Rheinwerk Computing series offers new and established professionals comprehensive guidance to enrich their skillsets and enhance their career prospects. Our publications are written by the leading experts in their fields. Each book is detailed and hands-on to help readers develop essential, practical skills that they can apply to their daily work.

Explore more of the Rheinwerk Computing library!

Benjamin Franz, Michaela Kauer-Franz

Usability and User Experience Design: The Comprehensive Guide to Data-Driven UX Design

2024, 673 pages, paperback and e-book
www.rheinwerk-computing.com/5778

Sebastian Springer

React: The Comprehensive Guide

2024, 676 pages, paperback and e-book
www.rheinwerk-computing.com/5705

Sebastian Springer

Node.js: The Comprehensive Guide

2022, 834 pages, paperback and e-book
www.rheinwerk-computing.com/5556

Christian Ullenboom

Spring Boot 3 and Spring Framework 6

2024, 934 pages, paperback and e-book
www.rheinwerk-computing.com/5764

Metin Karatas

Developing AI Applications: An Introduction

2024, 402 pages, paperback and e-book
www.rheinwerk-computing.com/5899

www.rheinwerk-computing.com

Kristian Köhler

Software Architecture and Design

The Practical Guide to Design Patterns

Editor Meagan White
Acquisitions Editor Hareem Shafi
German Edition Editor Lisa Helmus, Almut Poll
Translation Winema Language Services, Inc.
Copyeditor Yvette Chin
Cover Design Graham Geary
Photo Credits Shutterstock: 2325140695/© BINK0NTAN; iStockphoto: 157192883/© enot-poloskun
Layout Design Vera Brauner
Production Kelly O'Callaghan
Typesetting III-satz, Germany
Printed and bound in the United States of America, on paper from sustainable sources

ISBN 978-1-4932-2743-3
1st edition 2025

© 2025 by:
Rheinwerk Publishing, Inc.
2 Heritage Drive, Suite 305
Quincy, MA 02171
USA
info@rheinwerk-publishing.com
+1.781.228.5070

Represented in the E.U. by:
Rheinwerk Verlag GmbH
Rheinwerkallee 4
53227 Bonn
Germany
service@rheinwerk-verlag.de
+49 (0) 228 42150-0

Library of Congress Cataloging-in-Publication Control Number: 2025030592

All rights reserved. Neither this publication nor any part of it may be copied or reproduced in any form or by any means or translated into another language, without the prior consent of Rheinwerk Publishing.

Rheinwerk Publishing makes no warranties or representations with respect to the content hereof and specifically disclaims any implied warranties of merchantability or fitness for any particular purpose. Rheinwerk Publishing assumes no responsibility for any errors that may appear in this publication.

"Rheinwerk Publishing", "Rheinwerk Computing", and the Rheinwerk Publishing and Rheinwerk Computing logos are registered trademarks of Rheinwerk Verlag GmbH, Bonn, Germany.

All products mentioned in this book are registered or unregistered trademarks of their respective companies.

No part of this book may be used or reproduced in any manner for the purpose of training artificial intelligence technologies or systems. In accordance with Article 4(3) of the Digital Single Market Directive 2019/790, Rheinwerk Publishing, Inc. expressly reserves this work from text and data mining.

Contents at a Glance

1	Introduction	15
2	Principles of Good Software Design	63
3	Source Code and Documenting the Software Development	143
4	Software Patterns	197
5	Software Architecture, Styles, and Patterns	289
6	Communication Between Services	347
7	Patterns and Concepts for Distributed Applications	421

Contents

1 Introduction 15

1.1 Programming Paradigms 17
1.1.1 Structured Programming 18
1.1.2 Object-Oriented Programming 20
1.1.3 Functional Programming 22
1.1.4 Reactive Programming 25

1.2 What Are Design Patterns and How Did They Come About? 27

1.3 What Are Software Architecture and Software Design? 31
1.3.1 Tasks of a Software Architecture 33
1.3.2 Architectural Styles and Architectural Patterns 35

1.4 The Evolution of Software Development and Architecture 38
1.4.1 Class-Responsibility-Collaboration Cards 38
1.4.2 The "Gang of Four" Patterns and Their Structure 39
1.4.3 Clean Code 43
1.4.4 Component-Based Development 45
1.4.5 Core J2EE Patterns 50
1.4.6 Enterprise Integration Patterns 54
1.4.7 Well-Architected Cloud 60

2 Principles of Good Software Design 63

2.1 Basic Concepts of Object-Oriented Programming 64
2.1.1 Objects and Classes 65
2.1.2 Encapsulation 66
2.1.3 Abstraction 68
2.1.4 Inheritance 72
2.1.5 Polymorphism 73

2.2 Clean Code Principles 75
2.2.1 Identifier Names and Conventions in the Code 77
2.2.2 Defining Functions and Methods 89
2.2.3 Don't Repeat Yourself 104
2.2.4 Establishing Clear Boundaries Between Components 105
2.2.5 The Broken Windows Theory and the Boy Scout Rule in Software 107

2.3 SOLID Principles 108
2.3.1 The Single Responsibility Principle 109
2.3.2 The Open-Closed Principle 113
2.3.3 The Liskov Substitution Principle 119
2.3.4 The Interface Segregation Principle 123
2.3.5 The Dependency Inversion Principle and the Inversion of Control 128

2.4 Information Hiding 131

2.5 Inversion of Control and Dependency Injection 132

2.6 Separation of Concerns and Aspect Orientation 134

2.7 Quality Assurance with Unit Tests 137

3 Source Code and Documenting the Software Development 143

3.1 Comments in the Source Code 143
3.1.1 Documenting Interfaces Using Comments 145
3.1.2 Creating Expressive Code 149
3.1.3 Necessary and Meaningful Comments 150
3.1.4 Comments Not Needed 154

3.2 Documenting the Software Architecture 157
3.2.1 Documenting Quality Features 158
3.2.2 Rules for Good Software Architecture Documentation 159
3.2.3 arc42 for the Complete Documentation 162

3.3 Representing Software in Unified Modeling Language 169
3.3.1 Use Case Diagram 170
3.3.2 Class Diagram 171
3.3.3 Sequence Diagram 174
3.3.4 State Diagram 175
3.3.5 Component Diagram 177

3.4 C4 Model for Representing Software Architecture 180
3.4.1 System Context (Level 1 or C1) 184
3.4.2 Container (Level 2 or C2) 186
3.4.3 Component (Level 3 or C3) 187
3.4.4 Code (Level 4 or C4) 188

3.5 Doc-as-Code 188
3.5.1 AsciiDoc 189
3.5.2 PlantUML 191
3.5.3 Structurizr 194

4 Software Patterns   197

4.1 Factory Method   198
- 4.1.1 Problem and Motivation   198
- 4.1.2 Solution   200
- 4.1.3 Sample Solution   201
- 4.1.4 When To Use the Pattern   203
- 4.1.5 Consequences   204
- 4.1.6 Real-World Example in Open-Source Software   205

4.2 Builder   206
- 4.2.1 Problem and Motivation   206
- 4.2.2 Solution   208
- 4.2.3 Sample Solution   209
- 4.2.4 When To Use the Pattern   213
- 4.2.5 Consequence   214
- 4.2.6 Real-World Example in Open-Source Software   214

4.3 Strategy   216
- 4.3.1 Problem and Motivation   216
- 4.3.2 Solution   216
- 4.3.3 Sample Solution   217
- 4.3.4 When To Use the Pattern   220
- 4.3.5 Consequences   221
- 4.3.6 Real-World Example   222

4.4 Chain of Responsibility   223
- 4.4.1 Problem and Motivation   224
- 4.4.2 Solution   224
- 4.4.3 Sample Solution   226
- 4.4.4 When To Use the Pattern   228
- 4.4.5 Consequences   228
- 4.4.6 Real-World Example   229

4.5 Command   232
- 4.5.1 Problem and Motivation   232
- 4.5.2 Solution   233
- 4.5.3 Sample Solution   235
- 4.5.4 When To Use the Pattern   238
- 4.5.5 Consequences   239
- 4.5.6 Real-World Example   239

4.6 Observer   243
- 4.6.1 Problem and Motivation   243
- 4.6.2 Solution   244

	4.6.3	Sample Solution	245
	4.6.4	When To Use the Pattern	248
	4.6.5	Consequences	249
	4.6.6	Real-World Example	249
4.7	**Singleton**		**251**
	4.7.1	Problem and Motivation	251
	4.7.2	Solution	252
	4.7.3	Sample Solution	253
	4.7.4	When To Use the Pattern	255
	4.7.5	Consequences	256
	4.7.6	Real-World Example	258
4.8	**Adapter/Wrapper**		**259**
	4.8.1	Problem and Motivation	259
	4.8.2	Solution	260
	4.8.3	Sample Solution	261
	4.8.4	When To Use the Pattern	264
	4.8.5	Consequences	265
	4.8.6	Real-World Example	265
4.9	**Iterator**		**268**
	4.9.1	Problem and Motivation	269
	4.9.2	Solution	269
	4.9.3	Sample Solution	271
	4.9.4	When To Use the Pattern	273
	4.9.5	Consequences	274
	4.9.6	Real-World Example	274
4.10	**Composite**		**276**
	4.10.1	Problem and Motivation	276
	4.10.2	Solution	277
	4.10.3	Sample Solution	278
	4.10.4	When To Use the Pattern	281
	4.10.5	Consequences	281
	4.10.6	Real-World Example	281
4.11	**The Concept of Anti-Patterns**		**283**
	4.11.1	Big Ball of Mud	283
	4.11.2	God Object	284
	4.11.3	Spaghetti Code	285
	4.11.4	Reinventing the Wheel	286
	4.11.5	Cargo Cult Programming	287

5　Software Architecture, Styles, and Patterns　289

5.1	The Role of the Software Architect	290
5.2	Software Architecture Styles	292
	5.2.1　Client-Server Architecture	293
	5.2.2　Layered Architecture and Service Layers	294
	5.2.3　Event-Driven Architecture	299
	5.2.4　Microkernel Architecture or Plugin Architecture	304
	5.2.5　Microservices	307
5.3	Styles for Application Organization and Code Structure	310
	5.3.1　Domain-Driven Design	311
	5.3.2　Strategic and Tactical Designs	315
	5.3.3　Hexagonal Architecture/Ports and Adapters	315
	5.3.4　Clean Architecture	320
5.4	Patterns for the Support of Architectural Styles	324
	5.4.1　Model View Controller Pattern	324
	5.4.2　Model View ViewModel Pattern	329
	5.4.3　Data Transfer Objects	335
	5.4.4　Remote Facade Pattern	339

6　Communication Between Services　347

6.1	Styles of Application Communication	349
	6.1.1　Synchronous Communication	350
	6.1.2　Asynchronous Communication and Messaging	351
	6.1.3　Streaming	353
6.2	Resilience Patterns	356
	6.2.1　Error Propagation	357
	6.2.2　Subdivision of the Resilience Patterns	358
	6.2.3　Timeout Pattern	361
	6.2.4　Retry Pattern	366
	6.2.5　Circuit Breaker Pattern	372
	6.2.6　Bulkhead Pattern	378
	6.2.7　Steady State Pattern	383
6.3	Messaging Patterns	388
	6.3.1　Messaging Concepts	388
	6.3.2　Messaging Channel Patterns	389
	6.3.3　Message Construction Patterns	395
	6.3.4　Messaging Endpoint Pattern	402

Contents

6.4 Patterns for Interface Versioning 411
  6.4.1 Endpoint for Version Pattern 415
  6.4.2 Referencing Message Pattern 415
  6.4.3 Self-Contained Message Pattern 417
  6.4.4 Message with Referencing Metadata 418
  6.4.5 Message with Self-Describing Metadata 420

7 Patterns and Concepts for Distributed Applications 421

7.1 Consistency 422
7.2 The CAP Theorem 423
7.3 The PACELC Theorem 424
7.4 Eventual Consistency 425
7.5 Stateless Architecture Pattern 428
  7.5.1 Problem and Motivation 428
  7.5.2 Solution 429
  7.5.3 Sample Solution 431
  7.5.4 When To Use the Pattern 433
  7.5.5 Consequences 433
7.6 Database per Service Pattern 434
  7.6.1 Problem and Motivation 434
  7.6.2 Solution 434
  7.6.3 Sample Solution 435
  7.6.4 When To Use the Pattern 436
  7.6.5 Consequences 436
7.7 Optimistic Locking Pattern 437
  7.7.1 Problem and Motivation 437
  7.7.2 Solution 438
  7.7.3 Sample Solution 441
  7.7.4 When To Use the Pattern 443
  7.7.5 Consequences 443
  7.7.6 Pessimistic Locking 444
7.8 Saga Pattern: The Distributed Transactions Pattern 446
  7.8.1 Problem and Motivation 446
  7.8.2 Solution 447
  7.8.3 Sample Solution 447
  7.8.4 When To Use the Pattern 449
  7.8.5 Consequences 449

7.9	**Transactional Outbox Pattern**	450
	7.9.1 Problem and Motivation	450
	7.9.2 Solution	451
	7.9.3 Sample Solution	452
	7.9.4 When To Use the Pattern	454
	7.9.5 Consequences	455
7.10	**Event Sourcing Pattern**	455
	7.10.1 Problem and Motivation	455
	7.10.2 Solution	456
	7.10.3 Sample Solution	457
	7.10.4 When To Use the Pattern	460
	7.10.5 Consequences	461
7.11	**Command Query Responsibility Segregation Pattern**	461
	7.11.1 Problem and Motivation	461
	7.11.2 Solution	463
	7.11.3 Sample Solution	464
	7.11.4 When To Use the Pattern	467
	7.11.5 Consequences	467
7.12	**Distributed Tracing Pattern**	467
	7.12.1 Problem and Motivation	468
	7.12.2 Solution	468
	7.12.3 Sample Solution	470
	7.12.4 When To Use the Pattern	476
	7.12.5 Consequences	476

The Author	479
Index	481

Chapter 1
Introduction

Software development can look back on a long and eventful history characterized by diverse developments and a constant stream of new and fundamental ideas and concepts. The aim of these developments has always been to create better software. This chapter highlights the most important programming paradigms as well as various types of development approaches and influences that have significantly shaped today's software development ecosystem and will probably continue to impact the future.

Software development has a fascinating history and evolution to reflect upon. After Konrad Ernst Otto Zuse presented the world's first freely programmable, functional computer in 1941, increasingly complex software has been created.

Initially, software was developed as machine-readable binary code and a single unit together with the corresponding hardware. This development in machine language was time consuming and—due to the ever-increasing complexity—unfortunately quite error prone.

Over time, more sophisticated programming languages emerged that allowed developers to create source code in a more comprehensible form. Significant milestones were the invention of *Fortran (Formula Translation)* in the mid-1950s and the subsequent publication of *COBOL (Common Business-Oriented Language)* and *Lisp (List Processing)* in 1959.

The principles and concepts developed at that time are still valid and in use today—such as the *Backus-Naur form (BNR)*, which can be used to represent the syntax of a programming language in a *context-free grammar*.

> **Backus-Naur Form and Context-Free Grammar**
>
> In computer science, a *context-free grammar (CFG)* is a special type of grammar for describing the syntax of formal languages. The rules that define the structure of the language thus don't depend on the context in which they are used.
>
> CFGs arose from the need to describe recursive and hierarchical structures precisely, both in linguistics and in computer science. Noam Chomsky's work laid the theoretical

> foundation in the early 1950s, and computer scientists such as John Backus and Peter Naur ensured that CFGs became useful tools for the specification of programming languages.

At that time, the rapid progress of computers and their hardware repeatedly led to the further development of existing programming languages or to the creation of new, adapted languages through which specific problems could be solved more efficiently.

However, even with these new *higher programming languages*, the first *software crisis* in the 1960s could not be prevented: The costs for software development exceeded the costs of the hardware. Errors became more frequent, deadlines could not be met, and projects failed.

This crisis gave rise to the idea of *software engineering* in the 1970s, with the aim of establishing software development as an engineering process that follows clear methods, uses certain tools, and upholds standards. New concepts also emerged during this time, such as the *object-oriented programming* paradigm and new process models such as the *waterfall model*.

All these developments led to a plethora of programming languages and programming styles or *programming paradigms*, which became increasingly consolidated and standardized in the 1980s.

For example, Bjarne Stroustrup introduced the *C++* language in 1983, which is an object-oriented extension of the *C* language developed and widely used for Unix in 1972. *Agile methods* were developed, described, and used.

The rapid growth of the internet in the 1990s was a major challenge for the software industry. However, the internet created a new basis for the development of software systems, and software was often developed specifically for the internet environment.

With *Java, Sun Microsystems* introduced a new, object-oriented programming language in 1995, which has enjoyed great popularity ever since and has certainly contributed to the popularity of object-oriented programming.

In addition to the many new programming languages and concepts, standardization and quality assurance in software development became increasingly important. For this reason, in the late 1980s, the concept of standardizing proven solutions for frequently occurring problems in software development, called *patterns* or *design patterns*, emerged within software engineering, thus enabling efficient and uniform development processes.

Since then, tried and tested solutions have been summarized and described in various pattern catalogs. The book *Design Patterns: Elements of Reusable Object-Oriented Software* by the *Gang of Four (GoF)* represents a milestone in this respect and is certainly one of the best-known and most widely used books on the subject of software design.

Software development has changed and evolved continuously since its beginnings. New approaches, further fundamental concepts, and even additional programming languages have been developed to solve a wide variety of challenges optimally.

These further developments have not only led to new approaches and methods in software development, but also to the emergence of innovative technologies, which in turn has had lasting impact on the progress of software development. One example is microservices, a *programming style* that only became usable and popular with the introduction of container technologies such as Docker.

In the following sections, we'll describe various programming paradigms, the concepts behind design patterns, and influential developments or evolutions in software development in more detail. These concepts usually continue to form the basis for current products, technologies, or frameworks.

Ever since the earlier software crisis, it has been clear that one goal of software development and software design should always be to create easy-to-use, maintainable, and reusable code that helps to solve a problem. Developed systems should remain maintainable and expandable in the long term at reasonable costs.

Recurring, recognizable patterns during development make source code easier to understand these criteria can be better met. Repetitive structures and procedures are not only easier to create and document; they are also easier for a reader to grasp and understand.

Accordingly, various programming paradigms and proposed solutions or design patterns have been established at different levels of abstraction, which are presented in the individual chapters of this book. Only with knowledge of the basic concepts and procedures can you create and maintain clean software solutions.

However, software development consists of more than the creation of source code by developers. The solutions to be created or implementations created must be described and documented in an adequate way. Only in this way can software be created sustainably and later understood and maintained. This topic is covered throughout this book, especially in Chapter 3.

1.1 Programming Paradigms

Programming paradigms are the fundamental styles of programming. Independent of programming languages, programming paradigms define a basic concept or approach to structuring and organizing source code or software.

A programming paradigm influences developers in terms of how they approach problems, design algorithms, or implement code. Each paradigm has its own strengths and weaknesses.

The following sections present the best-known and most common styles.

1.1.1 Structured Programming

Structured programming is part of the *imperative programming* paradigm, in which a sequence of statements is defined and executed in order to achieve the desired end state.

As a further development of basic imperative programming, structured programming breaks down programs into a logical, tree-like structure and requires a restriction to basic control structures at the lowest code level.

This breakdown and structuring of software enables *recursive decomposition*: Larger problems to be solved are summarized in larger, higher-level functions and then subdivided into smaller, lower-level functions. This approach of dividing into smaller subordinate units can be continued almost indefinitely and is referred to as *functional decomposition*. This task often forms the basis for the development or design of a software architecture.

A structured subdivision of the code promotes possible reuse and improves maintainability, as errors can be localized more quickly.

Several well-known computer scientists laid the foundations for structured programming. For instance, in 1966, Corrado Böhm, an Italian theoretical computer scientist and computer pioneer, and his student Giuseppe Jacopini showed, in their *Böhm-Jacopini theorem*, that programs can be built upon just three control structures:

- Sequence (order of statements)
- Selection (or conditional branching)
- Iteration or repetition (loops)

Furthermore, the (global) jump statements often used until then can be dispensed with.

In 1968, Edsger Wybe Dijkstra published the now well-known essay "Go To Statement Considered Harmful," in which he writes:

> *"For a number of years I have been familiar with the observation that the quality of programmers is a decreasing function of the density of go to statements in the programs they produce."* —Edsger Dijkstra, 1968

In his investigations into goto statements, Dijkstra found that their use in some scenarios prevented modules from being subdivided and structured into smaller components. In other situations, however, subdivision was possible despite the use of goto.

Edsger Wybe Dijkstra (1930–2002)
Edsger Wybe Dijkstra was a Dutch computer scientist often regarded as the inventor of numerous software development principles. For his work, in 1972, he received the globally recognized *Turing Award*, which is awarded for outstanding contributions to the

development of computer science. As a matter of fact, the Turing Award is considered the highest honor in computer science.

One consequence of the knowledge gained for better subdivision of programs is the avoidance or—depending on the programming language—the limited and standardized use of jump statements, such as the `goto` statement.

Dijkstra's analysis showed, for example, that `goto` statements were in many cases equivalent to simple control structures or loops. In other words, these expressions could often be replaced by `if-then-else` or `do-while` statements.

Current programming languages do not support *absolute* jump statements, which jump to any point in the program code. The `goto` statement available in most languages is limited and does not allow *absolute* jump targets, only jumps to certain areas.

> **Examples of goto Statements in Modern Programming Languages**
>
> Modern programming languages generally include some kind of `goto` statement but with some variation:
>
> - Java: In Java, the `goto` keyword is reserved for the language, but no behavior is defined for it in the language specification. Accordingly, it cannot be used. With the existing `label` and `break` keywords, only limited "jumps" can be implemented in the program code.
> - JavaScript: This language does not provide a `goto` keyword. An implementation with `label` and `break` is limited (comparable to Java) and provides no "replacement" for a `goto` statement.
> - PHP: The PHP scripting language has a `goto` statement that cannot be used to jump without restrictions. The jump target must be in the same file and in the same context, which means jumping into or out of functions or methods is not possible.
> - Go: The Go language also has a `goto` statement. In this case as well, the jump targets can only be located within a function, and you cannot jump into code blocks.

All modern programming today is based on structured programming and allows a program to be divided into several components (e.g., functions, methods, or classes) to achieve better reusability, readability, and maintainability. Listing 1.1 shows an example.

```go
package main

import "fmt"

func extractedFunction() {
    fmt.Println("Extracted!")
}
```

```
func main() {

    //Subdivision
    extractedFunction()

    //Sequence
    fmt.Println("Start")
    fmt.Println("Hello world!")

    //Selection / conditional branching
    x := 5
    if x > 0 {
        fmt.Println("x is positive")
    } else {
        fmt.Println("x is not positive")
    }

    //Iteration
    for i := range 10 {
        fmt.Println(i)
    }
}
```

Listing 1.1 Structured Programming Example

1.1.2 Object-Oriented Programming

In the 1960s, Ole-Johan Dahl and Kristen Nygaard developed the *Simula* programming language, regarded as the predecessor of *object-oriented programming (OOP)*. Simula introduced concepts such as *classes*, *inheritance*, and *polymorphism*, which later became the fundamental principles of object-oriented programming.

In object-oriented programming, programs are organized on the logical basis of objects that combine data and methods. Communication between the individual objects takes place via the exchange of messages. In some definitions, object-oriented programming is also described as a way of mapping "the real world in the form of software."

The description of object-oriented programming varies greatly depending on the author, and even Alan Kay, one of the inventors of the first purely object-oriented programming language, *Smalltalk*, has adapted his definition of object orientation over time. In 1993, for example, Kay listed six aspects to describe object orientation in the context of Smalltalk:

- Everything is an object.
- Objects communicate by sending and receiving messages (which consist of objects).

- Objects have their own memory (structured as objects).
- Each object is an instance of a class (which must be an object).
- A class contains the behavior of all its instances (in the form of objects in a program list).
- To execute a program list, execution control is given to the first object, and the remaining is treated as its message.

In 2003, Kay placed more emphasis on message exchange with his new, adapted definition of object orientation:

"OOP to me means only messaging, local retention and protection and hiding of state-process, and extreme late-binding of all things." —Alan Kay, 2003

At the same time, the following definition of object orientation was created in 1999 with the International Organization for Standardization (ISO)/International Electrotechnical Commission (IEC) 2382-15 standard:

"Pertaining to a technique or a programming language that supports objects, classes, and inheritance." —ISO/IEC-2382-15 standard, 1999

If you take the definitions of object orientation together, some basic principles of object-oriented programming emerge:

- **Objects and classes**
  In object-oriented programming, data and methods are combined into logical objects and described by a class. A class represents the blueprint for similar instances or objects.

- **Encapsulation**
  The defined objects hide their internal states by only allowing access via a defined interface in the form of public methods. These methods are called via the receipt of corresponding messages.

- **Inheritance**
  Classes can be derived from other classes or inherit from them and thus take on the properties and behavior of this *base class*. Inheritance enables the reuse of code and the organization of classes in a hierarchy.

- **Abstraction**
  With the concept of abstraction, complex details of an implementation are hidden, and only relevant features and behavior of an object are visible and usable. Abstraction helps identify key features and promotes better structuring of software.

- **Polymorphism**
  The concepts of inheritance and abstraction allow algorithms to work with different object types without having to distinguish between them. As a result, derived objects can also be used instead of the base object, and the execution can therefore take on multiple forms (the very definition of "polymorphism").

Listing 1.2 shows an example of a class and an object in Java.

```
package com.sourcefellows.intro;

// Class description
public class Car {
    //Property encapsulated.
    //Property cannot be referenced outside the class.
    private int speed;
    //Abstraction of behavior.
    public void accelerate(int value) {
        speed += value;
        System.out.println("Accelerates to " + speed + " mph.");
    }
    public String getSpeed() {
        return String.format("%d mph", speed);
    }
    public static void main(String[] args) {
        //Object creation
        Car car = new Car();
        car.accelerate(10);
        System.out.println(car.getSpeed());
    }
}
```

Listing 1.2 Example of a Class and Object Orientation (Java)

Since object-oriented programming is a programming style, its use does not necessarily require an object-oriented programming language. The concepts can be transferred and applied to many languages, such as C, for example.

1.1.3 Functional Programming

Functional programming is a programming style that focuses on defining software mainly through the evaluation of functions.

The theoretical basis for functional programming is called *lambda calculus*, which was first used in the 1930s by Alonzo Church to define calculable functions clearly.

In contrast to the imperative programming style, in which internal states are changed within the program, functional programming does without these internal state changes. Possible side effects that could arise as a result of state changes are excluded.

In functional programming, programs are understood as combinations of functions. The individual functions accept parameters and generate a return that is uniquely and exclusively determined by their input values.

By nesting the functions, more complex structures can be created, and complete applications can be put together. Each function works independently and can be treated as a data object and passed as a parameter to another function, for example.

Example of Functional Programming

The Haskell example shown in Listing 1.3 illustrates the principle of functional programming.

```
-- Function that doubles a number (pure)
double :: Int -> Int
double x = x * 2

-- Function, which allows applying the
-- doubling to a list of numbers
-- (used higher function: 'map')
doubleList :: [Int] -> [Int]
doubleList xs = map double xs

-- Main program: Sample call
main :: IO ()
main = do
    let numbers = [1, 2, 3, 4, 5]

    let doubledNumbers = doubleList numbers
    putStrLn ("Original: " ++ show numbers)
    putStrLn ("Doubled: " ++ show doubledNumbers)
```

Listing 1.3 Example of Functional Programming

The example shows three function definitions:

- The first function, double, is a *pure function* that doubles the input value x and returns the corresponding result.
- The second function definition, doubleList, internally uses map, which is a *higher function* that applies a function to all elements contained in a list. The double function is used in the example, shown in Listing 1.4.
- The main function generates a list of integer values and doubles their values by calling the doubleList function. Both lists are then printed to the console.

One major advantage of functional programming is that the individual functions have no internal states (i.e., no attributes or fields) across a single call and therefore cannot be changed. This approach makes it easier to verify and optimize programs and functions. In addition, subexpressions of a program can, for example, be converted into a form that can be evaluated in parallel and also executed in parallel.

In many modern programming languages, what are called *lambda functions* can be implemented. These functions have a close relationship to functional programming and enable some of its basic principles. Lambda functions are essentially short-lived, anonymous functions that can be treated like data objects.

> **Lambda Expressions in Java**
>
> Java has supported lambda expressions since *Java Development Kit (JDK)* version 8 to better support functional programming.
>
> The example shown in Listing 1.4 illustrates how each element of a previously filled list can be output with a lambda expression.
>
> ```
> List<Integer> numbers = new ArrayList<>();
> numbers.add(5);
> numbers.add(9);
> numbers.forEach( (n) -> { System.out.println(n); } );
> ```
>
> **Listing 1.4** Sample Lambda Function
>
> In the last line of this example, a lambda function is passed as a parameter. The function has no status and always produces the same result for the same parameter.

Well-known representatives of predominantly functional programming languages include *Lisp, Clojure, Haskell, F#,* and *Scala*.

> **Example of Functional Programming Using Lisp**
>
> The example shown in Listing 1.5 illustrates the nesting of functions to calculate the area of two concentric circles with radii r1 and r2 in Lisp.
>
> ```
> (defun ringarea (r1 r2)
>     (flet ((sq (x)
>         (* x x)))
>     (* pi (- (sq r1)
>         (sq r2)))))
> ```
>
> **Listing 1.5** Example of Functional Programming (Lisp)
>
> In this example, the `ringarea` function is defined, which can be called with the corresponding parameters.
>
> Functions are called by placing the individual values in parentheses. If the helper function `sq`, which is also defined, is instead written separately, the result is a more readable expression:
>
> `((sq (x)(* x x)))`
>
> In this example, the expression reads, "Function sq is called with parameter x. The function reads: Multiply x by x."

1.1.4 Reactive Programming

The goal of *reactive programming* is to facilitate the handling of asynchronous and data-driven processes. Program parts are designed in such a way that they react to changes in data streams, and thus the control flow no longer needs explicit management.

Classic areas of application for reactive programming include, for example, user interfaces or the processing of real-time data.

In reactive programming, data is not viewed as static values but instead as a sequence of events that are continuously supplied by a *data source*, or what's called an *observable*. The resulting *data stream* is processed by one or multiple *operators*, as shown in Figure 1.1, until the data reaches a recipient, the *subscriber*.

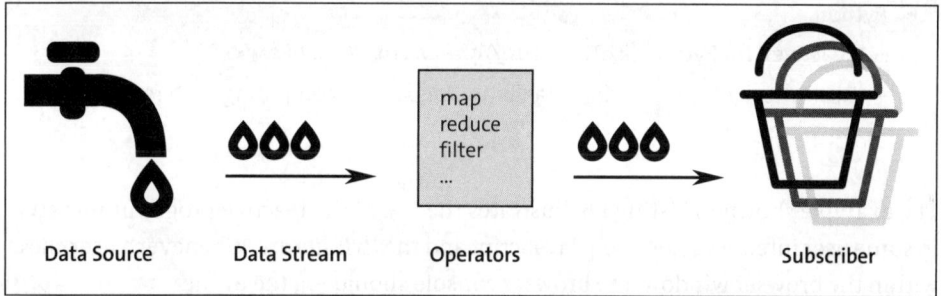

Figure 1.1 Data Flow in Reactive Programming

Operators play a key role in reactive programming and are responsible for manipulating, filtering, transforming, and combining multiple data streams. The result is declarative and modular code that is easy to understand and therefore easy to maintain. With this approach, the program flow is no longer explicitly controlled but triggered *by events*. The resulting application reacts to events or data generated by corresponding data sources.

As a rule, operators are implemented using functional programming to ensure flexible combinations and independent use, free of side effects. The individual operators are often executed asynchronously and in a non-blocking manner, thus enabling the efficient management of concurrent processes. Both complex and asynchronous systems, as well as event-driven systems, can maintain good responsiveness in this way.

In many modern programming languages, reactive programming is usually enabled via external libraries or made possible by existing language elements, such as `async` and `await` (in C#) or `go` and `chan` (in Golang).

> **Reactive Programming Using Libraries**
>
> Some examples of libraries that you can use to implement reactive programming include the following:

> - **JavaScript**
>   - RxJS: *https://rxjs.dev/*
>   - React: *https://react.dev/* (actually a UI framework, but it is based on reactive concepts)
> - **Java**
>   - Project Reactor: *https://projectreactor.io/*
>   - RxJava: *https://github.com/ReactiveX/RxJava*
>   - Akka Streams: *https://akka.io/*
> - **C#**
>   - Rx.Net: *https://github.com/dotnet/reactive*
> - **Python**
>   - ReactiveX for Python (RxPY): *https://github.com/ReactiveX/RxPY*
>   - Dask: *https://www.dask.org/* (a library for parallel computing with reactive concepts)

The example shown in Listing 1.6 illustrates the use of the reactive programming style in some user interface logic using JavaScript and the *RxJS library*. Whenever a user clicks within the browser window, the browser console should log the event.

Within the code, a data source (an *observable*) is created by calling `fromEvent(document, 'click')`, which feeds the JavaScript events `click`, which are registered on the entire document, into the corresponding data stream.

When the `map` operator is registered, the data that is available as a JavaScript event in the data stream is converted into a different, separate data format with only two attributes (x and y) and passed on.

At the end of this simple processing chain, a function is registered as a recipient or subscriber via the `subscribe` function, which then prints the data to the console.

```javascript
// Create observable from mouse clicks
const { fromEvent } = rxjs;
const { map } = rxjs.operators;

// Create observable for clicks on the document
const clicks = fromEvent(document, 'click');

// Transformation of the click events
const positions = clicks.pipe(
    map(event => ({ x: event.clientX, y: event.clientY }))
);
```

```
// Subscribe to the data stream (subscriber)
positions.subscribe(position => {
    console.log(`click at position: X=${position.x}, Y=${position.y}`);
});
```

Listing 1.6 Reactive Programming in JavaScript Using RxJS

1.2 What Are Design Patterns and How Did They Come About?

In 1977 and 1979, Christopher Alexander, a professor of architecture, and his colleagues at the Center for Environmental Structure in Berkley published the two books: *A Pattern Language* and *A Timeless Way of Building*. He himself described both books as connected parts of joint work.

In the first book, *A Timeless Way of Building*, Alexander describes the theory behind the creation and use of what he calls a *pattern language*. In the second book, *A Pattern Language*, he provides a concrete form of a language for describing architecture.

First, Christopher Alexander's aim was to enable anyone to use the model catalog, which had been developed over more than eight years of architectural work, to design or improve a city, a house, or an office. Secondly, he describes how anyone can develop their own language—either from completely new descriptions or by combining existing approaches. For *A Pattern Language*, he explicitly emphasized that the emphasis is on the word "A," i.e., it is only one form and that the compilation is not to be understood as complete and universally valid. Anyone should use the patterns, develop them further and share them themselves so that the architecture can continue to evolve.

For the sake of simplicity and ease of understanding, he proposed a uniform structure for the *patterns*, as he called the descriptions, and described 253 patterns on a total of 1220 pages in his work. Only successfully applied patterns were included.

Each of Alexander's *patterns* contained the following blocks in a fixed format:

- An *introductory image* illustrates the typical use of the pattern.
- An introductory section provides a *context description* describing where the pattern can be used and where a pattern might contribute to the implementation of a larger pattern.
- *Problem title and description*, the longest block, captures the problem to be solved, substantiated with empirical values to support the problem statement. In addition to the pure problem description, various fields in which the pattern could be applied are addressed.
- The *solution description* forms the core of a pattern and contains detailed information on how the solution can be used in the context described. This block also describes restrictions that might prevent a pattern from being used under certain conditions.

- The pattern is illustrated again in a *diagram*.
- Since each pattern has *relationships with other patterns*, these relationships are listed in the final section.

> **The Format of the Pattern Description**
>
> Christopher Alexander used strict formatting rules for the patterns. For example, three asterisks (***) separated the individual blocks of the patterns. No subheadings followed.
>
> The individual patterns were also "weighted" with star ratings to indicate how mature and how practicable a pattern was and whether there was still room for improvement.

Alexander divided the patterns in the resulting catalog into three subgroups. Each group refers to a level of abstraction in architecture.

The first set of patterns in the catalog described large-scale structures for entire cities and communities. The second category of patterns present solutions for the arrangement of houses, and the third group dealt with the construction of houses.

> *"The fact is, that we have written this book as a first step in the society-wide process by which people will gradually become conscious of their own pattern languages, and work to improve them."* —Christopher Alexander, A Pattern Language, 1977

For a long time, these descriptions and terms for patterns or pattern catalogs had nothing to do with software development. Ward Cunningham and Kent Beck first popularized the concept of patterns in software development in the 1980s. They worked together at Tektronix where they used the object-oriented programming language *Smalltalk*. At that time, they could not find good instructions for learning or writing down the associated object-oriented way of thinking or the use of terms from object orientation.

The book by Christopher Alexander, which was sent to Ward Cunningham by his student friend William Croft, gave Cunningham and Beck the idea of closing this gap by merging the world of architecture with the world of software development and adapting the concept of patterns and the pattern languages.

In September 1987, their presentation on *Using Pattern Languages for Object-Oriented Programs* at the OOPSLA (Object-Oriented Programming, Systems, Languages & Applications) conference in Orlando, made this idea more popular.

The presentation introduced Christopher Alexander's idea of a pattern language and its possible applications in software development. Also, part of the presentation were the first five numbered patterns for object-oriented user interface development:

1. Windows per task
2. Few panes per window
3. Standard panes

4. Short menus
5. Nouns and verbs

Cunningham and Beck already announced in the description of their presentation that they had identified and outlined a further 20 to 30 patterns and that they wanted to publish these in their own pattern language with around 100 to 150 patterns:

> "We have completed approximately ten patterns, have sketched out 20-30 more, and expect our finished pattern language to contain about 100-150 patterns. Our initial success using a pattern language for user interface design has left us quite enthusiastic about the possibilities for computer users designing and programming their own applications." —Kent Beck and Ward Cunningham, OOPSLA Conference, 1987

Subsequently, the concept of *design patterns* spread further in the world of software. In 1991, in Zurich, Erich Gamma laid the foundation for the book *Design Patterns* with his doctoral thesis entitled "Object-Oriented Software Development Using the example of ET++: Elements of Reusable Object-Oriented Software," coauthored with Richard Helm, Ralph Johnson, and John Vlissides in 1994.

This group of authors has since been called the *Gang of Four (GoF)* because of its size. The patterns listed in their book are therefore often called *GoF patterns*.

The Gang of Four

The *Gang of Four* consists of the following members:

- Erich Gamma
- Richard Helm
- John Vlissides
- Ralph Johnson

Patterns are grouped or classified according to their intent and area of application, in the same way as Christopher Alexander had done.

Even before the Gang of Four's book was published in 1994, another group met in August 1993 at the invitation of Kent Beck and Grady Booch in a mountain hut in Colorado to further explore the idea of design patterns. The participants of this meeting wanted to build upon the basic work of Erich Gamma's dissertation and develop a more reusable approach to generic patterns. Their work provided the basis for the later Gang of Four book.

This group also went down in history as the *Hillside Group*, in reference to the location of the meeting.

> **Hillside Group**
>
> The *Hillside Group* met in 1993 and consisted of the following individuals:
>
> - Ward Cunningham
> - Ralph Johnson
> - Ken Auer
> - Hal Hildebrand
> - Grady Booch
> - Kent Beck
> - Jim Coplien
>
> The Hillside Group is now registered as a non-profit educational organization that sponsors and organizes various conferences. It promotes the use of patterns and pattern languages so that knowledge can be recorded, analyzed, and shared easily. According to its own mission statement, the goal of the group is to improve the quality of life and society as a whole (*https://hillside.net*).

With the publication of *Design Patterns: Elements of Reusable Object-Oriented Software* by the Gang of Four, design patterns have become widespread in software development. This work is considered one of the most widely used and important books on the subject of software design.

Christopher Alexander's basic concepts for problem solving in architecture were subsequently adopted by various software developers who developed and published their own pattern languages.

Even though object-oriented languages were at the forefront of the development of design patterns in software development, the basic concept is not limited to this intent. Design patterns have always been used (and still are) to describe corresponding problem solutions in other contexts or languages.

Patterns always outline a successfully applied problem solution, and over time, four components have become established that every pattern description includes as a minimum:

- Name
- Problem
- Solution
- Consequences

Note that patterns do not represent a concrete form, they're "only" an abstract template with which the initially described problem can be solved.

> **Gang of Four Design Patterns and Java**
>
> In many people's minds, GoF patterns are closely associated with Java development. However, Java was not published until 1995/1996, that is, after the Gang of Four's book had been published.
>
> Their book therefore contains no Java examples, only C++ and Smalltalk examples.

Figure 1.2 shows the chronological sequence of this development.

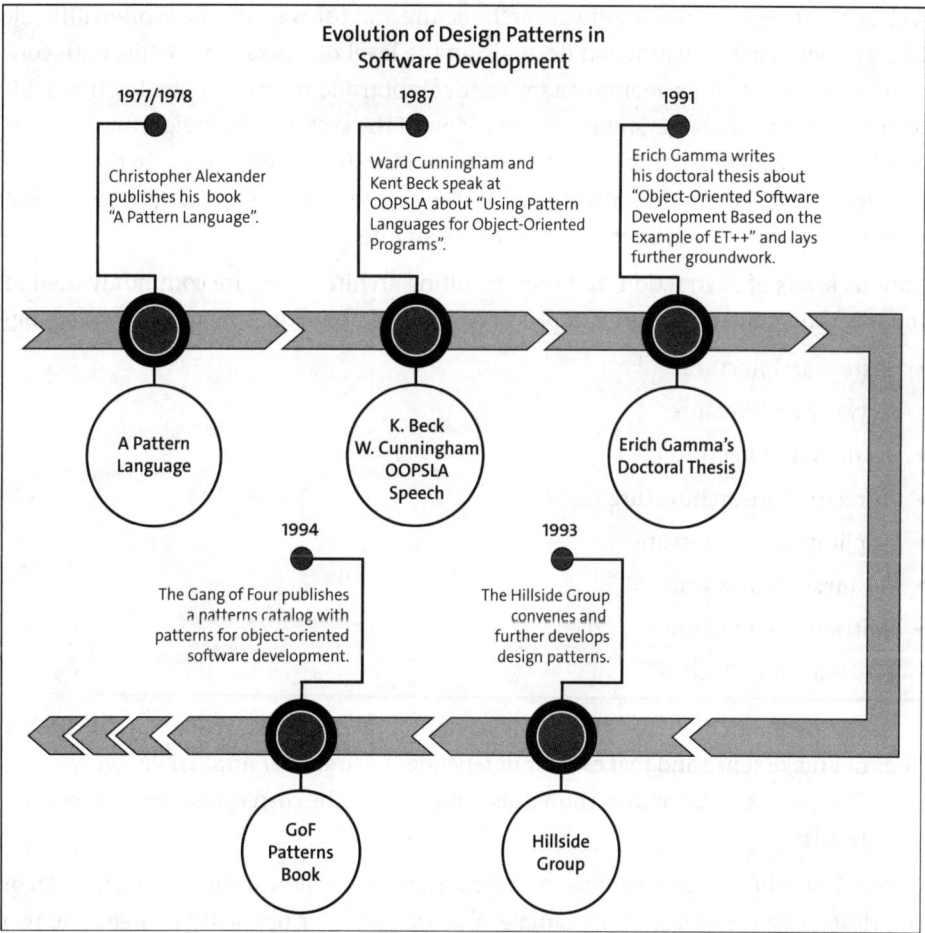

Figure 1.2 The Evolution of Design Patterns in Software Development

1.3 What Are Software Architecture and Software Design?

The emergence of the term *architecture* in the context of software development dates back to the 1960s, and the term as such still has no standardized definition. However, what the various definitions have in common is that they describe an architecture as an abstract representation of the basic structure of a software system, including its components and their interactions with each other.

Paul Clements co-authored a book called *Documenting Software Architectures: Views and Beyond*, which includes the following definition of software architecture:

> *"The software architecture of a computing system is the set of structures needed to reason about the system, which comprise software elements, relations among them, and properties of both."* —Paul Clements et al.

A clear distinction between software architecture and software design is often difficult to make because the distinction depends on the level of abstraction. While both concepts describe the components of a software solution and their relationships, the architecture focuses on a larger, more strategic view of the system. The design, on the other hand, makes concrete the basic framework provided by the architecture in the form of detailed specifications for individual modules and for their interactions. The level of abstraction used thus defines the focus of the architecture.

Various levels of abstraction, and their resulting architectures, are commonly used in the software industry. Some examples of common architectures include the following:

- System architecture
- Network architecture
- Security architecture
- Infrastructure architecture
- Application architecture
- Business architecture
- Software architecture
- Hardware architecture

Considering an architecture without considering the details described by the design does not make sense and makes their independent use almost impossible. Every architect who designs architectures should also think about the corresponding implementation details.

Robert C. Martin, author of various software design and architecture books and founder of the clean code movement, for example, also sees no clear boundary between the two terms and describes the interplay of the different levels of abstraction through the analogy of a house: A building architect designs the appearance of a house, but at the same time must not neglect certain functional details, such as the placement of sockets or water pipes. A bathroom without a water supply is just as useless as an application architecture without possible points for user interaction, for example, a web store that cannot be interacted with and from which no products can be ordered.

Defining a structure for software is important for creating higher-quality software. Software that lacks a formally defined architecture usually suffers unclear structures and is characterized by a strong coupling of the individual components. The interwoven structure of this application makes its maintenance and further development more difficult because changes in one area may have unforeseeable effects on other areas of the system.

However, software architectures should never be regarded as static. They must be able to develop further and adapt to new or changed requirements. The task of a software architect is to define these structures, maintain them over the course of a project, and modify them when necessary.

The software development ecosystem is constantly evolving, and new architectural approaches are emerging all the time. The next big thing in architecture is not, as one might expect, created on a whiteboard by architects, but instead emerge through an organic process in which existing concepts are recombined, adapted, or expanded. Software solutions, architectures, and framework conditions change, and projects must be kept flexible accordingly.

The architectural style of microservices, for example, would not have made sense in the 1990s, given the technologies available at the time, or would have been too difficult to implement. Only the development of container architectures and the increased popularity of open-source operating systems have made it possible to run multiple applications as entirely independent components on a single computer in a cost-effective and efficient manner. Expensive licensing and hardware costs were avoided, and new deployment scenarios were created.

Although developing innovative and individual solutions and adapting your own approach accordingly are important, most developers should understand, and be able to apply, basic software architecture approaches.

1.3.1 Tasks of a Software Architecture

As described earlier, software architectures define the fundamental framework and basic structure of an application and thus help to break down the complexity of the solution into more manageable parts. An architecture lays the foundation for the efficient development, maintenance, and scaling of an application. As a blueprint for the development of an entire software system, an architecture ensures that all components work together consistently and that overall requirements are met.

The software architect is responsible for defining and documenting the architecture of a system. They must consider the various aspects of an architecture, such as its scalability, security, performance, and maintainability, during the development process and include these aspects in architectural decisions. Chapter 3 describes how to document software architectures for easy sharing with others and for communicating about architectures.

A software architecture must fulfill multiple tasks, while documenting various aspects of the tasks undertaken:

- **Structuring**
  A software architecture breaks down complex systems into smaller, more manageable modules or application layers to enable a better organization and management of the system.

- **Design principles**
  The design principles define the basic guidelines and patterns that serve as a guide for all decisions made during the system development process.

- **Communication between modules**
  The architecture defines how the individual modules communicate with each other and how they exchange data.

- **Implementing requirements**
  Applications have both functional and non-functional requirements that must be considered and implemented by the software architecture.

- **Technological decisions**
  The technologies, programming languages, frameworks, and tools to be used must be defined as part of the architecture.

- **Adaptability**
  Good architectures enable a software system to adapt easily to new requirements and can be scaled well in terms of functionality. This capability facilitates the development, testing, and maintainability of the software.

- **Integration with other systems**
  Software architectures not only define the communication mechanisms within the application but also how the software integrates with other systems.

- **Documentation and communication**
  The documentation of a software architecture can serve to communicate between different stakeholder groups. Documentation can build a shared understanding of the system and the solutions it provides.

For Robert C. Martin, the goal of a good software architecture is to minimize the amount of personnel required to create and maintain a system:

> *"The goal of software architecture is to minimize the human resources required to build and maintain the required system."* —Robert C. Martin

The advantages of a solid and well-documented software architecture include the following:

- **Improved maintainability**
  A well-structured application with a clear architecture is easier to understand, and errors can be localized and rectified more easily.

- **Increased expandability**
  Clear structures and defined boundaries between the individual components make it easier to integrate new functions without impairing existing functionality and causing side effects.

- **Simplified collaboration**
  By dividing an application into more independent blocks, multiple teams can work simultaneously without getting in each other's way.

- **Improved reusability**
  Clear and well-considered interfaces between the individual components can increase reusability.

- **Better testability**
  Adherence to clear structures and clearly defined dependencies promotes the testability of the individual components. They can be tested in isolation and autonomously, whereby the external dependencies of the component can be replaced by what are called *mock objects* based on the existing interface specifications.

- **Higher security**
  A well-structured architecture prevents security gaps and better protects the system against attacks.

1.3.2 Architectural Styles and Architectural Patterns

In the early 1990s, American computer scientist and university professor Mary Shaw examined various software systems and, like Ward Cunningham and Kent Beck, recognized certain recurring architectural concepts in various applications even at a lower code level. She referred to these concepts as "elements of a design language for software architecture" or "design idioms" and thus laid the foundation for the description of software architectures.

Almost at the same time, Dewayne Perry and Alexander Wolf also adopted existing concepts from building architecture in 1992 in order to transfer them to the world of software. One of these concepts was what's called an *architectural style*. Like Mary Shaw, Perry and Wolf recognized some recurring design forms in the structure of software systems.

For building design, an architectural style describes a particular way or method of designing or constructing buildings. Typically, a style encompasses certain design principles, such as aesthetic features or construction methods characteristic of a particular era or region. Architectural styles can evolve and adapt to new technologies or approaches. Each style can be identified or classified by its own special characteristics.

For software systems, one goal was to find similar classifications and to identify systems based on the forms used or their characteristics. The structure or organization of the code and the interactions of various software components are often used as characteristics.

Four well-known architectural styles in software development are, for example, the following:

- Layered architecture: In a *layered architecture*, the software is organized in different layers, with each layer taking on a specific function. Common examples are *three-tier architectures* in which a distinction is made between the presentation, business logic, and data management layers.
- Client-server architecture: A *client-server architecture* divides an application into two main components: one component for presentation (as a user interface) and a central component running on a server that provides the business logic.
- Microservices: In this approach, an application is implemented using several smaller, independent services, each of which covers a specific business area. Each service can be developed, provisioned, or scaled independently.
- Event-driven architecture: This architectural style is based on the concept of event-driven programming. Components react to events that are generated by other components. This style does not use any synchronous, strongly coupled communication between the components.

Deciding which architectural style should be used depends on your software requirements. In larger software systems, you'll often find combinations of several styles because different aspects of various approaches might be used at different levels of the software.

In 1996, Frank Buschmann and his colleagues at Siemens established the logical missing link between software architectures and design patterns. Their *Pattern-Oriented Software Architecture, Volume 1: A System of Patterns* introduced the concept of *architectural patterns*. This concept spread quickly, and the book developed into a standard work for software architects, analogous to the Gang of Four book for programmers.

> **Books on Pattern-Oriented Software Architecture**
>
> After the original *Pattern-Oriented Software Architecture (POSA)* book, four extensions were later published that focused on different aspects of software architectures. The complete series consists of the following books:
>
> - Volume 1: A System of Patterns
> - Volume 2: Patterns for Concurrent and Networked Objects
> - Volume 3: Patterns for Resource Management
> - Volume 4: A Pattern Language for Distributed Computing
> - Volume 5: On Patterns and Pattern Languages

In their book, Buschmann and colleagues describe architectural patterns through a standardized description template to make patterns comparable. They also used a context-problem-solution description as a basis, which they extended by a few points:

- Name: Including alternative names, to identify the pattern.
- Example: This example represents the problem to be solved.
- Context: Description of the situations in which the pattern can be used.
- Problem: Description of the problem to be solved and a discussion of the necessary dependencies.
- Solution: A basic representation of the solution.
- Structure: A detailed representation of the structure of the pattern using corresponding diagrams. *Class-responsibility-collaboration (CRC)* cards and class diagrams from the *object-modeling technique (OMT)* are used. Both are tools for communicating and representing software design that have been replaced by the *Unified Modeling Language (UML)*.
- Dynamics: Typical scenarios that describe the runtime behavior of the pattern. In this case, diagrams like *Object Message Sequence Charts*, a predecessor of UML sequence diagrams, are used for visualization.
- Implementation: Guidelines, suggestions, and code fragments for implementing the pattern.
- Further aspects of the example: For the example representing the problem, further aspects about implementation are described, aspects that have not yet been dealt with in the solution, structure, dynamics, or implementation sections.
- Variants: A brief presentation of possible variations of the pattern.
- Known deployments: Examples from real-world systems to prove that this pattern has actually been successfully applied.
- Consequences: Advantages and disadvantages of using the pattern.
- See also: Cross-references to other related and similar patterns.

Pattern-Oriented Software Architecture (POSA), Volume 1 lists the following architectural patterns with this description template:

- Layers
- Blackboard
- Model-view-controller
- Microkernel
- Pipes and filters
- Broker
- Presentation abstraction control
- Reflection

The following design patterns were also described:

- Whole-part
- Proxy
- View handler
- Client-dispatcher-server
- Master-slave
- Command processor
- Forwarder-receiver
- Publisher-subscriber

1.4 The Evolution of Software Development and Architecture

Software development, software design, and software architecture have changed and evolved continuously since their beginnings. *Design patterns* continue to play a major role in today's software development. However, their focus or level of abstraction has increased significantly over time. Not only are they used to describe problem solutions for object-oriented programming languages but also solutions for system architectures, for example. After the first few books, some real hype buzzed around design patterns, and they were used for description and incorporated everywhere.

In parallel to design patterns, other new concepts, architectural styles, approaches, and methodologies have continuously emerged, all of which continue to influence today's software development further.

In the following sections, I present several well-known and widespread "evolutions," or further developments, in software design and software architecture that have had or still have a major influence. Describing them in more detail, new architectural patterns emerge in an organic process in which existing concepts are recombined, adapted, or expanded. For this reason, look at the origins and background of various topics makes sense.

1.4.1 Class-Responsibility-Collaboration Cards

With the increasing popularity of object-oriented software development, new means for communicating and presenting software design were also sought. At the late 1980s, Kent Beck and Ward Cunningham published *class-responsibility-collaboration (CRC) cards* as a visual tool for analysis and documentation that addressed precisely this problem.

CRC cards consist of three elements: the name of the class, information about its responsibilities, and information about its collaboration with other classes. The display format of the CRC cards is not fixed or predefined.

Even if CRC cards are almost irrelevant today, their development indicates that developers felt a meaningful representation of software design was necessary early on and that a structured and possibly graphical representation would provide better communication options.

The *Unified Modeling Language (UML)*, which is now almost universally used for modeling and representation, was developed under the umbrella of the *Object Management Group (OMG)* on the basis all the considerations regarding the graphical representation of software and its structure. It defines a standardized graphical notations systems for the visualization, specification, construction, and documentation of software and system architectures. We provide an overview of UML in Chapter 3, Section 3.3.

1.4.2 The "Gang of Four" Patterns and Their Structure

In Section 1.2, I mentioned the book *Design Patterns: Elements of Reusable Object-Oriented Software* by the Gang of Four. Since its publication in 1994, this book has been regarded as one of the standard works on software development and has contributed greatly to the spread of design patterns.

Erich Gamma further laid the foundation with his dissertation, in which he documented and catalogued common reusable solution patterns. Many of the patterns listed in the book were already part of his dissertation and were expanded and supplemented with the help of the other authors.

The challenge at the time was to find a description format with which the patterns could be presented in such a way that other developers who had not yet encountered them could also understand them and recognize their relevance. It was not just about describing "how" something can be done, but above all about the "why."

Recognizing your own patterns is usually easier in real life than documenting them in such a way that other developers understand the meaning behind them and can successfully adapt the patterns for their own projects.

The GoF book is based on the structure and format of the design patterns proposed by Christopher Alexander. However, the four main elements of a pattern description (name, problem, solution, consequences) have been expanded into a more comprehensive template:

> *"Design patterns are not about designs such as linked lists and hash tables that can be encoded in classes and reused as is. Nor are they complex, domain-specific designs for an entire application or subsystem. The design patterns in this book are descriptions of communicating objects and classes that are customized to solve a general design problem in a particular context."* —GoF

I will present the components of the Gang of Four pattern template in more detail since I use the template for further pattern descriptions throughout this book.

The template consists of the following elements listed in the following sections.

Pattern Name and Classification

The pattern name should clearly describe the problem, the consequences, and the solution in one or two words. The name should be clear and distinctive so that the design can be clearly described and referenced in the documentation or in communications with colleagues. In addition, the name is intended to support the development of a design terminology.

As described earlier, the Gang of Four introduced a classification of patterns, whereby two criteria were distinguished: intent and applicability.

The intent of a pattern should describe and demonstrate the task of the pattern. Accordingly, the GoF distinguishes between the following categories of patterns:

- Creational patterns: These patterns are used in the object creation process.
- Structural patterns: These patterns deal with the composition of objects or classes.
- Behavioral patterns: These patterns deal with the interaction of multiple objects or classes.

Two criteria were used to determine applicability: class or object.

> **Objects and Classes**
> If you're not yet familiar with the concepts of *object* and *class*, we recommend referring to Chapter 2, Section 2.1.

This resulted in the matrix shown in Table 1.1 into which the individual patterns can be divided:

Applicability	Intent: Generation	Intent: Structure	Intent: Behavior
Class	Factory method	Adapter	▶ Interpreter ▶ Template method
Object	▶ Abstract factory ▶ Builder ▶ Prototype ▶ Singleton	▶ Adapter ▶ Bridge ▶ Composite ▶ Decorator ▶ Facade ▶ Flyweight ▶ Proxy	▶ Chain of responsibility ▶ Command ▶ Iterator ▶ Mediator ▶ Memento ▶ Observer ▶ State ▶ Strategy ▶ Visitor

Table 1.1 Classification of GoF Patterns by Criteria

Each GoF pattern has a corresponding classification with an applicability and an intent, indicated by its name. The *composite pattern*, for example, is included in the *object-structure patterns*.

> **Pattern Names**
>
> In their book, the Gang of Four describe finding a good name for a pattern as the biggest challenge. Some patterns were also renamed during the work on the catalog to make the name more meaningful. For example, the pattern known today as *decorator* was previously called *wrapper* and the *glue pattern* later became *facade*.

Intent

The intent section in a GoF template basically corresponds to the problem description of Christopher Alexander's pattern term. The question as to which problem or design problem the pattern solves should be answered briefly and concisely here.

Other Well-Known Names

As I mentioned in the box, a pattern can be known by several names. Other well-known names of the pattern are listed here.

Motivation

Within the motivation description, the design problem is illustrated by a sample scenario, and it is shown how it can be solved by the pattern. The scenario should help to better understand the abstract description in the following sections.

Applicability

This section of the description explains when it makes sense to use the pattern. Negative examples help by highlighting design decisions that would cause problems if the pattern were not applied. They thus illustrate the benefits and necessity of the pattern for specific development situations.

Structure

The structure of the pattern is outlined with the help of the structure section of the GoF template. The authors use three types of presentation to illustrate the elements and their relationships:

- Class diagram
- Object diagram
- Interaction diagram

However, only the class diagram is mandatory for all patterns.

The Gang of Four used the *object-modeling technique (OMT)*, *Objectory*, and the *Booch method* to create the diagrams. All of them together are precursors of the *UML*, which had not yet been defined at that time.

Components Involved

Each pattern consists of one or more components, which are named here in this template. In addition to the names, the relationships between the components and their specific responsibilities are also explained. The aim is to give readers a clear understanding of the structure and functionality of the pattern.

Interaction

This shows how the classes or objects involved interact during execution and which specific tasks they take on.

Consequences

This section describes the effects and possible trade-offs of the pattern. Both positive and negative consequences are shown in order to create a decision-making basis for an application. This should enable the reader to assess whether the advantages of the pattern outweigh the potential disadvantages and whether it therefore makes sense to use the pattern.

Implementation

The abstract description of the implementation contains tips for implementing the pattern and describes considerations and possible challenges. In some cases, language-specific features are also highlighted, which must be observed.

Sample Code

Since C++ and Smalltalk were common object-oriented languages at the time the GoF book was published, the sample code fragments presented in the book, which illustrate the implementation of the patterns, are written in these languages.

The patterns are always templates that you have to adapt yourself, which means you'll not find any complete implementations here, but only smaller blocks of code with the most important implementation details that characterize the pattern.

Known Use Cases

Since patterns are always proven solution descriptions, this section shows two successfully implemented use cases from real systems in different domains that demonstrate the success of the patterns.

The Gang of Four refers here, for example, to earlier publications or existing libraries in which the pattern was implemented and discovered.

Cross-References

In this section, links to other patterns are established and both related and different patterns are presented. The description emphasizes the corresponding differences. In addition, possibilities are shown for combining the current pattern with other patterns.

Design Principles

In addition to the pattern catalog and the corresponding descriptions of the patterns, the GoF book contains general design guidelines and notes on implementations for object-oriented software.

These *design principles* are often not associated with the GoF book and are not as extensive and elaborate as the patterns.

Example:

"Favor object composition over class inheritance." —Gang of Four

1.4.3 Clean Code

With the increasing spread and use of object-oriented programming, more and more design guidelines were formulated, based on which software could be better structured and developed more efficiently and expressively. These idioms or design principles provided the foundations or basic structures according to which the design patterns were built.

In contrast to the design patterns, the *clean code* approaches do not describe major solutions for specific problems, but rather the sensible use of programming languages, in particular object-oriented languages.

In his book *Advanced C++: Programming Styles and Idioms*, James Coplien published a catalog of such idioms for C++ and a series of programming styles in 1991, which were and still are adapted to other languages.

One of its described idioms, *Handle/Body*, for example, deals with the separation between the public interface and the implementation in order to reduce the complexity of an interface and hide private details of an implementation.

These overarching design principles, techniques and practices are primarily aimed at making the source code more readable, understandable, and maintainable:

"Any fool can write code that a computer can understand. Good programmers write code that humans can understand." —Martin Fowler

In 2008, Robert C. Martin published his book *Clean Code: A Handbook of Agile Software Craftsmanship* and thus significantly coined the term *clean code*. This has since become a synonym for software quality in the software industry and a standard for the maintainability of systems. The principles are now considered "good practice" in development.

> **Robert C. Martin alias Uncle Bob**
>
> Robert C. Martin, also known as Uncle Bob, has significantly shaped the concept of *clean code* with his books, publications and lectures and is internationally recognized as an expert.
>
> As a cosignatory of the Agile Manifesto and member of the Software Craftsmanship movement, he tries to make software development perceivable as an independent craft and to further professionalize it.
>
> The following list is an excerpt from the books Robert C. Martin has already published:
> - 1995: Designing Object-Oriented C++ Applications
> - 2002: Agile Software Development. Principles, Patterns and Practices
> - 2008: Clean Code. A Handbook of Agile Software Craftsmanship
> - 2011: Clean Coder. A Code of Conduct for Professional Programmers
> - 2017: Clean Architecture: A Craftsman's Guide to Software Structure and Design

Martin describes several important principles and practices in his book:

- Meaningful names: For variables, functions, or classes, you should define meaningful, clear names that can be easily understood and used by others.
- Functions with exactly one task: Functions or methods should fulfill exactly one clear, specific purpose. They should be written as briefly and concisely as possible.
- Avoidance of duplication: Redundancies in the code should be avoided as much as possible; otherwise, maintainability could be negatively affected.
- Meaningful comments: The source code should be as self-explanatory as possible so that no comments are needed to understand it.
- Uniform formatting rules: Source code should be formatted uniformly within a project.
- Testability: Source code must be tested, and the tests must also be compact, clear, and maintainable.

In addition to these rather fine-grained considerations within the source code, Uncle Bob is also dedicated to codifying the basic structure of classes or systems. For the organization of classes and their interactions, he described several principles, such as the following:

- Single responsibility principle
- Dependency injection
- Cross-cutting concerns

His thoughts and approaches were published in the book *Agile Software Development. Principles, Patterns, and Practices*, in which he compiled and presented various design principles for the first time. To encapsulate these principles, the now well-known and widespread *SOLID* acronym was created in 2004. SOLID stands for:

- **S**ingle responsibility principle
- **O**pen-closed principle
- **L**iskov substitution principle (LSP)
- **I**nterface segregation principle
- **D**ependency inversion principle

These principles, as well as others, and their practical use are described in more detail in Chapter 2, Section 2.3.

The clean code movement, also called the clean architecture approach, goes hand in hand with the use of design patterns. Some principles, such as the *single responsibility principle*, provide the basis for patterns. However, they can also be used independently—possibly with an adjusted focus. Compliance with the *open-closed principle* not only helps with the implementation of a class, for example, but can also be useful when defining a public HTTP interface.

These principles are quite general but can improve your solution's quality at various points, not just at the source code level.

1.4.4 Component-Based Development

In the 1990s, many new applications were created using object-oriented concepts. In most cases, however, these solutions were implemented as larger coherent blocks and only featured limited reusability of their components.

As a result, what are called *component models* became increasingly important in libraries and frameworks with the goal of developing reusable, flexible, and scalable components that can be used in several applications or implemented with less effort.

The best-known component models include the following:

- Microsoft's *Component Object Model (COM)*
- The Object Management Group (OMG)'s *Common Object Request Broker Architecture (CORBA)*
- Sun Microsystems's *Java Enterprise JavaBeans*
- *Java applets*

These component models continued the modularization trend in software systems. What were previously rather monolithic systems have been split into several, more independent components for the purpose of reuse. These individual components can be developed, tested, and maintained independently of each other.

Each component model describes the requirements for the component itself as well as for the environment in which the component should be executed. For components, these requirements usually include defining interfaces or metadata that must be made available, as shown in Figure 1.3. These requirements are often referred to as *contracts* that must be fulfilled.

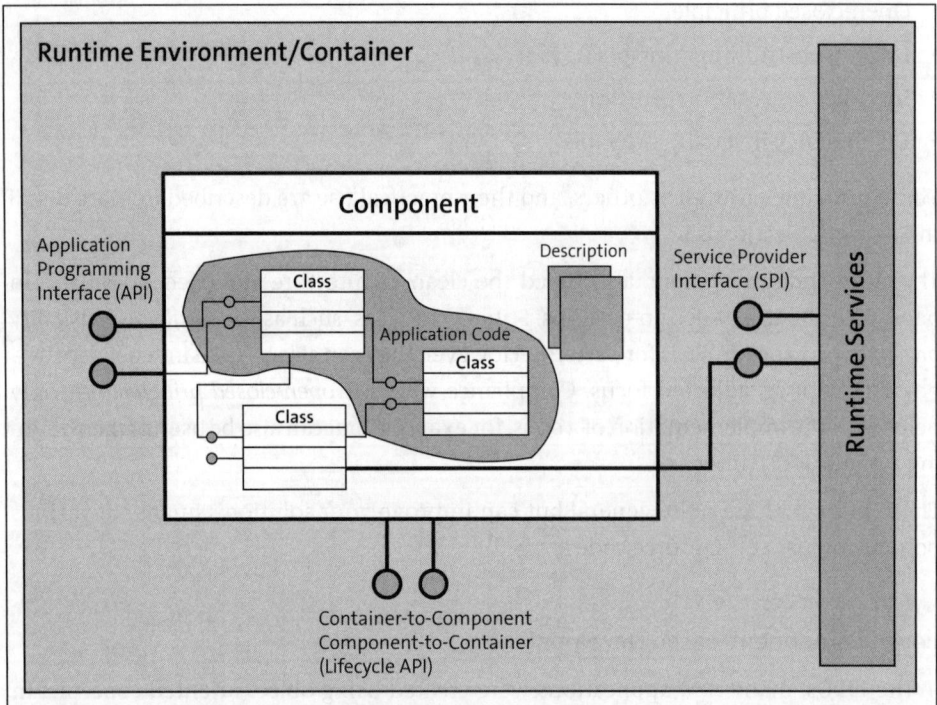

Figure 1.3 Individual Parts of a Component

An important point is that no individual component can no run on its own, nor can it be executed without a *runtime environment* or a *container*. Most models assume a predefined life cycle for the components. Thus, in this case, the runtime environment takes over the task of instantiating or destroying the respective component, for example. Installing the component in a runtime environment is referred to as its *deployment*.

Some component models are described as specifications, and manufacturers have the option of implementing this specification to create their own implementations. Java, for example, has the *web container specification*, which defines a web component's runtime environment. The *Jakarta EE Technology Compatibility Kit (TCK)* then checks whether the specification (i.e., the defined contract) is fulfilled.

If the contract has been fulfilled, the manufacturer may designate its implementation as Jakarta EE-compatible. *Apache Tomcat* or *Jetty* are examples of implementations of the *Jakarta servlet specification*, which is part of Jakarta EE.

With *Apache TomEE*, the *Apache Software Foundation* also provides a distribution of Apache Tomcat with additional components that together implement the *Jakarta EE 9.1 Web Profile specification*.

Components are usually supplied as a unit and consist of the following elements:

- A separate application code, which consists of one or more classes
- Exported interfaces of the component
- Descriptive information about the component so that it can be executed in a runtime environment

A basic distinction can be made between three types of exported interfaces:

- **Application Programming Interfaces (APIs)**

  Business logic is implemented within a component either via one class or multiple classes. The *application programming interface (API)* then makes this business logic available to other components and to enable interaction.

- **Service provider interfaces (SPIs)**

  Runtime environments provide shared functionality, such as the connection to a database connection pool or a transaction manager. These functionalities can also be called by the components in the environment via an interface, called a *service provider interface (SPI)*.

  In addition to using an SPI, components themselves can implement interfaces and make them available to other components. You can thus make your runtime environment components yourself.

- **Container interfaces**

  This third type of interface is used for communication with the runtime environment. Component models usually define a *life cycle* for the components to be executed, which defines, for example, when a new instance should be created. In most cases, the components must implement an interface specified by the environment.

In our next example, a simple Hello World program, a web component is implemented to better illustrate a current kind of component-based development, referred to as a *servlet* in *Jakarta EE*. The component should be installed or deployed in a runtime environment and respond to HTTP requests.

What Are Java Servlets?

Java servlets are web components for generating dynamic content. They are executed within a runtime environment, called a *web container*, and deliver a corresponding response to a request (request/response). Servlets are standardized as part of the Jakarta EE specification (see also Section 1.4.4).

To install a servlet within a runtime environment (deployment), the servlet is provided as part of a web application. Provisioning is achieved via a defined directory structure, which is often packed into a separate archive, a *Web Application Archive (WAR)*.

To fulfill the servlet component contract, which states that the component must implement the servlet interface, the Hello World component extends the abstract HTTPServlet class, which implements the corresponding interface. Thus, the class does not need to implement all the methods of the servlet interface and remains much more clear. Incidentally, this approach is also recommended by the specification. Figure 1.4 shows the class hierarchy of the HttpServlet class.

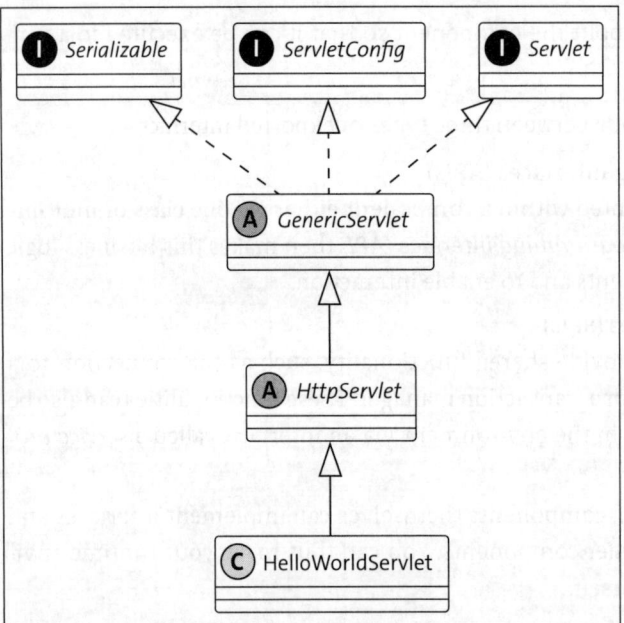

Figure 1.4 HelloWorldServlet Class Diagram

The resulting source code for implementing a web component is clearly laid out. For this example, only the doGet method needs to be implemented with the corresponding parameters, as shown in Listing 1.7.

```
package com.sourcefellows.javaee.helloworld;

import jakarta.servlet.annotation.WebServlet;
import jakarta.servlet.http.HttpServlet;
import jakarta.servlet.http.HttpServletRequest;
import jakarta.servlet.http.HttpServletResponse;

import java.io.IOException;
import java.io.PrintWriter;

@WebServlet("/HelloWorld")
public class HelloWorldServlet extends HttpServlet {
```

```
    @Override
    protected void doGet(HttpServletRequest req,
                HttpServletResponse resp)
                throws IOException {

        try(var writer = resp.getWriter()) {
            writer.println("Hello World!");
        }
    }

}
```
Listing 1.7 HelloWorldServlet Example (Java)

The parameters required and transferred by the component contract are used by the component implementation to communicate with the runtime environment. In this example, for instance, the output is written using a transferred `PrintWriter`.

In the Jakarta EE specification, metainformation, such as the name under which the servlet should be accessible, can be specified directly in the class as an annotation. In our example, we've used the `@WebServlet` annotation with a corresponding parameter.

The `HelloWorldServlet` class is already complete. The class can be compiled and deployed within a WAR archive in a compatible web container. This class cannot run on its own; it can only be used within a container.

After deployment, the container or runtime environment takes control of the component, and the installed component can be called as shown in Figure 1.5.

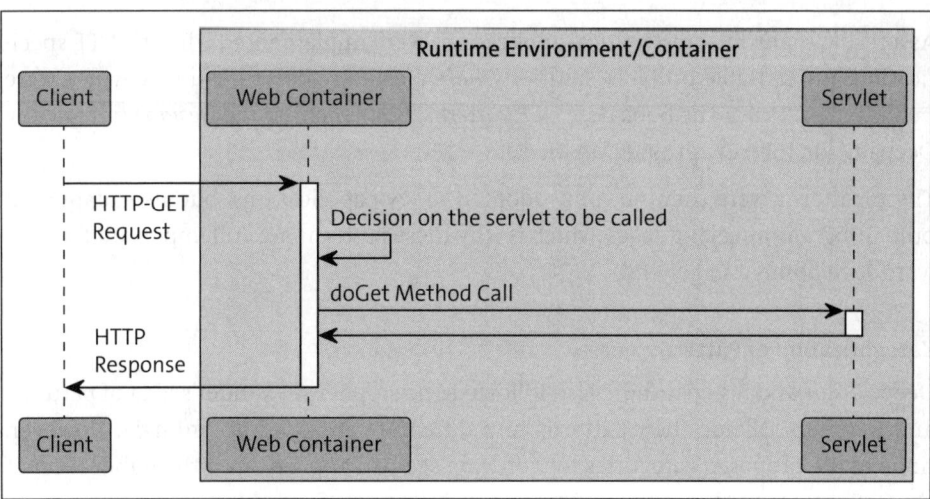

Figure 1.5 Sequence Diagram of a Servlet Call

1.4.5 Core J2EE Patterns

On January 23, 1996, Sun Microsystems released *Java 1.0*, another object-oriented programming language. Java was intended to be simpler and more universal than C/C++ and was characterized by its platform independence. Bytecode generated by the compiler could be run within a platform-dependent, provided *virtual machine (VM)*. Java applications no longer require laborious translation for multiple platforms.

These aspects certainly contributed to Java's increasing popularity, and Java was soon used to create business-critical applications.

As a result, many distributed Java-based applications were created, and the standard library was expanded to include further specifications and components in the form of classes and objects for corresponding development.

This development finally led to these individual standards being brought together and coordinated under a common umbrella in 1999: The *J2EE specification* was born.

> **From J2EE to Java EE and Jakarta EE**
>
> Sun Microsystems developed the *Java 2 Platform, Enterprise Edition (J2EE)* in 1999. J2EE was a compilation of several independent specifications from the Java Enterprise environment. Sun's aim was to bring development together under one roof and coordinate it by creating a common standard. Server providers could identify their *application servers* as compatible with a *J2EE specification* and applications were compatible between the servers (at least according to the advertising).
>
> J2EE was renamed *Java Enterprise Edition (Java EE)*, with the introduction of version 1.5 in 2006 and then the introduction of *Jakarta EE* in 2019.

As a result, many business-critical applications were implemented with the J2EE specification, and reusable problem solutions were created as design patterns, which were published in 2001 as the book *Core J2EE Patterns: Best Practices and Design Strategies* by Deepak Alur, John Crupi, and Dan Malks.

These patterns were then not only adopted in Java applications, but also adapted in other programming languages, which is why these patterns are still important for software development in general.

Categorization of Patterns

Deepak Alur and his coauthors also followed Christopher Alexander's idea of patterns and wanted to divide their patterns into different categories accordingly. However, none of the proposed categories for software seemed applied since the publication of the GoF pattern book seemed to fit.

Up until then, there had been:

- Design patterns
- Architecture patterns
- Analysis patterns
- Creational patterns
- Structural patterns
- Behavioral patterns

In their view, each of their patterns was located between the design and architecture patterns, which is why they created their own *J2EE pattern* category.

In the end, they assigned each of the 15 patterns from the first version of the book to one of the three logical architectural layers of a three-tier architecture:

- Presentation
- Business
- Integration

> **Three-Tier Architecture**
>
> A three-tier architecture is a common layer model in which an application is divided into three logical layers:
>
> - Presentation layer
> - Application layer
> - Persistence or integration layer
>
> The presentation layer is traditionally responsible for processing the data for the purpose of display or output. The application layer contains the actual business logic, and the persistence or integration layer is where the business data is stored or passed on to external systems.
>
> For decades, this three-tier architecture was the predominant architecture for business-critical, server-based applications.

With the second edition of the *Core J2EE Pattern* book from 2003, six additional patterns were added, bringing the total number of patterns to 21:

- **Presentation**
    - Intercepting filter
    - Application controller (from version 2)
    - Context object (from version 2)
    - Dispatcher view
    - Front controller

- Composite view
- View helper
- Service to worker

▶ **Integration**
- Service activator
- Web service broker (from version 2)
- Domain store (from version 2)
- Data access object

▶ **Business**
- Application service (from version 2)
- Service locator
- Transfer object
- Value list handler
- Transfer object assembler
- Business delegate
- Session facade
- Business object (from version 2)
- Composite entity

> **Sample Application: Pet Store**
>
> With the J2EE 1.3 specification, a complete sample application was created for the first time, with which the J2EE patterns were demonstrated in an executable application. The so-called *pet store* was created as part of the *BluePrint Program* at Sun Microsystems.
>
> The idea of a providing a complete sample application for technologies, concepts, or frameworks has spread since then. You can always find complete pet stores that present a framework, for example.

Pattern Description

No fundamentally new description template was used in the J2EE patterns. The authors largely worked based on the Gang of Four template, which they adapted only slightly. For example, the description of different variants of a conversion has been added. The following structure was used:

▶ Problem
▶ Prerequisites
▶ Solution
▶ Participants and responsibilities

- Different variants of implementation/strategies
- Consequences
- Sample code/implementation
- Cross-references

Criticism of J2EE: Spring Arose

The J2EE specification and the Core J2EE patterns published for it were subject to a great deal of criticism. The solution approaches were often considered far too complex and far too technically oriented. The required amount of code led to the complexity of the applications, which subsequently had to be maintained and mastered.

The *Enterprise Java Beans (EJB)* specification with its various types of *entity beans* triggered the most criticism. Basically, using a framework or specification was supposed to keep complexity out of the application logic. However, the opposite was often the case, and applications became unnecessarily complex and overloaded with many technical details. Critics have noted that J2EE applications are quickly *over-engineered*.

In his 2002 book *Expert One-on-One J2EE Design and Development*, Rod Johnson presented how, from his point of view, J2EE applications can be usefully structured and developed. He argued in favor of (re)orienting oneself to the existing object-oriented approaches that were available through already widespread design patterns and not allowing application and object design to be strictly dictated by a technical specification like J2EE or EJB. He questioned not all specifications, but individual J2EE specifications. In his opinion, the use of the EJB specification was particularly questionable, and the strict linking of the terms J2EE and EJB was often a mistake.

A second book, *Expert One-on-One J2EE Development without EJB*, which Rod Johnson published in 2004 with Jürgen Holler, provided alternative development approaches for J2EE-based Java applications—without using the EJB specification, as the name suggests.

To illustrate their approaches and demonstrate their practicality, the Johnson and Holler used a framework created in 2003 from code examples in the first book, called the *Spring Framework*.

The rest is history. Now, we can't imagine the Java world without the Spring Framework, and its concepts and approaches have also influenced further J2EE development. With the later renaming of J2EE to Java EE and later Jakarta EE, Sun certainly wanted to get rid of the negative connotations of the name.

Meanwhile, the relationship between Jakarta EE and the Spring Framework can be characterized as both competitive and complementary within the Java ecosystem. Both frameworks have different strengths and objectives. Spring provides an alternative, flexible ecosystem, while Jakarta EE acts as a standardized platform.

1.4.6 Enterprise Integration Patterns

The focus of the first design patterns in software development was on the internal structure and creation of software. This focus made sense, but in real life, software is rarely used on an isolated system. Usually, software must interact with other systems.

In 2004, Gregor Hohpe, Bobby Woolf, and colleagues published the book *Enterprise Integration Patterns*, in which they broadened their focus and compiled design patterns for the integration of various software systems. The book is a joint effort by a number of authors who had already started their own pattern catalogs, now incorporated into this work.

> **Contributors to Enterprise Integration Patterns**
>
> The book *Enterprise Integration Patterns* is the result of the work of many contributors. Gregor Hohpe and Bobby Woolf are the official authors, and they were supported by the following coauthors:
>
> - Kyle Brown
> - Conrad F. D'Cruz
> - Martin Fowler
> - Sean Neville
> - Michael J. Rettig
> - Jonathan Simon

From their point of view, the same fundamental challenges always arise during the integration of software systems that run in multiple environments, challenges such as the following:

- Networks are unreliable.
- Networks are slow.
- Applications are different.
- Changes are the daily norm.

When integrating applications, data must usually be transferred from one computer to another. Even in more modern system landscapes, these computers may only be virtualized environments, and the potential for error increases many times over compared to a local method call. Network access may be slow, data may be filtered, or a connection may fail completely. The number of components involved in a transmission is enormous, and each component represents a potential source of error.

Once the data has been successfully transferred, integration solutions must ensure that the data is also technically understood. The applications concerned may use different technologies, such as programming languages, operating systems, or data formats. Applications are different, and you may need to transform the data.

However, the technically correct transmission and the appropriate technical transformation of the data does not necessarily guarantee that all applications can handle and process that data. Software changes, even changes to only one part of the system, should not affect the other components involved. Dependencies must also be minimized as well.

These challenges were addressed by different developers with different solutions:

- File exchange
- Shared database
- Remote procedure calls
- Messaging

The first part of the *Enterprise Integration Patterns* book discusses the differences between, as well as the relative advantages and disadvantages of, these integration types. The second part of the book then describes individual integration styles and finally ends with a comprehensive catalog of patterns for an asynchronous messaging-based communication style. This is the main focus of the book, and these patterns have also contributed to its lasting importance.

The purpose of most patterns in *Enterprise Integration Patterns* is to create a system that is as loosely coupled as possible while maintaining high degrees of flexibility and stability.

Many of these patterns provide the basis for communications in distributed environments, for example, in the cloud. Message transmission or event-based communication is ubiquitous, and these patterns have become even more important with the advent of cloud environments.

> **Why Are Enterprise Integration Patterns Useful?**
>
> For the first time, the *Enterprise Integration Patterns* catalog dealt with patterns for asynchronous communication and with concurrent processing on a large scale.
>
> In today's cloud environments, integration and communication between applications still play a central role, and you'll find certain patterns again and again in various products or in the documentation provided by cloud providers. Enterprise integration patterns are omnipresent.
>
> Some patterns can also be adapted for concurrent processing within just one application.

The individual patterns are presented with a description template containing the following items. However, the descriptions are not given in a table or list, but rather in prose. Each pattern is described with the following information:

- The *pattern name* corresponds to GoF semantics. This unique, easy-to-memorize name describes the task of the pattern and can be used for communication.
- Most patterns are also represented by a small *graphic*, as shown in Figure 1.6. In addition to the text, this graphic is intended to create a visual language to support communication through images, popular in software development. Hohpe and Woolf developed special image components for their book, which were also intended to illustrate possible combinations of individual parts.

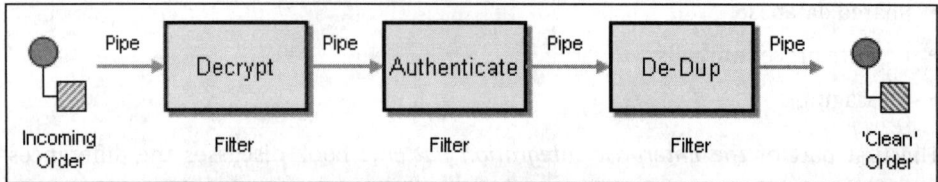

Figure 1.6 Sample Graphic for Enterprise Integration Patterns (Source: https://enterpriseintegrationpatterns.com/)

- The *context* describes the framework in which the pattern can be used and what problems can be solved by the pattern. Related patterns that are potentially already in use are usually listed.
- The *problem* describes the challenge you're facing in the form of a question. In contrast to other pattern catalogs, these authors limited themselves to a single sentence describing whether the pattern is important for their own work. For example:

How does a requester who has received an answer know for which request this is the answer? (The problem of the correlation identifier pattern.)

- Each pattern description *reinforces* the pattern's approach by presenting further approaches that could be used to solve the problem. The alternatives are intended to highlight the added value of this solution, as they are not expedient.
- A description of the *solution* that can be used to tackle the problem is provided.
- Each solution description is supplemented by a short, corresponding *sketch*, which is also about quickly identifying the solution. Ideally, it should be possible to recognize each pattern by looking at the name and the sketch.
- In the *results* section, the details of an application and the effects of the pattern are clarified. Any new challenges that may arise are mentioned.
- *Additional patterns* are then named that could or must be considered after the current pattern has been used.
- Technical problems or variants are discussed in the *sidebars*. This section is separate from the actual description.
- The final section presents *examples* for the use of the pattern.

The 683 pages of the book contain a total of 61 enterprise integration patterns, namely, the following:

- **Messaging**
  - Message channel
  - Pipes and filters
  - Message translator
  - Message
  - Message router
  - Message endpoint
- **Message channel**
  - Point-to-point channel
  - Datatype channel
  - Dead letter channel
  - Channel adapter
  - Message bus
  - Publish-subscribe channel
  - Invalid message channel
  - Guaranteed delivery
  - Messaging bridge
- **Message construction**
  - Command message
  - Event message
  - Return address
  - Message sequence
  - Format indicator
  - Document message
  - Request-reply
  - Correlation identifier
  - Message expiration
- **Message router**
  - Content-based router
  - Dynamic router
  - Splitter
  - Resequencer
  - Scatter-gather

- Message filter
- Recipient list
- Aggregator
- Composed message processor
- Routing slip
- Message broker

▶ **Message transformation**
- Envelope wrapper
- Content filter
- Normalizer
- Content enricher
- Claim check
- Canonical data model

▶ **Message endpoint**
- Messaging gateway
- Transactional client
- Event-driven consumer
- Message dispatcher
- Durable subscriber
- Service activator
- Messaging mapper
- Polling consumer
- Competing consumers
- Selective consumer
- Idempotent receiver

▶ **System management**
- Control bus
- Wire tap
- Message store
- Test message
- Detour
- Message history
- Smart proxy
- Channel purger

> **Products with Enterprise Integration Patterns**
>
> Implementations for using enterprise integration patterns are already built into some products. The following open-source products support the patterns directly:
> - Spring Integration
> - Apache Camel
> - RedHat Fuse
> - Mule ESB

The following example from the *Spring Integration* documentation shows a simple XML-based configuration for processing incoming web service messages. Some patterns contained in the catalog are used, such as the following:

- Gateway, through the `gateway` element
- Message channel, through the `input-channel` attribute
- Normalizer, all messages are normalized internally
- Content enricher, through the `header-enricher` element

```xml
<!-- Web Service -->
<int:gateway id="wsGateway" service-interface="foo.TempConverter"
    default-request-channel="viaWebService" />

<int:chain id="wsChain" input-channel="viaWebService">
  <int:transformer
      expression="'&lt;FahrenheitToCelsius xmlns=&quot;https://www.w3-schools.com/xml/&quot;&gt;&lt;Fahrenheit&gt;XXX&lt;/Fahrenheit&gt;&lt;/FahrenheitToCelsius&gt;'.replace('XXX', payload.toString())" />
  <int-ws:header-enricher>
      <int-ws:soap-action value="https://www.w3schools.com/xml/FahrenheitToCelsius"/>
  </int-ws:header-enricher>
  <int-ws:outbound-gateway
      uri="https://www.w3schools.com/xml/tempconvert.asmx"/>
  <int-xml:xpath-transformer
      xpath-expression="/*[local-name()='FahrenheitToCelsiusResponse']/*[local-name()='FahrenheitToCelsiusResult']"/>
</int:chain>
```

Listing 1.8 Spring Integration: Example of Data Processing

1.4.7 Well-Architected Cloud

In the late 1990s and early 2000s, Amazon faced a challenge: expanding its rapidly growing online retail platform while at the same time reducing operating costs. Amazon invested heavily in the expansion of its IT infrastructure and the development of internal software solutions. In addition to new software solutions, this investment has resulted in technologies and tools for virtualizing servers and automating operating processes.

In 2002, Amazon launched the *Amazon Web Services Developer Program*, which gave developers access to Amazon.com's resources to develop and operate applications in the Amazon infrastructure. The environment known today as the *cloud* was invented.

The official launch of *Amazon Web Services (AWS)* as an independent business segment in 2006 underlined the importance of the cloud for Amazon. AWS opened up the environment to other customers. In 2008 and 2010, Google and Microsoft followed suit with their own cloud offerings, making their own resources available to customers, as shown in Figure 1.7.

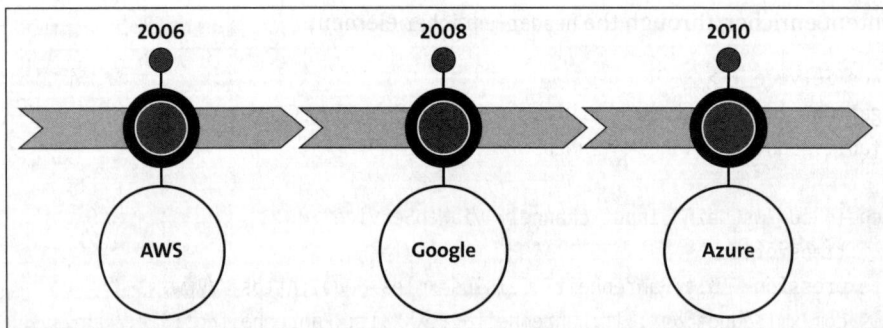

Figure 1.7 Emergence of the Major Cloud Providers

The business model of marketing computer infrastructure has developed steadily since and has become a major line of business for providers.

While many customer-specific and proprietary applications from cloud providers were successfully migrated to the corresponding environment or newly developed for it, new challenges arose in some implementations. Both the large cloud providers and their customers sometimes experienced difficulties during implementation, and further development and adjustments were necessary to processes and working methods.

In 2015, AWS published a collection of best practices and principles for the design and implementation of cloud applications and infrastructures, coining the term *well-architected cloud*. Other providers have also recognized the need for solution descriptions and thus have published their own descriptions of how the new challenges of cloud computing can be mastered.

> **Well-Architected Cloud Instructions from Cloud Providers**
>
> All major cloud providers have published their own collections of best practices for development in their environments. These collections can be accessed at the following URLs:
>
> - *https://aws.amazon.com/en/architecture/well-architected/*
> - *https://learn.microsoft.com/en-us/azure/well-architected/*
> - *https://cloud.google.com/architecture/framework?hl=en*

The guidelines defined for AWS are currently based on six *pillars* and describe solutions in the following areas:

- Operational excellence, with a focus on operational processes and procedures and their continuous optimization
- Security, with a focus on the protection of information and systems
- Reliability, with a focus on ensuring operational availability and recovery from failures
- Performance and efficiency, with a focus on optimizing the use of resources, monitoring performance, and maintaining efficiency
- Cost optimization, with a focus on the management and reduction of operating costs
- Sustainability, with a focus on minimizing the environmental impacts

A well-designed cloud architecture and software development are closely linked. A well-structured cloud deployment provides a stable, secure, and scalable basis for software development, while good software development practices help to optimize the use of the cloud.

With certain software development methods, for example the following, you can support a well-architected cloud and make optimum use of it:

- Modularization or microservices architecture: By developing an application in smaller units, updates, extensions, and deployments can be carried out much more quickly thus allowing the environment to be used more flexibly.
- Continuous integration/continuous deployment (CI/CD): An automated build, test, and deployment pipeline accelerates the development cycle and supports the improvement process.
- Containerization: The packaging of complete applications with their dependencies in containers simplifies their provision and operation.
- DevOps practices: With practices such as *infrastructure-as-code*, the consistency and reproducibility of environments can be increased, thereby increasing the reliability of the environment, among other things. In addition, sensible monitoring and logging solutions allow problems to be quickly localized and resolved.

- Scaling: If applications are designed to be stateless, their instances can be scaled quickly and adapted to the required load or reliability requirements.

A well-architected cloud provides the following advantages for software development:

- Scalability: The cloud provides the option of dynamically adding or removing new resources. This feature enables reacting quickly to changing requirements possible and setting up new development or test environments, for example, which can increase the speed and reliability of software releases.
- Cost efficiency: Optimum use and dynamic adjustment of the required resources can minimize unnecessary costs when billing is based on consumption. Most cloud providers provide tools for monitoring and optimizing costs.
- Reliability: Redundantly designed systems and automated failover mechanisms can ensure the availability of applications even in the event of disruptions.
- Security: Cloud providers usually offer integrated security functions that developers can use in their own applications. These features include identity and access management, encryption, and network security groups.

Chapter 2
Principles of Good Software Design

Good software design plays a key role in making software sustainable and keeping it usable and maintainable in the long term. This chapter highlights various concepts and principles for sound, effective software design and shows you how to build and structure your software in an intelligent way.

Software development is a tool for overcoming challenges and problems. Software development should not be used to create a pure functionality as an end in and of itself, but instead to develop solutions for specific business needs and requirements.

The purpose of software development is to create error-free, maintainable, and expandable products that meet user requirements and that can be developed and maintained with reasonable investment in terms of time, resources, and costs.

A sound software design and a structured and systematic approach form a solid foundation for a successful software project. Both ensure stability, scalability, and maintainability of a product over time.

In real life, however, a business unit's desire for quick solutions often clashes with the actual possible development speed by the software development department, which strives for high-quality and reliable results. Compromises are made that may seem practicable in the short term but can lead to serious problems in the long term, as subsequent corrections are often not made or are associated with considerable challenges and costs.

For example, some projects start with the rapid implementation of superficial features that might give both management and end customers a false impression of the development speed of individual functions. This rapid development is often achieved at the expense of compromises in terms of quality, functionality, or maintainability. In many cases, some stakeholders argue that the compromises will be resolved later.

Dave LeBlanc, a software developer whose ironic mock law became famous via Robert C. Martin's *Clean Code* book, puts it in the following way:

> "Later equals never." —The "Dave LeBlanc Law"

In other words, if management has been falsely led to believe that new features are developed more quickly than in reality, further compromises might later be made to continue to meet unrealistic expectations, and the quality of the software will increasingly suffer.

The software crisis in the 1960s, and many other failed projects of the past, clearly demonstrate that the quality of the code plays a decisive role in the success of a software project—not only at the start in the development phase, but above all during the expansion and maintenance of the software.

Good software design is a continuous process that is not only created in the code itself but that extends to various levels. The design of a robust and efficient system begins with the requirements, and in the concept phases, the design is taken further by architecture-related decisions and finally implemented in the source code.

Efficient and error-free software development is a key challenge today. Poor and inadequately structured code acts like a car's brake block that slows down the development process and leads to frustration among developers (and everyone else involved in the project). This frustration often results from a lack of comprehensibility of the code, which also encourages errors.

Clean code and good software design are therefore the keys to accelerating software development. Clear structures, clear formatting, and well-documented functions make your code easier to understand and modify if necessary.

In the following sections, I will introduce you to several concepts and principles underlying good software design at the code level. These principles are designed to help you create software that is robust, maintainable, scalable, and easy to understand.

2.1 Basic Concepts of Object-Oriented Programming

As software projects became increasingly complex, new concepts and approaches were sought back in the 1960s to make software more maintainable and reusable. As a result, *object-oriented programming (OOP)* was developed, as described earlier in Chapter 1. OOP is based on a few simple concepts, which I will present in more detail in the following sections:

- Objects and classes
- Encapsulation
- Abstraction
- Inheritance
- Polymorphism

In object-oriented programming languages, these concepts are directly supported by their corresponding language constructs or keywords. For example, the *Java, TypeScript, C#, Python,* and *C++* languages, as well as others, use the class keyword to introduce a class definition.

However, these concepts and ideas can also be used in other programming languages or software architectures to improve maintainability. Examples include the concept of encapsulation or abstraction.

2.1.1 Objects and Classes

In object-oriented programming, logically related data and functions are combined into objects and described by a *class*, which represents the blueprint for similar objects. Objects are therefore instances of a class.

> **Object, Class, and Instance**
> Sometimes, these terms are not clearly differentiated from one another and might be used ambiguously. In this book, however, the term *object* always stands for a concrete instance (i.e., a characteristic or an example) of a class.

To describe object orientation, parallels can be drawn in the real world: Let's define a class such as Person. Each person has a name and age properties as well as a common behavior, for example, being able to greet other people.

You can have several instances (or objects) of this Person class within an application, each of which is created and configured differently. Each individual instance has their own names and their own ages, and yet they were all created using the same blueprint. To create these objects, a class contains the instructions for creating their instances. Most programming languages provide a what's called a *constructor* for this purpose; a constructor is a special function of the class. Functions that are closely linked to a class are referred to as its *methods*.

As mentioned earlier, class definitions in programming languages in which object-oriented programming is possible are usually introduced by the class keyword. In the Python example shown in Listing 2.1, for example, the Person class is defined with name and age properties. Such a property is also referred to as an *attribute* or a *field*. In addition, our example includes the greet function, which outputs a text with attributes of an object and is therefore bound to the class (via what's called a self reference). As a result, the greet function a method of the class.

```
class Person:
    # Constructor
    def __init__(self, name, age):
        self.name = name   # Attribute
        self.age = age     # Attribute

    # Method
    def greet(self):
        print(f"Hi, my name is {self.name}, I am {self.age} years old.")
```

Listing 2.1 Example Class Definition (Python)

In addition to the greet method, the class has a constructor that you can use to create new instances of the class. In Python, this constructor is always referred to as __init__.

Now, let's use the class we've created, as shown in Listing 2.2.

```
# Create objects using a constructor
person1 = Person("Alice", 30)
person2 = Person("Bob", 25)

# Call the defined method
person1.greet()   # Output: Hi, my name is Alice, I am 30 years old.
person2.greet()   # Output: Hi, my name is Bob, I am 25 years old.
```

Listing 2.2 Using Our Person Class (Python)

For Java, the example is analogous, as shown in Listing 2.3.

```
public class Person {
    public String name;
    public int age;

    public Person(String name, int age) {
        this.name = name;
        this.age = age;
    }

    public void greet() {
        System.out.printf("Hi,… %s, I am %d years old.\n", name, age);
    }
}
```

Listing 2.3 Example Class Definition (Java)

2.1.2 Encapsulation

Encapsulation describes a mechanism by which the data and the defined methods of an object are protected against unauthorized access and manipulation from the outside.

Objects hide their internal states by only allowing access via a defined interface in the form of public methods. The purpose of encapsulation is to ensure the integrity of an object's data and to conceal its own implementation from external access.

In addition, the internal structure of encapsulated objects can be changed without necessarily having to adapt the external interface. This feature increases the maintainability of your code as well as the reusability of the objects since objects can be treated as independent and autonomous units.

2.1 Basic Concepts of Object-Oriented Programming

In most programming languages, encapsulation is achieved by using *access modifiers* such as public, private, or protected.

Listing 2.4 shows a Java example in which the guestCount field of the Party class is protected against direct access from outside. The field has the access modifier private and can therefore only be set or read/displayed using the newInviteParty and showPartySize methods. In this case, the information "Number of party guests" is protected against unauthorized access and changes.

```java
public class Party {
    private int guestCount;

    public void inviteNewGuests(int newGuests) {
        guestCount = guestCount + newGuests;
    }

    public void showPartySize() {
        System.out.printf("There are %d guests\n", guestCount);
    }
}
```

Listing 2.4 Example Encapsulation (Java)

Bypassing Encapsulation

Even though many object-oriented programming languages support the concept of encapsulation through access modifiers, this encapsulation of objects is not usually strong or absolute.

In Java, for example, you can use *reflection* to bypass encapsulation and change the attributes of an object without an access method.

For example, the *Spring Framework* uses reflection to set references to other objects directly. The code shown in Listing 2.5 bypasses the encapsulation of the object.

```java
@Service
public class BadDatabaseAccountService {
    @Autowired //Encapsulation is bypassed
    private Assessor assessor;
    public BadDatabaseAccountService() {
        // The "assessor" attribute should be passed as a parameter!
    }
}
```

Listing 2.5 Removing Encapsulation in the Spring Framework

For this reason, I also recommend that Spring applications only set references via external interfaces, such as through the constructor. The code shown in Listing 2.6 illustrates this approach.

```
@Service
public class GoodDatabaseAccountService {
    private Assessor assessor;
    @Autowired
    public GoodDatabaseAccountService(Assessor assessor) {
        this.assessor = assessor;
    }
}
```

Listing 2.6 A Better Solution with Encapsulation in the Spring Framework

2.1.3 Abstraction

As with objects and classes, a comparison with the real world can be helpful to describe *abstraction*: Let's suppose you want to drive a car. How the engine works is simply not relevant to you. You interact with the abstract user interface—steering wheel, pedals, and gear stick—to simplify the complex mechanisms in the engine compartment and transmission.

Abstraction in software development works according to the same principle. The complex details of an implementation are hidden by abstract layers, and only the relevant features and behaviors of an object become visible and usable. Abstraction helps identify key features and promotes better structuring of software.

In object-oriented programming languages, abstraction is achieved by using *interfaces* or *abstract classes*. Both define a generally valid interface that callers can use and for which specific implementations exist. The abstract interfaces describe the behaviors that are implemented by concrete classes.

Interface

An *interface* defines a contract that all implementing classes must fulfill. The example shown in Figure 2.1 illustrates an (abstract) interface definition Car in a Unified Modeling Language (UML) representation. Our interface definition is comprised of four methods and defines the general behavior of a car.

> **Unified Modeling Language Representation**
>
> The *Unified Modeling Language (UML)* defines a standardized graphical notation for the visualization, specification, construction, and documentation of software and system architectures. If you have little experience with the UML syntax, we provide an overview in Chapter 3, Section 3.3.

2.1 Basic Concepts of Object-Oriented Programming

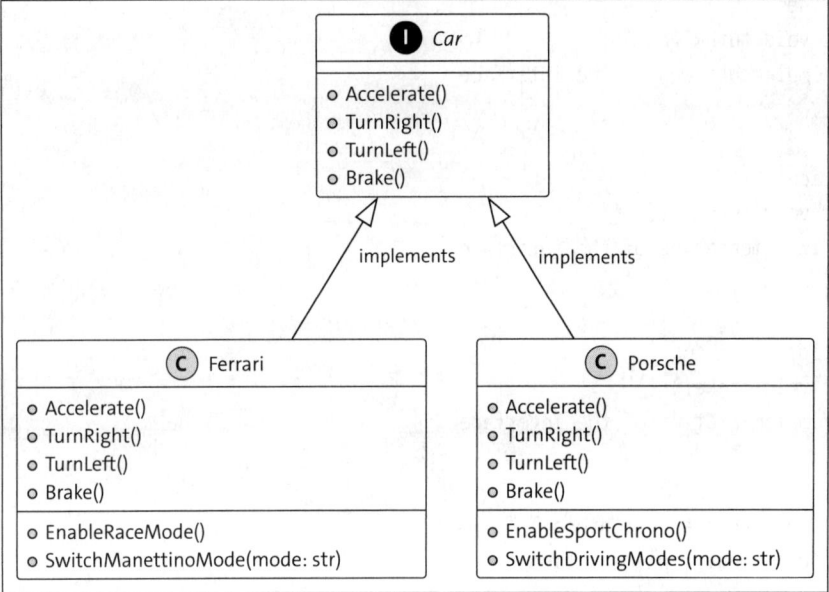

Figure 2.1 Abstract Types Using of a Car

The Ferrari and Porsche classes implement the Car interface and must therefore realize the four defined methods. Each class in the example has additional specific functionality in the form of additional methods.

Let's focus now on the engine we mentioned earlier. Further details of the specific implementations are not relevant for callers: You can accelerate, turn, and brake the car implementations.

The implementation of the Car interface can be created in Java, as shown in Listing 2.7.

```java
interface Car {
    void accelerate();
    void turnRight();
    void turnLeft();
    void brake();
}
```

Listing 2.7 Abstract Definition of the Car Interface (Java)

A concrete implementation of the interface is shown in Listing 2.8.

```java
public class Porsche implements Car {
    @Override
    public void accelerate() {
        //Implementation of the interface
        ...
    }
```

```java
    @Override
    public void turnRight() {
        //Implementation of the interface
        ...
    }
    @Override
    public void turnLeft() {
        //Implementation of the interface
        ...
    }
    @Override
    public void brake() {
        //Implementation of the interface
        ...
    }
    public void EnableSportChrono() {
        //Additional implementation
        ...
    }
    public void SwitchDrivingModes(string mode) {
        //Additional implementation
        ...
    }
}
```

Listing 2.8 Concrete Implementation of the Porsche Class (Java)

Abstract Class

Abstract classes also describe a generally valid behavior. However, unlike interfaces, abstract classes can also implement methods and have fields or instance variables.

Figure 2.2 shows a UML diagram for an abstract class named Animal. This class has an abstract method called makeSound and a concrete method called sleep. Derived classes must therefore implement the makeSound method and can continue to use the functionality of the sleep method.

Listing 2.9 and Listing 2.10 show the corresponding implementation in Java. Listing 2.11 shows how the concrete implementation, Dog, can be used.

```java
abstract class Animal {
    protected String name;

    public Animal(String name) {
        this.name = name;
    }
```

```
    public abstract void makeSound(); // Abstract method

    public void sleep() { // Concrete method
        System.out.println(name + " sleeps...");
    }
}
```

Listing 2.9 Abstract Class (Java)

```
class Dog extends Animal {
    public Dog(String name) {
        super(name);
    }

    @Override
    public void makeSound() {
        System.out.println(name + " barks: Woof!");
    }
}
```

Listing 2.10 Concrete Implementation of an Abstract Class (Java)

```
Dog fido = new Dog("Fido");
fido.sleep();
fido.makeSound();
```

Listing 2.11 Application of the Concrete Dog Type (Java)

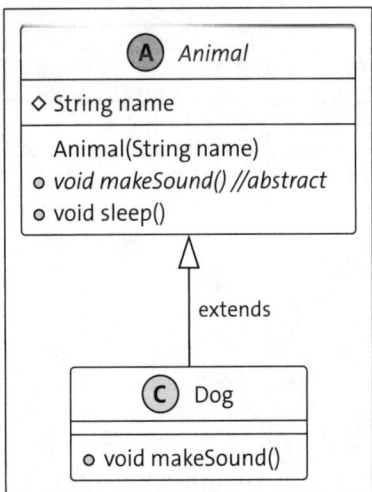

Figure 2.2 UML Diagram of an Abstract Class

Effects of Abstraction

The use of abstraction can make your software more simple because the calling code can focus exclusively on the lower complexity of the abstraction layer. The caller won't need to deal with the details of the individual implementations.

In addition to the source code level, abstraction arises in many areas of software development, such as the following:

- Application programming interfaces (APIs) provide an abstract interface for interacting with other software components. Internal details of the called component do not need to be known for a call.
- Libraries often provide simple abstract interfaces or objects for complex issues that are implemented within the library.
- Frameworks provide ready-made structures and abstractions for certain software applications.

2.1.4 Inheritance

In software development, *inheritance* is a special form of abstraction and defines special relationships between classes. The *subclass* or *derived class* adopts the properties and behaviors in the form of methods from another class, which is called the *superclass* or *base class*.

As with inheritance, abstraction is often best illustrated with an analogy from the real world: Imagine you have a base class called Animal, shown in Figure 2.3, that has the properties Name and Age and a Run method, for example.

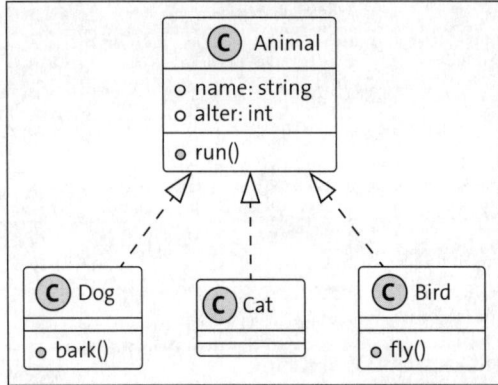

Figure 2.3 Inheritance through the Animal Class

You can then also define the subclasses Dog, Cat, and Bird, each of which inherits from the base class Animal and thus adopts the properties and methods specified in the superclass. Thus, due to inheritance, a dog could already run and already has a name and an age. In addition to the transferred properties and methods, you can define your own specific

2.1 Basic Concepts of Object-Oriented Programming

properties or methods for each derived class. A dog could also bark, while a bird could fly.

With inheritance, properties and methods are passed on to the derived classes and thus don't need to be implemented in the derived class again. The functionality is already implicitly available.

In Java (as in many other object-oriented programming languages), class inheritance is expressed through the extends keyword. Listing 2.12 shows the implementation of the example shown in Figure 2.3.

The run method is defined within the Animal base class, and Animal passes the method on to the Dog class for which it can be called.

```
class Animal {
    public void run() {
        System.out.println("Animal runs away");
    }
}
class Dog extends Animal {
    public void bark() {
        System.out.println("Dog barks");
    }
}
public class Main {
    public static void main(String[] args) {
        Dog d = new dog();
        d.run();
        d.bark();
    }
}
```

Listing 2.12 Example of Inheritance (Java)

Inheritance enables the reuse of code by automatically passing on the functionality of a base class to its derived classes. The same functionality doesn't need to be rewritten again and again, and thus development effort is considerably reduced. At the same time, inheritance provides flexibility for extensions: Subclasses can add new functions without changing the original base class. This principle enables a hierarchical organization of the code, which modularizes the software and thus makes your code better structured and easier to understand.

2.1.5 Polymorphism

Polymorphism (Greek for "many-shaped" or "versatility") is another fundamental principle of object-oriented programming that enables objects of different classes that have a common interface to be handled in a uniform way. Polymorphism allows, for example, objects of different types to be stored in a variable and to trigger different actions when called, depending on the class or object type.

In this case, too, the polymorphism should be described with an analogy to the real world: Imagine a parking enforcement officer handing out tickets to illegally parked vehicles in a no-parking zone. The person goes from vehicle to vehicle, notes the license plate number, and issues a parking ticket. They always work on one vehicle at a time, and it is irrelevant to them which specific version of a vehicle is currently in front of them. Whether motorcycle, car, sports car, or truck—the person only notes the license plate number, which must be legibly attached to the vehicle.

In terms of software development, you can, for example, create a Vehicle interface that defines an interface for reading the license plate number. The EnforcementOfficer class has a currentVehicle variable with the Vehicle interface, which is filled in with the specific instance of a vehicle. For the EnforcementOfficer class, which specific implementation is used doesn't matter as long as it provides the interface for reading a license plate and has a corresponding implementation. Each class can have its own implementation. The corresponding implementation is executed at runtime.

This *versatility* of the currentVehicle variable illustrates the principle of polymorphism its name. Figure 2.4 shows a corresponding class diagram.

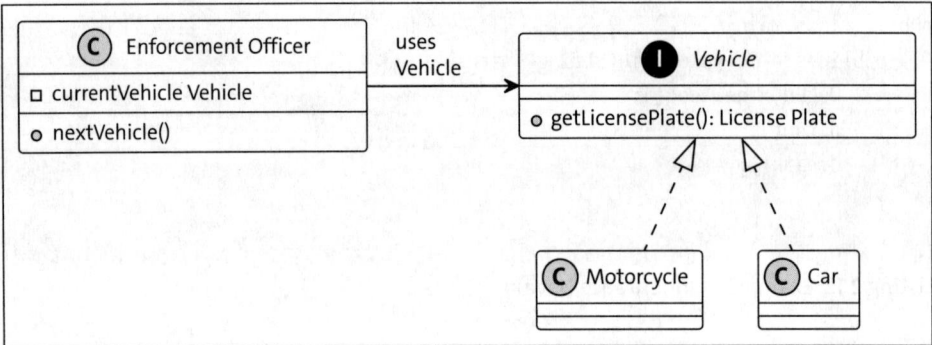

Figure 2.4 Class Diagram for Polymorphism

If several methods with the same *method signature* are implemented at different levels in a class hierarchy, which of the methods is used for an object is only determined at runtime. This process is referred to as *dynamic binding*.

> **Method Signature**
>
> A unique identifier for a method, a *method signature* consists of a name for the method and a list of its parameters, including their data types and order. A method signature enables a clear distinction between different methods, especially when *overloading* and *overriding*:

> - Overloading refers to the possibility of implementing multiple methods with the same name but different parameter lists within a class.
> - Overriding refers to the process of redefining a method of a base class within a derived class.

Listing 2.13 shows the important parts of an implementation for our current example, in Java.

```java
interface Vehicle {
    LicensePlate getLicensePlate();
}

public class EnforcementOfficer {
    Vehicle currentVehicle; //Interface type!

    void nextVehicle() {
        //...
        LicensePlate plat = this.currentVehicle.getLicensePlate();
        //...
    }
}
```

Listing 2.13 Example of Polymorphism (Java)

2.2 Clean Code Principles

Comprehensible, expressive, and readable source code is the basis for maintainable code and the foundation for good software design.

In the 1970s, various programmers and software architects compiled several guidelines and best practices to help write and structure source code more clearly. These *clean code principles* have been adapted time and again over the years and have continued to evolve ever since.

The development of the principles was significantly influenced by the following people:

- *Robert C. Martin* (Uncle Bob): One of the most important representatives of the clean code movement. With his books and especially with his book *Clean Code*, published in 2009, he shaped the modern development of the term and presented many central principles in it.
- *Kent Beck*: As one of the founders of *Extreme Programming (XP)*, he was already of the opinion in the late 1990s that software should always be improved. One of his goals was easily modifiable and maintainable source code.

- *Fred Brooks*: He published his book *The Mythical Man-Month* in 1975, which also had an influence on the clean code movement. Brooks emphasized the importance of good documentation and modularity in software development.
- *James Coplien*: In 1991 he published his book *Advanced C++: Programming Styles and Idioms*, in which he published the sensible use of the object-oriented programming language C++ as a procedure catalog. The programming styles he described were subsequently adapted to many other languages.

The principles of clean code are not rigid rules, but rather guidelines for better programming practice. Overall, they are based on experience from many successful software projects and are intended to help make code more readable, facilitate troubleshooting, and reduce maintenance costs. The clean code principles are still an important part of modern programming today and serve as the basis for many other development methods and guidelines.

Well-structured and readable code is easier to understand. This applies to the creator as well as to other developers who work with the code over time. Usually, code is read more often than it is written during the life cycle of software. Studies show that software maintenance, for example, accounts for around 60% to 80% of life cycle costs.

If you invest an extra 10 or 20 minutes in creating a piece of code that can later be better understood by other developers, the extra time will quickly pay for itself. If, for example, an error in the production system is localized and rectified more quickly as a result, it may not be necessary to carry out time-consuming corrections to inconsistent production data or the downtime of a production line can be prevented.

If you think about good software design and apply the clean code principles, this usually pays off after a short time, as you can minimize maintenance costs in this way and thus save money over the longer life cycle of a software.

In 2008, Thom Holwerda provided what he considered to be the universal quality metric for good source code in a humorous comic strip: *The only valid measurement of code quality*: WTFs/minute (see *https://www.osnews.com/story/19266/wtfsm/*): The more often other developers swear about the code, the more incomprehensible and therefore worse it is in their opinion.

It is often easy to identify poor and incomprehensible code, and it is easy to complain about it. However, it is much more difficult to produce expressive and legible code yourself!

The following sections present various practices that will help you to produce more expressive, more readable, and therefore more maintainable code. However, the practices presented will not be the only means for better code. It is also important to read and understand the code of other developers and to adopt proven, successful patterns from it.

Ultimately, learning best practices and reading good code is an effective way to improve your own coding skills in the long term. And in the end, there are three things that really make your code better: practice, practice, practice.

2.2.1 Identifier Names and Conventions in the Code

The naming of identifiers is an important part of software development, as it significantly influences the readability and comprehensibility of the code. Unambiguous names avoid confusion and misinterpretation, while descriptive names make the meaning of the respective element quickly recognizable.

If you take a closer look at the source code, you'll notice that the selected identifiers of functions, classes, methods, etc. make up the majority of the code base. In the *Software Quality Journal*, for example, Florian Deissenboeck and Markus Pizka published an article entitled "Concise and Consistent Naming" in 2006, in which they determined a percentage of 70% for the proportion of identifiers in the code and clearly deduced from this that the chosen identifier names have a major influence on the comprehensibility of the source code.

The challenges and difficulties of naming identifiers in source code are often referred to in the industry with the humorous statement; *Naming Things: The Hardest Problem in Software Engineering* (or in slightly different words). Tom Benner even published a book of the same name in 2023, in which he discusses the challenges in more detail.

Often quoted in this context is the statement by Phil Karlton, a well-known principal developer who worked at Xerox PARC and Netscape, among others. He sees naming as only the second most difficult thing to implement:

> *"There are only two hard things in computer science: cache invalidation and naming things."* —Phil Karlton

However, there is a grain of truth in these humorous statements. The naming of identifiers in the source code is one of the strongest levels of abstraction in programming. John Ousterhout, a pioneer of software development and inventor of the *Tcl* language, put it in a nutshell:

> *"[T]he best names are those that focus attention on what is most important about the underlying entity, while omitting details that are less important."*
> —John Ousterhout

Good names therefore concentrate on the most important features of the objects being named. Four basic principles should guide every name you choose:

- Comprehensible
- Short and concise
- Consistent
- Distinguishable

Comprehensible Names

A name should help the reader to understand as much as possible the concept or purpose it describes.

Without additional information or context, the variable name c for an instance of the Car class, for example, is difficult to understand. It gets even less transparent when generic names such as item or incorrect names for types of this Car class such as bicycle are used. In these cases, it becomes almost impossible for people who have to read such code to recognize a concept or purpose of an identifier! The example shown in Listing 2.14 illustrates how difficult it can be to recognize, in later code, what was stored earlier in the foo variable.

```
Car foo = new Car();
```

Listing 2.14 An Incomprehensible Variable Name (Java)

Good names are memorable and explain what is being represented. If a reader already knows the concept behind the designated object, the name should also be familiar. If a reader is unfamiliar with the concept, the name should convey the correct technical terminology.

The example shown in Listing 2.15 features the car variable, from which it is clear that the referenced object is an instance or the "concept" Car. The variable describes a car and should be named as such. This link makes it clear in the source code what the referenced object is.

```
Car car = new Car();
```

Listing 2.15 Comprehensible Identifier Names (Java)

Every name used for an identifier should appear in a dictionary, if possible, whether a general, commonly used dictionary such as the Merriam-Webster or a domain-specific glossary, which should always be included in a domain-driven design.

Common terms are much easier for a reader to understand and remember than invented terms that are not used in communications, for example, between the specialist department and the development department. The example shown in Listing 2.16 demonstrate some corresponding names for variables.

```
// Poor examples
o, org, people_group, guys

// Better example
organization, computer
```

Listing 2.16 Using Names from Dictionaries!

In addition to the terms and names for technical objects, domain-independent concepts are usually also used in software. These concepts include, for example, abstract data types, algorithms, general or technical concepts, and design patterns. You should use the generally accepted and established names for these objects or concepts and not invent your own names.

If standardized definitions of terms (such as units for time or physical units) or commonly recognized names (e.g., technical terms from computer science for networks or databases) are available, you should use them.

Working with an entry in an SQL-based database will most likely be more intuitive and easier to grasp for most developers if you work with a `row` rather than an `entry`, as databases are row-oriented, and this concept is ubiquitous. The same applies to actions that are carried out with the database: SQL contains the `delete` command. For this reason, you should name its corresponding methods according to these terms and concepts, as shown in Listing 2.17.

```
// Poor example
database.removeEntry(…);

// Better example
database.deleteRow(…);
```

Listing 2.17 Deriving Names from Established Terms and Concepts!

One of the most important rules when finding names is to use existing, common terms. Don't invent new words!

As I've already mentioned several times, good naming conventions make it easier to understand the software. For this reason, you should also pay attention to the correct spellings of names as well as use correct grammar.

Listing 2.18 shows an example of the incorrect plural form for a variable name, which can cause confusion in later code passages because the reader may not assume that only one organization is being used.

```
// Poor example
organizations := repository.GetAll()[0]

// Better example
organization := repository.GetAll()[0]
```

Listing 2.18 Incorrect Pluralization of Variable Names (Go)

In real life, an assignment of identifier types to grammatical functions, for example, in Table 2.1, has become established for the implementation of clearer grammar.

Type	Grammatical Function	Example
Class	▶ Noun ▶ Nominalizations	`Organization` `PaymentMethod`

Variables	▶ Noun
	▶ Combination of verb and noun
		`name`
		`age`
		`isValid`
		`succeeded`

Methods	▶ Verbs
	▶ Combination of verb and noun
		`validate`
		`findAll`
		`canHandle`
		`isValid`
		`ParseTime`

Interfaces	▶ Noun
	▶ Nominalizations
	▶ Adjectives
		`Handler`
		`DataInput`
		`Calculator`
		`Formatter`
		`Runnable`

Table 2.1 Assignment of Identifier Types to Grammatical Functions

The example shown in Listing 2.19 illustrates the use of a different grammatical function that makes the code more difficult to understand. The `Cancellation` method of the `transport` object would be better if named `CancelRequest`, which makes it clear the method cancels a request for this transport object. The `IsCancelled` method is then used to check whether the `Request` object has actually been canceled.

```
// Poor example
transport.Cancellation(request)
cancelled := request.Cancel()
```

```
// Better example
transport.CancelRequest(request)
cancelled := request.IsCancelled()
```

Listing 2.19 Using Grammatical Functions (Go)

Some libraries, especially older ones, often contain interfaces for specifying time spans, such as timeouts, for example. Unfortunately, they often have no indication of the corresponding time unit to be used. For example, Listing 2.20 shows the configuration for *Apache HTTPClient* version 4.3.

```
// Poor example
int timeout = 5000;
RequestConfig config = RequestConfig.custom()
  .setConnectTimeout(timeout)
  .setConnectionRequestTimeout(timeout)
  .setSocketTimeout(timeout).build();
CloseableHttpClient client =
  HttpClientBuilder.create().setDefaultRequestConfig(config).build();
```

Listing 2.20 Apache HTTPClient Version 4.3 Code Example (Java)

Based on the number 5000, you might possibly assume that we are using milliseconds, but it is certainly not clear to the reader at first.

Names should have a corresponding unit as part of the name so that its meaning is clear. Let's adapt our earlier example, as shown in Listing 2.21.

```
int timeoutInMilliseconds = 5000;
RequestConfig config = RequestConfig.custom()
  .setConnectTimeoutInMilliseconds(timeoutInMilliseconds)
  .setConnectionRequestTimeoutInMilliseconds(timeoutInMilliseconds)
  .setSocketTimeoutInMilliseconds(timeoutInMilliseconds).build();
```

Listing 2.21 Extension of the HTTPClient (Java)

The variable has been renamed `timeoutInMilliseconds`, and the methods have been extended by the `InMilliseconds` addition. In this way, the units are known. The additional effort required for longer identifier names should certainly be accepted for better readability.

Units in APIs: A New Version of the HTTPClient

Usually, newer APIs already contain interfaces with a suitable unit. The example of the HTTPClient looks clearer in version 5.x:

```
ConnectionConfig connConfig = ConnectionConfig.custom()
    .setConnectTimeout(timeout, TimeUnit.MILLISECONDS)
    .setSocketTimeout(timeout, TimeUnit.MILLISECONDS)
    .build();
```

A second parameter is used to transfer the corresponding time unit, which is another way to formulate APIs more clearly.

In Go, for example, the `time.Duration` type was defined for the transfer of time periods in the standard library. This special type contains a value with an associated unit. The transfer of a timeout value as a `duration`, for example, therefore always contains a value with an associated unit:

```
context.WithTimeout(ctx, 2 * time.Second)
```

As mentioned earlier, names should be comprehensible and preferably come from dictionaries. Abbreviations or single characters as names can make code more difficult to understand.

However, you can use abbreviations or common single character names if they are actually common and well-known terms, as shown in Listing 2.22. However, if a reader does not know an abbreviation, for example, recognizing the intention behind a code block can be difficult. Abbreviations should therefore be avoided wherever possible.

```
// Possibly problematic abbreviations
ap   //for AccessPoint (?) AccountPeriod (?) ...

// Known abbreviations
url, dns, id
```

Listing 2.22 Example Abbreviations

The situation is similar for single character names. The letters i and j are often used for loop variables, a widespread and common practice.

If the context of the variable's use is larger than a manageable code block and the reader can no longer see the area in which the variable is used all at once, using a descriptive name is important here too: As soon as the reader has to translate the names into their own mental model, the identifiers used are not intuitive enough and should be adapted.

The use of detailed names creates an additional advantage in the code base: It is easier to search for them. Thus, new developers can more easily find their way around the code or navigate to a relevant section more quickly.

In general, all names of identifiers should always be aligned with the reader and have no ambiguities. A name should generally only be assigned once. Jokes or puns must never be included in names! The meaning of a method called goldenCoconut is unclear without contextual knowledge (or if you don't have the same sense of humor as the author) and unnecessarily leads to confusion.

Uncle Bob puts it quite clearly:

"Say what you mean. Mean what you say." —Uncle Bob

Short and Concise Names

As explained in the previous section, names should be as comprehensible as possible and tailored to the audience. On the other hand, the names should not be too long or take up too much space—neither in the source code nor in the reader's mind.

Studies have shown that longer names are more difficult to grasp, memorize, and process. Accordingly, they should be short and concise.

A good name conveys a large amount of information with a small number of letters:

> *"The more you say, the less people remember."* —François Fénelon

Finding the right level of abstraction is essential. The name should not be too abstract. As a result, it would provide little or no relevant information for the target audience. On the other hand, the name should not be too specific and thus provide too much and therefore irrelevant information for the audience.

A comparison with the real world should make this clear: Which of these sentences would you use to introduce yourself (assuming all statements apply)?

1. Hi, I am human.
2. Hi, I ride a bike.
3. Hi, I am a mountain biker and ride a Lux Trail CF7.

With the second sentence, you might identify yourself as a cyclist in a group of athletes. The other two variants would either provide the target audience with too little or too much information in this context. So which level of abstraction you use depends on the audience.

The same applies to programming: Names should be chosen in such a way that they clearly communicate the objective or function of the named object, rather than indicating technical details of its implementation.

Listing 2.23 shows an example with several options for different, alternative method names for the `TaxCalculator` type. The method should calculate the taxes for a product.

```go
type TaxCalculator struct{}

func (tc *TaxCalculator) Process(p Product) {}
func (tc *TaxCalculator) CalculateTax(p Product) {}
func (tc *TaxCalculator) ProductPriceMultipleTaxValue(p Product) {}
func (tc *TaxCalculator) MultiplyValues(p Product) {}
```

Listing 2.23 Selecting a Name in an Implementation (Go)

The first method name, `Process`, seems unsuitable because it is too abstract and does not provide the reader with any information about the action performed or intended. The name `MultiplyValues` is on the other extreme. This name is too specific and describes exactly the internal structure of the method, which basically multiplies the price of a product by a tax rate.

In this example, the name `CalculateTax` best describes the intention or context of the method without publishing its implementation details.

You should therefore check each name assignment to ensure that a name does not publish too many implementation details and is not too abstract. The intention behind an identifier should always be clear.

In some companies, style guides prescribe prefixes or suffixes for identifier names. Class attributes must have the prefix m_, for example, as it is a *member attribute*, or an I should be used as a prefix for interface names. From the point of view of clean code or the principle of short and concise names, such prefixes are not necessary or even disruptive. Current development environments display the corresponding types, and code should generally be structured in such a way that there are no endless stretches of code from which it is no longer clear what is happening and what is hidden behind an identifier.

Classes, methods, or functions should remain so straightforward that prefixes or suffixes and other metadata (e.g., type information) for identifiers are unnecessary.

Name components that do not significantly improve the readability of the names and thus the source code should always be omitted, as shown in Listing 2.24.

```
// Poor example
user.deleteNow();

//Better example
user.delete();
```

Listing 2.24 Superfluous Name Components

In the example shown in Listing 2.24, the addition of Now does add any significant value. The fact that the user is deleted when the method is called is the expected behavior—as long as all other methods do not work asynchronously.

Consistent Names

In addition to the structure of names and their components, their uniform use and a uniform format are important.

If the following rules are observed, consistent names are created:

- Each concept is given exactly one name.
- Similar concepts are named similarly.
- Similar names have a similar structure.

If the same concepts are used in different places in an application, they should also be labeled in the same way. This approach makes your code easier to understand overall, as the recognition of code passages or procedures is increased. Searching for similarities is also easier.

Regardless of the programming language, design patterns provide a standardized language for describing concepts and solutions. Their use, and the use of the uniform names defined there, makes sense and provides an advantage in that the patterns are recognized and recorded more quickly and that an application can be better understood. Listing 2.25 shows an example.

```
// Poor example
CarServiceCreator

// Better example with a known name
CarServiceFactory
```
Listing 2.25 Using the Names of Design Patterns

Every programming language or framework also follows some generally accepted naming conventions, naming patterns, or concepts. Incorporating these conventions into your own code base makes sense and enables other developers to find their way around your code more quickly to gain a deeper understanding of your application. Deviations from the familiar patterns of a language or framework often lead to confusion and misunderstandings among developers.

In the *Java Servlet* specification, for example, the term *filter* is used to describe a component that can influence the flow of HTTP requests and responses (an example of aspect-oriented programming is discussed further in Section 2.6). In Go, the term *middleware* is used for this approach. If you work in a corresponding environment, its commonly used terms should be used.

In the example shown in Listing 2.26, we created a Go http.HandlerFunc named myservlet, which is based on the *Java servlet* specification. Using servlet leads to confusion and thus should be avoided! A name like myHandlerFunc would be better.

```
func main() {
   http.HandleFunc("/", myservlet)
   fmt.Println(http.ListenAndServe(":8081", nil))
}

func myservlet(writer http.ResponseWriter, request *http.Request) {
   fmt.Fprint(writer, "Hello World")
}
```
Listing 2.26 Mixed Concepts Leading to Confusion!

> **Rethink Your Approach When Switching Programming Languages**
>
> Developers often keep their preexisting concepts when changing programming languages and ignore new concepts of the new language or framework are statements, such as "Ah, this code comes from a [insert different programming language here] developer. What is being done here? Why is it like that? But we do things differently here!"
>
> While you may have good reasons for switching to a new programming language, you should familiarize yourself with the new language and its concepts and conventions without neglecting clean code principles.

Some common concepts also include technical terms and the naming of types or attributes. Values with the same meaning must also be named the same in different places.

Listing 2.27 shows two Go `structs` representing a container ship and a sailing ship. Both ships have fields for indicating the home port. Although the fields have the same meaning, the two types have different names: `homeport` and `homeharbor`. A reader may not immediately realize that this is the same information. The same applies to the different names for ships, for instance, `ship` versus `vessel`. While perhaps technically correct, it also potentially leads to confusion.

```go
type ContainerShip struct {
    homeport string
}
type SailingVessel struct {
    homeharbor string
}
```

Listing 2.27 Inconsistent Code with Synonyms (Go)

A more uniform and easier to read conversion is shown in Listing 2.28.

```go
type ContainerShip struct {
    homeport string
}
type SailingShip struct {
    homeport string
}
```

Listing 2.28 Uniform Designations (Go)

The naming of methods or other generic concepts is subject to the same consistency rules: The same or similar concepts should be given the same or similar names.

As shown in Listing 2.29, two methods are defined for two different types, each of which starts a loading or unloading process of a ship (`Run` and `Start`). The fact that synonymous verbs were used for the method names does not clearly indicate that the concepts are analogous.

```
var unloader internal.ShipUnloader
var onboarder internal.ShipOnboarder

onboarder.Run()
unloader.Start()
```

Listing 2.29 Inconsistent Method Names (Java)

By aligning the method names, you reduce potential confusion for new employees. Listing 2.30 shows how to implement this alignment. Table 2.2 contains more examples.

```
var unloader internal.ShipUnloader
var onboarder internal.ShipOnboarder

onboarder.Start()
unloader.Start()
```

Listing 2.30 Consistent Naming for Identical Concepts (Java)

No Consistency	Consistent Names
Differences in grammar	
▶ `ShipService`	▶ `ShipService`
▶ `ContainersService`	▶ `ContainerService`
Antonyms (words with opposite meanings)	
▶ `Open()` and `Kill()`	▶ `Open()` and `Close()`
▶ `Increase()` and `Lower()`	▶ `Increase()` and `Decrease()`
▶ `Acquire()` and `GetRidOff()`	▶ `Acquire()` and `Release()`
▶ `Create()` and `Clean()`	▶ `Create()` and `Destroy()`
Synonyms (words that have the same meaning)	
▶ `Fetch()`, `Get()`, `Load()`, `Find()`	▶ `Fetch()`

Table 2.2 More Examples of Consistent Naming

Creating a *style guide* and automated review for style using what's called a *linter* can help minimize deviations from conventions or known concepts.

> **Style Guides and Linters**
>
> A code style guide is a collection of guidelines and conventions designed to ensure that your code is formatted and structured in a uniform manner. Following a style guide should increase readability and thus lead to a more efficient collaboration within the team, organization, or project.
>
> By using a *linter*, source code can be automatically analyzed for potential errors, security vulnerabilities or style problems, or deviations from the style guide using static code analysis. As part of a build pipeline, a linter ensures that the established rules are adhered to.
>
> Many popular code style guides are available, such as *PEP 8* for Python or the *Google Java Style Guide*, which you can use as a basis for your own style guide. These guides already contain common conventions, and thus only minimal adjustments would be necessary.
>
> Some functionalities in development environments that format code automatically and ensure compliance with a corresponding style guide can also be used for the uniform formatting of the source code. A corresponding configuration can be exchanged and set

> up between the developers. For Go, for example, the Go team already provides formatting tools, which makes a separate configuration unnecessary.

Distinctive Names

To ensure that names for identifiers are comprehensible and concise, they should also not be ambiguous or interchangeable. The meaning of a name should always be clear.

For this reason, words with context-dependent meanings are often unsuitable for names. For example, the word "bank" can be understood as a financial institution or as the land alongside a river or stream. The English language features many words that have different meanings depending on whether they are used as a verb or a noun, which can cause confusion when used in programming. Table 2.3 shows some well-known examples.

Word	Meaning as a Verb	Meaning as a Noun
record	to register	the data record
list	to enumerate or itemize	the list
check	to examine	the check
map	to represent something as a map	a drawing or picture of an area

Table 2.3 Risk of Interchangeable English Words

Good names are easy to distinguish, and you should run no risk of interchanging them with each other. Caution is also advised with words that differ from each other only in individual characters, such as in the following examples:

- *affect* versus *effect*
- *principal* versus *principle*

Programming languages generally distinguish between uppercase and lowercase in the source code. For the compiler, differences in uppercase and lowercase letters or similar-looking characters are not a problem, but for the reader, capitalization can be challenging, as shown in Listing 2.31.

```
var onboarder internal.ShipOnboarder
var onBoarder internal.ShipOnboarder
var onBOarder internal.ShipOnboarder
var onB0arder internal.ShipOnboarder
```

Listing 2.31 Difficult-to-Read Source Code Due to Only Minimal Differences

2.2.2 Defining Functions and Methods

Structured development is essential to design applications efficiently and clearly. The smallest unit of this structure is formed by functions or methods that specifically execute individual tasks and thus divide the code into smaller, reusable blocks.

The readability and comprehensibility of the implementation of those methods and functions is essential. Functions and methods should be implemented in such a way that every reader can quickly and clearly recognize their purpose. A straightforward communication of the intention or task is crucial in order to avoid misunderstandings and ensure traceability for all parties involved.

A few basic principles and rules serve as a basis and orientation for implementation. Functions and methods should fulfill the following prerequisites:

- Names should be meaningful.
- Names should be short, concise, and clear.
- A function or method should fulfill exactly one task.
- A function or method should be on exactly one level of abstraction.
- A function or method should have no side effects.
- A function or method should have suitable parameters.
- A function or method should be user-friendly and "foolproof."

Meaningful Names

I have already covered this topic in detail in the previous section. With methods, an important requirement is that the content of the method should already be clear when reading the name, without having to study the method itself.

Short and Concise

The ideal length of a function or method depends on multiple factors, such as the context or the complexity of the task. No general rule exists for specifying the length beyond which a function or method is considered too long. However, as length increases, internal complexity will also increase, and fully grasping the complete range of functions can be difficult.

Accordingly, your goal should be to implement methods as briefly and concisely as possible in order to maintain readability and comprehensibility:

> "The object programs that live best and longest are those with short methods."
> —Martin Fowler, *Refactoring: Improving the Design of Existing Code*

For example, in his book *Extreme Programming*, Kent Beck suggests a method length of a few lines to a maximum of a few dozen lines. Length specifications such as "maximum one screen page" are often given as a rough guide.

Functions and methods that are formulated briefly and concisely concentrate on their core functionality and can also be reused more easily than complex or detailed implementations in other contexts.

To achieve the goal of clear and concise methods and functions, we recommend the techniques and procedures published by Martin Fowler and other authors in his book *Refactoring: Improving the Design of Existing Code* as a guide.

The goal of *refactoring* actually aims to optimize existing code and make it easier to understand. The authors refer to *code smells* which serve as indicators of potential issues in source code that should be revised accordingly. For each of these "smells," they provide techniques and a suitable target image of how the code should be structured. The target image and the corresponding techniques can serve as guidelines for creating methods.

The following tricks can help keep methods and functions as short and clear as possible:

- Use a method instead of documentation
- Write short block statements
- Maintain low nesting depth

Let's consider these suggestions more closely: Within methods, logical blocks are often created during programming to tackle a complex task in a structured way. If you extract this logic into separate methods, shorter and clearer methods will automatically be created, as shown in Figure 2.5.

As the example shown in Listing 2.32 and Listing 2.33 illustrates, an existing comment above a logical block can indicate that code can and should be extracted into a separate method.

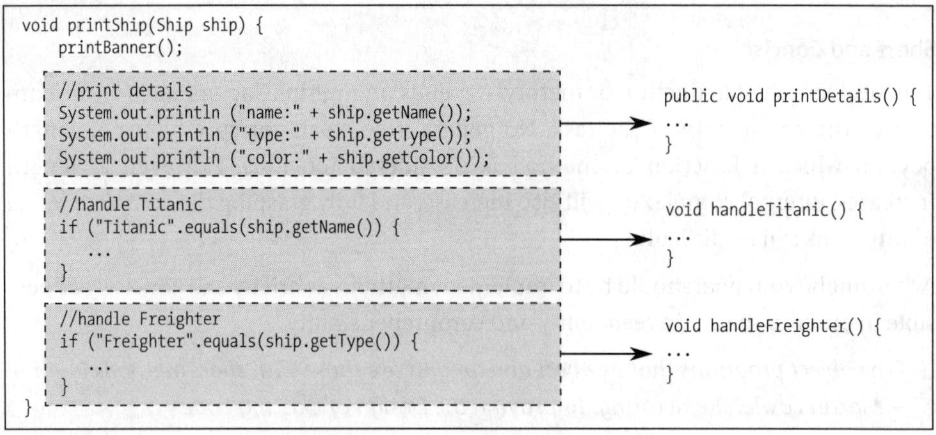

```
void printShip(Ship ship) {
    printBanner();
    //print details
    System.out.println ("name:" + ship.getName());
    System.out.println ("type:" + ship.getType());
    System.out.println ("color:" + ship.getColor());

    //handle Titanic
    if ("Titanic".equals(ship.getName()) {
        ...
    }

    //handle Freighter
    if ("Freighter".equals(ship.getType()) {
        ...
    }
}
```

```
public void printDetails() {
    ...
}

void handleTitanic() {
    ...
}

void handleFreighter() {
    ...
}
```

Figure 2.5 Converting Blocks into Methods

With a reasonably chosen method name, you can also avoid creating documentation for the block, and the resulting code becomes clearer and easier to understand. The reader no longer needs to analyze the entire method line by line to understand its functionality. A corresponding method name should not describe *how* something is done in the method, but *what* is implemented in the method.

```java
void printShip(Ship ship) {
    printBanner();

    //print details
    System.out.println ("name:" + ship.getName());
    System.out.println ("type:" + ship.getType());
    System.out.println ("color:" + ship.getColor());
}
```

Listing 2.32 Long Method with a Comment Above a Block (Java)

The `printShip` method can be shorter and clearer after a corresponding conversion, as shown in Listing 2.33.

```java
void printShip(Ship ship) {
    printBanner();
    printShipDetails(ship);
}
private void printShipDetails(Ship ship) {
    System.out.println("name:" + ship.getName());
    System.out.println("type:" + ship.getType());
    System.out.println("color:" + ship.getColor());
}
```

Listing 2.33 Extracted Block of a Method (Java)

Martin Fowler refers to the extraction of such a logical code block into a separate method as *extract method*.

Not only can longer methods quickly become confusing. High nesting depth and long nested blocks often lead to confusing code. Listing 2.34 shows a cautionary example. Most developers will find it difficult to quickly grasp the task of the code block in this example.

```java
public static void main(String[] args) {
    int[][] data = new int[3][3];
    for (int i = 0; i < data.length; i++) {
        if (data[i][0] % 2 == 0) { // Decision 1
            for (int j = 1; j < data[i].length; j++) {
                if (data[i][j] > 5) { // Decision 2
                    // Nesting 1
                    for (int k = 0; k < data.length; k++) {
```

```java
                        if (data[k][0] == i + 1) { // Decision 3
                            // Nesting 2
                            for (int l = 1; l < data[k].length; l++) {
                                if (data[k][l] < 4) { // Decision 4
                                    System.out.println("found: " + data[i][j]);
                                    return; // Exit the loops
                                }
                            }
                        }
                    }
                }
            }
        }
    }
}
```

Listing 2.34 Difficulty Reading Deeply Nested Code (Java)

During development, the following rules of thumb can help you to keep your code clean:

▶ Allow condition blocks to be only a few lines long. If possible, execute a method call to an extracted method.

▶ Set up a maximum nesting depth of 1 or 2 levels.

In combination with the *extract method* approach, the implemented search logic shown in Listing 2.34 can be converted into the code shown in Listing 2.35.

```java
int[][] matrix = Example.data;

Predicate<Integer> greaterThanFive = i -> i > 5;
Predicate<Integer> smallerThanFour = i -> i < 4;

List<Integer> rowIdx = rowIndexesWithValuesInFirstColumn(matrix, Example2::isEven);
for (int idx : rowIdx) {
    var currentRow = matrix[idx];

    List<Integer> indexesOfValues =
                columnIndexesOfValues(currentRow, greaterThanFive);
    if (indexesOfValues.isEmpty()) {
        continue;
    }

    Predicate<Integer> currentIndexPlusOne = i -> i == idx + 1;
    if (!containsRowWithValues(matrix, currentIndexPlusOne, smallerThanFour)) {
```

```
        continue;
    }

    var indexToUse = indexesOfValues.getFirst();
    System.out.println("Data point found: " + currentRow[indexToUse]);
    return;

}
```
Listing 2.35 Possible Customization of a Nested Block (Java)

Extracting methods does not necessarily mean that the code becomes slower, as is often claimed. Today's compilers optimize the calls of methods and usually carry out the opposite refactoring internally, in what's called the *inline method*.

Your goal should always be readable and clear code. Do not perform any questionable optimizations (i.e., anything you're not sure is actually an optimization).

Exactly One Task, One Level of Abstraction

In well-structured, readable code, each method has a clearly defined task that is implemented by one or more statements within the method. Each of these statements must be at the same level of abstraction. Levels of abstraction can be compared to the zoom levels of a camera: You can focus on different details or keep a broader overview. Mixing different levels of abstraction within a method leads to confusion for the reader and makes it difficult to distinguish between central concepts and details of the method.

The example shown in Listing 2.36 mixes different levels of abstraction. First, an XML string is read and an SQL statement is created at a high level of detail. Secondly, a tax calculation is performed at a business level.

```
public void parseDocumentAndCalculatePriceAndStoreInDB(String data) {
    int beginIdx = data.indexOf("<tax-value>");
    int endIdx = data.indexOf("</tax-value>");
    String taxValueString = data.substring(beginIdx + 11, endIdx);
    BigDecimal taxValue = new BigDecimal(taxValueString);
    beginIdx = data.indexOf("<price-value>");
    endIdx = data.indexOf("</price-value>");
    String productPriceString = data.substring(beginIdx + 13, endIdx);
    BigDecimal productPrice = new BigDecimal(productPriceString);
    String price = String.format("%.2f", taxValue.multiply(productPrice));
    String sqlString = "UPDATE products SET price =";
    sqlString += price;
    sqlString += " WHERE id = ";
    beginIdx = data.indexOf("<product-id>");
    endIdx = data.indexOf("</product-id>");
```

```
    sqlString += data.substring(beginIdx + 12, endIdx);
    Database.execute(sqlString);
}
```

Listing 2.36 Different Levels of Abstraction (Java)

Overall, the code is quite confusing, and even the name of the method (parseDocumentAndCalculatePriceAndStoreInDB) indicates that several tasks and levels have been mixed here. Method names that contain the word And already indicate that more than one clearly defined task has been implemented.

To find the correct or uniform level of abstraction within a method, Robert C. Martin described what he called the *stepdown rule*. Methods should be read or written from top to bottom at a uniform level of detail, and each method called represents another, deeper level of abstraction with details. Uncle Bob also recommends sorting methods within a file according to their abstraction level. In his opinion, the more higher-level concepts or algorithms are included, the higher up in a file the methods should be included.

A possible solution for our previous example is shown in Listing 2.37. The logic and details for parsing the transferred string and for saving the data have been extracted, and the name of the method (updateProductGrossPrice) clearly indicates the task of the method. The Product class was also introduced to bring together the data of a product.

```java
public void updateProductGrossPrice(String data) {
    Product product = Product.fromString(data);
    BigDecimal grossPrice =
            product.getNetPrice().multiply(product.getTaxValue());
    Database.updateGrossPrice(product, grossPrice);
}
```

Listing 2.37 Extracting Details to Separate Abstraction Levels (Java)

Listing 2.38 shows an excerpt of some extracted parsing logic, which in this example has been included directly in the Product class within the static fromString method.

```java
//Not yet complete and perfect example
public class Product {
    private BigDecimal taxValue;
    private BigDecimal netPrice;

    public Product(BigDecimal taxValue, BigDecimal netPrice) {
        this.taxValue = taxValue;
        this.netPrice = netPrice;
    }
```

```java
    public static Product fromString(String data) {
        int beginIdx = data.indexOf("<tax-value>");
        int endIdx = data.indexOf("</tax-value>");
        String taxValueString = data.substring(beginIdx + 11, endIdx);
        BigDecimal taxValue = new BigDecimal(taxValueString);
        beginIdx = data.indexOf("<price-value>");
        endIdx = data.indexOf("</price-value>");
        String productPriceString = data.substring(beginIdx + 13, endIdx);
        BigDecimal productPrice = new BigDecimal(productPriceString);
        return new Product(taxValue, productPrice);
    }
    ...
}
```

Listing 2.38 Excerpt from the Product Class with Parsing Logic (Java)

> **Parsing Logic Within the Class?**
>
> In this example, it would also be possible and make a lot of sense to extract the parsing logic, which is implemented within the fromString method, to a separate class. The factory pattern, for example, which is described in detail in Chapter 4, can be used for this purpose. It can be used for creating objects.

Methods should either change the state of an object or provide data or information about the object, but never both. For a caller, for example, the method from Listing 2.39 can be confusing and lead to errors.

```java
public boolean setTaxValue(BigDecimal taxValue) {
    if (taxValue.signum() == -1) {
        return false;
    }
    this.taxValue = taxValue;
    return true;
}
```

Listing 2.39 Two Tasks in One Method (Java)

In this example, a user of the interface cannot recognize from the method signature that a return value must be evaluated in order to determine that the value has not been accepted. Listing 2.40 shows the incorrect use and the correct use of the method since only one variant is checked as to whether the transferred value is valid.

```java
Product product = new Product();
product.setTaxValue(new BigDecimal(-1));
```

```
// or "correctly"
boolean valid = product.setTaxValue(new BegDecimal(-1));
if (!valid) {
    ...
}
```

Listing 2.40 Correctly and Incorrectly Calling the Set Method (Java)

The division into two methods makes sense in this example. One method could check the validity of the value, and the second method would only be used to set the new value. Listing 2.41 shows the application of the customized interface.

```
if (isValidTax(val)) {
   product.setTaxValue(val);
}
```

Listing 2.41 Division into Two Methods (Java)

The principle of splitting a task into separate action and query methods is often referred to as the *command query separation* principle.

To sum up, to create better-structured methods, you should keep the following points in mind:

- Find the main task or responsibility of the method.
- Do not mix tasks within a method.
- A method should only be on one level of abstraction. Move details to other methods.
- Method names containing And are a *code smell* and should therefore be avoided.
- Methods should either execute an action or answer requests.

No Side Effects

The tasks and activities of a method must be clearly defined and communicated. This clarity can be achieved via the documentation or via a meaningful method name. Hidden, undocumented side effects can cause confusion, make errors difficult to catch, and create unwanted dependencies between different parts of your code.

Unwanted and unexpected behaviors when calling a method might arise, for example, by the following undocumented internal actions:

- Unexpected changes to fields in your own class
- Calls or changes to global resources
- Unexpected changes to transferred parameters

The example shown in Listing 2.42 illustrates how some actions may be unexpected for the caller.

```java
public boolean isValid(Product product) {
    if (product.getId() == null) {
        // Change to parameter!
        product.setId(String.valueOf(Math.random()));
        return false;
    }
    if (...) {
        return false;
    }
    // Unexpected change to a global resource
    SessionCache.initWithProduct(product);
    return true;
}
```

Listing 2.42 A Method with Side Effects (Java)

The `isValid` method name suggests that the method checks the passed `Product` object for validity. However, if the `product` does not have an `Id`, then the ID will be set with a new, random value:

`product.setId(String.valueOf(Math.random()));`

In this example, the globally defined and accessible `SessionCache` will also be initialized with the `product` within the code block if the check is successful. This approach, too, can lead to confusing behavior for the caller and may force it to call the API in a fixed order. The global `SessionCache` may need to be initialized at a fixed time, and calling the `isValid` method is the only way to do this.

These temporal and implicit, unexpected dependencies, which are caused by unwanted side effects, are confusing to callers. Dependencies can be minimized and made more transparent by clearly defining the task of a method and adhering to the levels of abstraction I have already described.

Renaming the `isValid` method to `validateAndInitSession` makes the side effect of the dependency more visible but violates the *single responsibility principle*, which states that each method should only have one task or better: one reason for a change.

One possible solution could be to remove the setting of the `Id` and the call of the `SessionCache` from the method and to create a new enclosing method with a more suitable name. The task of the new method would then be to initialize a new session with a valid product. Accordingly, its name could be `initializeSessionWithValidProduct`, for example.

Suitable Parameters

Methods and functions work with data. They analyze this data, calculate new information from it, or change its structure. Parameters and fields are used to transfer data to methods. In object-oriented languages, fields that represent the properties of an object can be used, while in other programming languages parameters are used to pass data directly to a method.

However, each additional parameter of a method increases its complexity and makes the method more difficult to read, more incomprehensible, and more complex to test. Ideally, methods should therefore have no parameters or as few parameters as possible. You should avoid methods with three or more parameters altogether if possible or only use them sparingly and with good reason.

In the following sections, we'll provide a few tips to keep the parameters of methods clear and small in number.

Methods with One Parameter

Methods with only one parameter should be used for questions or for the transformation of one parameter into another. For example, the following method clearly answers a question with one parameter: `boolean isValid(myObject)`. A method with the `File openFile("myFile.txt")` signature, for example, transforms a string parameter into a corresponding file handle and returns it.

As mentioned, it makes sense for each method to process either a task or a query, but never both at the same time. This applies in particular in the context of a parameter transfer. Always make a clear statement as to which task or query a method should perform.

If a method only has one `boolean` parameter, this method can and should be completely replaced by two corresponding, more readable methods, as shown in Listing 2.43.

```
// Boolean parameter only
void run(boolean full);

// replace with
void runSingle();
void runFull();
```

Listing 2.43 Avoiding Single Boolean Parameters (Java)

Methods with Multiple Parameters

If methods have more than one parameter, they tend to become more confusing—especially if the order of the parameters plays a role. The example shown in Listing 2.44 illustrates a corresponding problem from the *JUnit* test framework, that is, a method with two parameters of the same data type.

```
// Assert that expected and actual are equal.
static void assertEquals(int expected, int actual)
```

Listing 2.44 Example of the JUnit API (Java)

This method compares the two transferred values to see whether they are the same and, if there is a discrepancy, reports a test error with information about which value was expected and what value was actually transferred. Without prior knowledge or the API documentation at hand, recognizing which parameter is the expected value versus the actual value for the test comparison can be difficult when using the method in this example.

Meaningful Method Names

Meaningful method names that clearly describe the task of the method and the parameters with their potential sequence can make your source code easier to understand. For methods with one parameter, for example, verb/noun combinations can be formed, as shown in Listing 2.45. For methods with multiple parameters, a useful approach is to use the parameter names as part of the method's name.

```
// Method with one parameter
String serialize(Person person);

// Method with one parameter
String serializePerson(Person person);
```

Listing 2.45 Using a Parameter Name as Part of the Method's Name (Java)

In this example from JUnit, for instance, the `assertExpectedEqualsActual` name could have been used to clearly communicate the order of the parameters and save the reader from having to jump to the API.

> **An Alternative Solution for More Readable JUnit Test Code**
>
> To create more readable test code with *JUnit*, you can use popular external libraries, such as *AssertJ* or *Hamcrest*, in addition to using a custom API. The libraries have a more modern approach and a more readable API.
>
> If you use the *Hamcrest* library and its *matcher*, you can formulate test comparisons, for example, in the following way:
>
> `assertThat(calculator.subtract(4, 1), is(equalTo(3)));`
>
> If you use AssertJ, a test for comparison can be expressed in the following way:
>
> `assertThat(frodo).isNotEqualTo(sauron);`

As I have already mentioned, the complexity of methods increases with an increasing number of parameters. To keep the number of parameters as manageable as possible, we recommend refactoring according to Martin Fowler's book *Refactoring: Improving the Design of Existing Code*. The *long parameter list* code smell can be cleaned up using the following refactoring approaches, for example:

- Replace a parameter with a method
- Introduce a parameter object
- Preserve a whole object

I would like to briefly present the "target states" of these refactoring techniques next. These techniques can become the basis for clean development and prevent *code smells* from arising.

Replace a Parameter with a Method

If a value is determined by a method call and then passed as a parameter to a method, this call can be moved to the called method. Listing 2.46 shows an example.

```
price := 1.20
taxValue := TaxValue()
grossPrice := CalculateGrossPrice(price, taxValue)
```

Listing 2.46 Returning the Value of a Method Call as a Parameter (Go)

The call of the `TaxValue` method, which determines the tax rate, can be moved to the `CalculateGrossPrice` method itself, as shown in Listing 2.47.

```
price := 1.20
grossPrice := CalculateGrossPrice(price)
```

Listing 2.47 Method Call Moved to the Method (Go)

This technique is only possible if the following prerequisites are fulfilled:

- The transferred value is not required multiple times in the calling method, and multiple calls would lead to different values.
- The parameter value of the moved method is the same each time it is called or, ideally, the method has no parameters and always returns the same value.
- The called method has a reference to the method to be called.

Introduce a Parameter Object

If multiple logically related parameters are passed to a method, these parameters can be combined into a single parameter object.

Listing 2.48 shows an example of an `UpdatePosition` method in which the values for the length and width of the position are passed and stored separately.

```
type Ship struct {
    name string
    lat  float64
    lon  float64
}
```

```go
func (s *Ship) UpdatePosition(lat, lon float64) {
    s.lat = lat
    s.lon = lon
}
```

Listing 2.48 Changing a Position Using Latitude and Longitude (Go)

Alternatively, the individual values can be combined into a summarizing parameter object and passed as a single object to the method. In the code shown in Listing 2.49, I have adapted the example with a `Position` parameter object, which is also used internally to store the data.

```go
type Position struct {
    Lat float64
    Lon float64
}

type Ship struct {
    name     string
    position Position
}

func (s *Ship) UpdatePosition(pos Position) {
    s.position = pos
}
```

Listing 2.49 Introduction of a Parameter Object (Go)

If the parameter object is also used as a parameter in various methods, the class interface becomes more consistent overall and therefore easier to understand.

Code that previously had to be implemented in the individual methods can now be implemented and shared in the parameter object.

Preserve the Whole Object

For methods requiring several values that all originate from the same object, consider passing the complete object as a parameter instead of its individual values.

The example shown in Listing 2.50 demonstrates the `calculateDistance` method, which calculates the distance between two points in kilometers. The parameters used are the latitude and altitude of the start and destination points, which are passed individually as parameters.

```go
type Ship struct {
    name                 string
    position             Position
    traveledDistanceInKm int
```

2 Principles of Good Software Design

```go
}

func (s *Ship) UpdatePosition(pos Position) {
    s.position = pos
    if s.position.Lat == 0 {
        return
    }
    s.traveledDistanceInKm += calculateDistance(s.position.Lat,
                                                s.position.Lon,
                                                pos.Lat,
                                                pos.Lon)
}

//Too many individual parameters! Room for improvement!
func calculateDistance(startLat, startLon, endLat, endLon float64) int {
    start := geodist.Coord{Lat: startLat, Lon: startLon}
    target := geodist.Coord{Lat: endLat, Lon: endLon}

    _, km, _ := geodist.VincentyDistance(start, target)
    return int(km)
}
```

Listing 2.50 Single Value Transfer to a Method That Could Be Improved (Go)

The parameters of the method come in pairs from the Position class and should not be passed individually after *Preserve Whole Object refactoring* but instead passed as an entire object. Listing 2.51 shows an adapted variant with only two parameters, which is much clearer.

```go
//Better variant
func calculateDistance(startPos Position, endPos Position) int {
    start := geodist.Coord{Lat: startPos.Lat, Lon: startPos.Lon}
    target := geodist.Coord{Lat: endPos.Lat, Lon: endPos.Lon}

    _, km, _ := geodist.VincentyDistance(start, target)
    return int(km)
}
```

Listing 2.51 Using Position Objects for Parameter Transfer (Go)

This example can be adapted even further: For example, the Movement struct could be used to combine the two Position objects into one Movement, as shown in Listing 2.52.

```go
type Movement struct {
    from Position
    to   Position
```

```go
}
func calculateDistanceWithMovement(movement Movement) int {
    start := geodist.Coord{Lat: movement.from.Lat,
                           Lon: movement.from.Lon}
    target := geodist.Coord{Lat: movement.to.Lat,
                            Lon: movement.to.Lon}
    _, km, _ := geodist.VincentyDistance(start, target)
    return int(km)
}
```

Listing 2.52 Introducing the Movement Type (Go)

A further step towards better code readability could be to move the method for calculating the distance between the points to the Movement type. The corresponding refactoring is referred to as *Move Method* refactoring and results in the code shown in Listing 2.53.

```go
type Movement struct {
    from Position
    to   Position
}

func NewMovement(fromPosition Position, toPosition Position) *Movement {
    return &Movement{from: fromPosition, to: toPosition}
}}

func (m *Movement) calculateDistanceinKm() int {
    start := geodist.Coord{Lat: m.from.Lat, Lon: m.from.Lon}
    target := geodist.Coord{Lat: m.to.Lat, Lon: m.to.Lon}

    _, km, _ := geodist.VincentyDistance(start, target)
    return int(km)
}

type Ship struct {
    name                string
    position            Position
    traveledDistanceInKm int
}

func (s *Ship) UpdatePosition(pos Position) {
    s.position = pos
    if s.position.Lat == 0 {
        return
    }
```

```
        s.traveledDistanceInKm +=
            NewMovement(s.position, pos).calculateDistanceinKm()
}
```
Listing 2.53 Move Method Refactoring of the Distance Calculation (Go)

2.2.3 Don't Repeat Yourself

The *don't repeat yourself (DRY)* principle is a fundamental concept in programming, which states that code duplication should be avoided wherever possible. If the same code is used multiple times, that code should be maintained centrally in one place and made reusable.

Without repetition, code becomes more compact, more readable, and easier to maintain since changes, extensions, or necessary bug fixes only need to be made in one place and since blocks are easier to capture by combining and storing them in methods. Centralized and reusable code also contributes to greater speed in software development, as the same task doesn't need to be implemented in every project or module. Robert C. Martin agrees:

> *"Duplication may be the root of all evil in software."* —Robert C. Martin

Repetition in the code can be minimized using the following techniques and approaches:

- Functions or methods: The extraction of logic in functions or methods is the basic framework for reuse.
- Objects and abstraction: With the help of classes, data and functionality can be summarized and used as a blueprint for objects. Abstractions also make it possible to combine the functionality of multiple classes and thus promote reusability—even with calling code.
- Modules or libraries: The extraction of functionality in modules or libraries enables other applications to use existing implementations. This saves development work.
- Configurations: If a configuration is extracted from the code, code can be adapted more flexibly to different scenarios.
- Extension mechanisms: Classes that can be extended via derivations or so-called plugins enable better reuse.
- Templates: When generating HTML pages or data reports, for example, general structures can be defined and filled differently using placeholders.

Do not overuse DRY: Code should not be made unnecessarily complex for the sake of reusing a few lines, and you should not extract or summarize each functionality in separate classes. Every time code is extracted, a dependency is created between modules or classes, which in turn can also cause problems.

The example shown in Figure 2.6 illustrates a potential problem: An application uses `Library Y` library in version 1. `Library Y` in turn uses `Library X`, also in version 1.

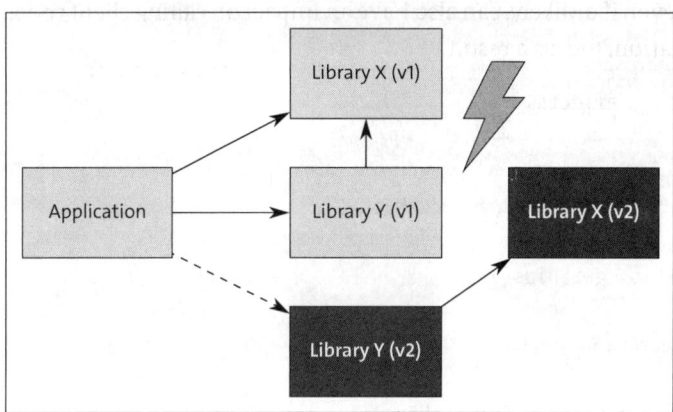

Figure 2.6 Potential Dependency Issues with Updates

If an update of `Library X` to version 2 becomes necessary but the library can only be used with `Library Y` in a new version 2, an application may require adaptation to a degree not previously considered since neither incompatible version can be used at the same time. This incompatibility leads to higher development costs and potentially more errors in areas that are not directly affected by the update.

Some final advice from Rob Pike, who became known as a codeveloper of Go and a long-time researcher at Bell Labs and Google:

> "A little copying is better than a little dependency." —Rob Pike

2.2.4 Establishing Clear Boundaries Between Components

In most cases, we don't develop an entire software system ourselves but instead rely on external libraries to solve specific tasks efficiently. A crucial step is that these libraries, regardless of their origin, must be integrated cleanly and correctly into your own software so you can continue to guarantee the quality and stability of your software.

Each integration of an interface of an external module inevitably leads to a closer coupling and dependency between the two modules involved. If you do not control these dependent interfaces yourself, dependency problems may arise, such as unwanted side effects or your own application being controlled by the external dependencies. Therefore, establishing appropriate boundaries or barriers to external dependencies makes sense.

If an external data type, such as the `java.util.Map` interface shown in Listing 2.54, is returned as return type and is therefore part of your own exported interface (`getShips` and `setShips`), the user of the class also gains access to the complete interface. Therefore, the user of this class could, in this example, directly create ships or completely clear the map using the `clear` method. In this case, the internal details of the `Harbor` class are not protected against external access.

Changes to the interface, even if unlikely, can also have an impact on calling client code, which then require adaptation, too, as a result.

```java
// Example of possible side effects
public class Harbor {

    private Map<String, Ship> ships = new HashMap<>();

    public Map<String, Ship> getShips() {
        return ships;
    }

    public void setShips(Map<String, Ship> ships) {
        this.ships = ships;
    }
}
```
Listing 2.54 Unintentional External Dependency in an Interface (Java)

Listing 2.55 shows unintentional access to the exported `Map` interface and the resulting possible call of the `clear` method.

```java
Ship titanic = new Ship("Titanic", "Cruise ship", "blue");
Map<String, Ship> ships = new HashMap<>();
ships.put("Titanic", titanic);

Harbor shouthampton = new Harbor();
shouthampton.setShips(ships);

// Unwanted access!
shouthampton.getShips().clear();
```
Listing 2.55 Published Interface Causing Errors (Java)

The problem of unwanted access is compounded by the extensive interface. Interface providers are usually interested in offering the broadest possible generic interface, with many methods, so that their implementation can be used in different areas of applications. Clients, on the other hand, are usually only interested in a small section of the interface (see also the *interface segregation principle*). This conflict of interest can be resolved by defining a separate interface for each client.

For our current example, as shown in Listing 2.56, the new `Ships` class can be created, which encapsulates access to the `Map` interface and provides a separate interface optimized for the application. The internal management of all ships is still carried out via the external Map dependency, but the interface for your own clients is defined by you, and internal details are hidden from callers. Access can no longer be made directly via the Map interface. In addition, the effects of changes to the external library are minimized because this dependency is now only an internal detail and is no longer exported.

```java
public class Ships {

    private final Map<String, Ship> ships = new HashMap<>();

    public void addShip(Ship ship) {
        ships.put(ship.getName(), ship);
    }
    public Ship getShip(String name) {
        return ships.get(name);
    }
}
```

Listing 2.56 Ships Class Encapsulating the External Map Interface (Java)

By encapsulating external libraries or modules in their own classes, you can create clear distinctions between internal and external components. These implementations, often referred to as *abstraction boundaries* or *crack stoppers*, prevent the propagation of forced changes that might be triggered by external libraries.

2.2.5 The Broken Windows Theory and the Boy Scout Rule in Software

In 1982, James Q. Wilson and George L. Kelling advanced the *broken windows theory*, which originated in the field of criminology. This theory states that visible damage to an environment, such as a broken window, is a sign of neglect and crime and that these signs can lead to further offenses. Experiments and research have shown that people pay less attention to neglected or damaged items than to well-maintained ones. Things that were already damaged continued to be destroyed much more quickly.

In software development, the broken window theory can mean that poor design, a small bug, or some inaccurate documentation may seem insignificant at first, but over time, these small errors lead to more serious problems, such as larger bugs or a deteriorating software architecture, as one may feel that another shortcoming or bad architectural decision is no longer a big deal. Ignoring small errors or shortcomings can give team members the impression that quality and maintenance are not important. This lax attitude results in more bugs, and more poorly structured software components, and the whole project becomes a mess. Constantly correcting errors that arise in this way can greatly slow down development and take focus away from new functions.

One tenet of the worldwide Scouting movement, founded by Robert Baden Powell, is that you should leave the world a better place than you found it.

This rule can be observed and applied in a variety of ways during programming, such as the following:

- Address inadequacies directly or fix them immediately (e.g., typos, missing documentation, missing tests, and so on)
- Foster a culture of *continuous improvement* for constant further development
- Enable early error detection through highly automated test coverage
- Perform regular *code reviews*
- Ensure architecture-related decisions are transparent

The broken windows theory demonstrates the importance of quality assurance throughout the entire software development process. It is better to fix minor problems quickly than to allow them to multiply over time, resulting in major problems later. This kind of foresight is also a key to good software design.

2.3 SOLID Principles

The acronym SOLID refers to five principles of software design compiled and popularized by Robert C. Martin.

In 2002, Martin's published the book *Agile Software Development: Principles, Patterns, and Practices* in which he provided several design principles for agile software. Two years later, in 2004, Michael Feathers, another author in the software sector, emailed Martin that, if the guidelines were rearranged, their initial letters would result in the SOLID acronym. Martin then rearranged the order of the principles, and the well-known acronym was born.

The individual principles Martin described were based on the work of many others as well as on the experience that "Uncle Bob" himself had gained in various software projects.

SOLID principles are intended to develop stable, maintainable, and expandable software and thus contribute to a longer service life for applications. However, the use of these principles is not only useful at the source code level, as Martin had originally conceived: In the age of microservices and distributed applications, these principles must also be interpreted and used for system architectures and system designs.

The five principles are:

- Single responsibility principle (SRP)
- Open-closed principle (OCP)
- Liskov substitution principle (LSP)
- Interface segregation principle (ISP)
- Dependency inversion principle (DIP)

The following sections describe these principles in more detail and explain their use through examples.

2.3.1 The Single Responsibility Principle

The *single responsibility principle* expresses the fact that only one responsible client, stakeholder, or group should exist for a module. In other words:

> "A module should be responsible to one, and only one, actor." —Robert C. Martin

Only this one group of people with the same interests can initiate changes to this module or be the reason for a change.

The term *module* is quite broad for the single responsibility principle. This term might refer to an individual source code file or to a larger, but still coherent, collection of functions and data.

The term *cohesion* is often used to describe the interconnection among individual parts—in this case, between functions and data. Figuratively speaking, the responsibility for a module is to serve as the adhesive that represents or controls the cohesion of the individual parts.

> "This principle is about people." —Robert C. Martin

Let's illustrate this principle using an example, the Product class, as shown in Figure 2.7. This class includes the CalculateTax and GetHTMLRepresentation methods as well as the WebStore and Accounting actors.

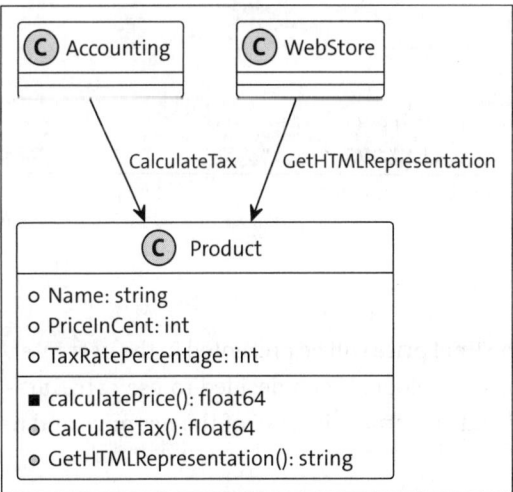

Figure 2.7 Problems Arising Without the Single Responsibility Principle

The CalculateTax method was defined by the Accounting actor and is only called by the latter. The WebStore actor is responsible for the specification and implementation of the GetHTMLRepresentation method and has expanded the Product class accordingly.

With its structure, the Product class violates the single responsibility principle since multiple actors share responsibility for the class and can initiate adjustments. Combining the source code of multiple actors into one class results in a *tight coupling* between actors that are otherwise meant to be independent.

If changes are later required for the WebShop, errors may arise in the logic of another actor due to possibly sharing existing source code.

In this example, the inner private calculatePrice method determines a price from both methods, as shown in Listing 2.57. In the HTML interface of the store, the price is displayed next to the tax rate, and the accounting team uses the value as the basis for calculating the overall price.

```go
type Product struct {
    Name               string
    PriceInCent        int
    TaxRatePercentage  int
}

func (p *Product) calculatePriceInCent() int {
    // extensive logic!
    return p.PriceInCent
}

func (p *Product) CalculateTax() int {
    return p.calculatePriceInCent()/100 + p.TaxRatePercentage*p.TaxRatePercentage
}

func (p *Product) GetHTMLRepresentation() string {
    return fmt.Sprintf("<div>%d cent plus %d %% VAT</div>",
        p.calculatePriceInCent(), p.TaxRatePercentage)
}
```

Listing 2.57 A Shared Function (Go)

By expanding the store interface, only the final price will be presented in the user interface with immediate effect. The web store developers then decided to use calculatePrice to determine the overall price for end customers instead of the net price and to display this directly.

The tests carried out by the store team are successful. However, the developers did not realize that the adjustment also had an impact on the accounting team. In its logic, value-added tax is added twice in the tax calculation because the tax is calculated on the basis of the calculatePrice method, which no longer provides a net price. At this stage, deploying the application without further testing would be fatal.

The more code is managed by several teams together, the more likely that changes made by one team will impact another. Even if the changes made by the two teams affect different functionalities, several teams might be working in parallel on the same source code file. Then, conflicts may arise in the version management system during the *merging* of the various changes. If these conflicts are not cleared up correctly, even more errors will occur.

The single responsibility principle can help solve this problem. Several implementations are possible that differ only in the distribution of the data and the functionality of the classes concerned. In all variants, however, the logic required by each actor is extracted into separate components and can be maintained there separately.

Complete Separation of Data and Functionality

Figure 2.8 shows an implementation in which the data (i.e., the attributes) of the Product class are completely separated from its functionality.

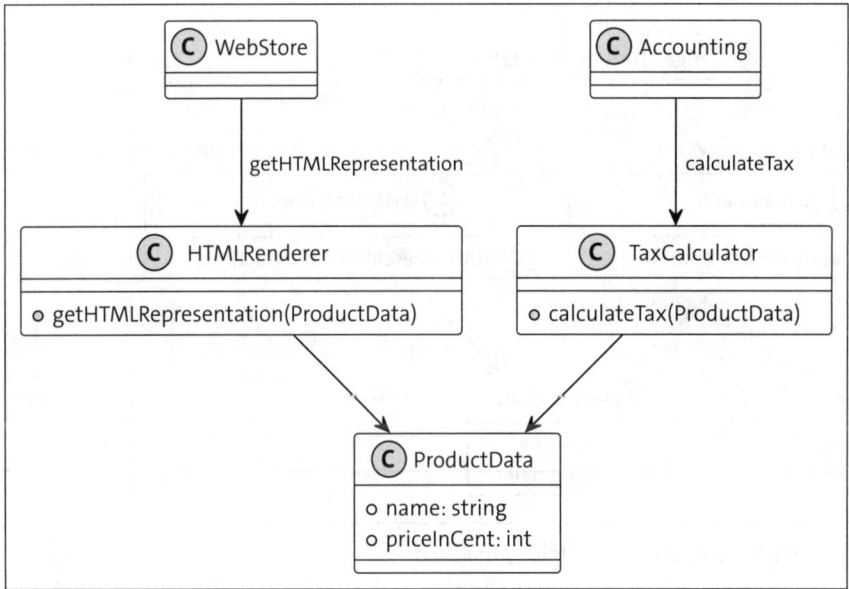

Figure 2.8 Example of the Single Responsibility Principle

The logic of each actor is implemented in a separate class, and both implementations access the ProductData class, which acts as a data container.

Each actor is responsible for its own class, and changes to any actor only have a local effect on their own subarea.

One disadvantage of implementing this solution is that the web store developers and the accounting developers must each know their corresponding data and logic classes. In our example, the WebStore must explicitly know, instantiate, and use the HTMLRender and ProductData classes.

2 Principles of Good Software Design

Separation of Functionalities and the Use of a Facade

As an alternative to a complete separation and disclosure of the details of the Product class, what's called a *facade* can be included, as shown in Figure 2.9. This element contains code for instantiating the data and actor logic classes as well as for redirecting the calls to the corresponding methods in the logic classes. With this approach, the data is still stored in the separate ProductData class, but this class no longer needs to be known to the caller.

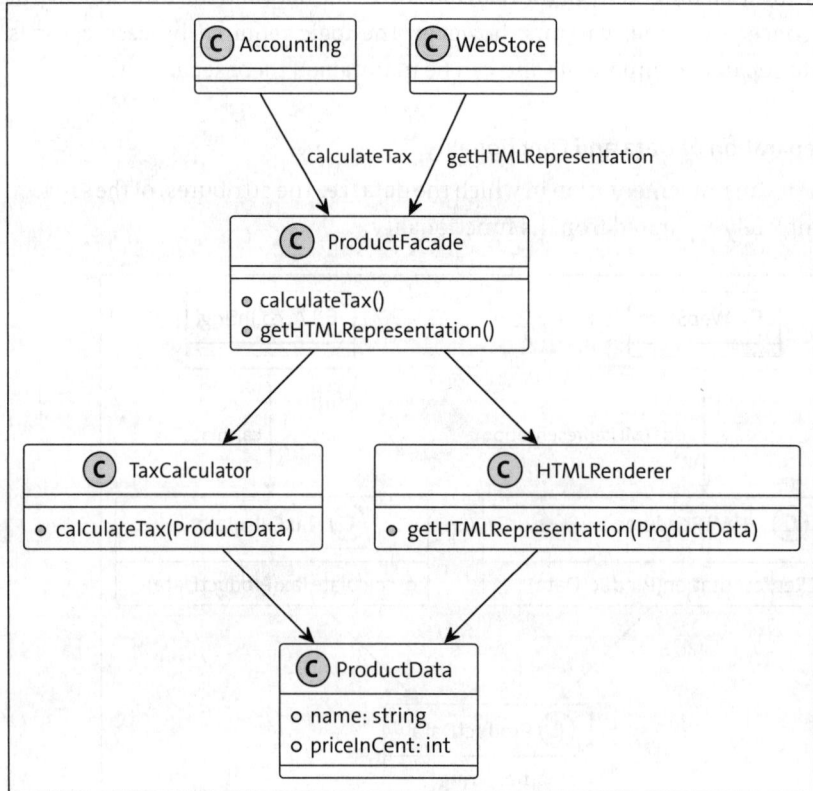

Figure 2.9 Single Responsibility Principle with Facade

One advantage of this approach is the simpler interface for the WebStore and Accounting classes, whose responsible developers now don't need to deal with the details of instantiating or using the data and logic classes. For these developers, the interface remains easy to use.

Data Object as a Facade to Separate the Functionality

A third implementation variant is to use the data object itself as a facade and thus also create the possibility of implementing important functionality within this class. Figure 2.10 shows a corresponding class diagram.

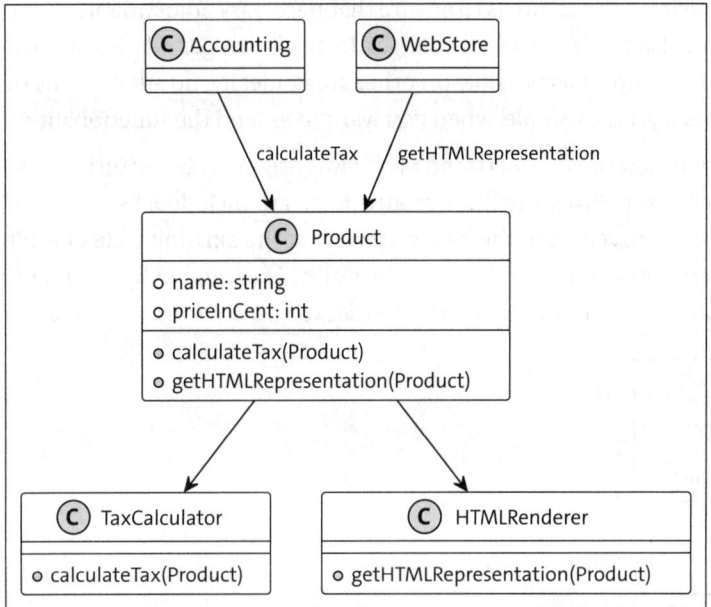

Figure 2.10 Single Responsibility Principle with Data Object as Facade

With this approach, the `WebStore` and `Accounting` classes can continue to work with a simple interface, and changes to the extracted logics of the individual actors do not affect each other.

Advantages of the Single Responsibility Principle

For each implementation variant, the logic is implemented in a separate module or class, thus *encapsulating* the functionality accordingly. This encapsulation improves the readability and maintainability of the resulting code. Errors within the logic classes, for example, can be identified more easily and rectified accordingly.

If functionality is clearly separated and a clear responsibility is defined, it is possible to reuse a logic class more easily in other application areas.

2.3.2   The Open-Closed Principle

The *open-closed principle*, published as early as 1988 by Bertrand Meyer in his book *Object-Oriented Software Construction*, expresses the idea that software should be written in such a way that it can be extended without affecting existing functionality.

> "A software artifact should be open for extension but closed for modification."
> —Bertrand Meyer

Rarely is software developed, completed, and then adapted in a single step, never to be touched again. The requirements placed on software will change over time, and adjustments often become necessary.

2 Principles of Good Software Design

The goal of a good software architecture is to ensure that necessary adjustments to the system do not repeatedly lead to far-reaching changes. Instead, changes relate to a small sub-area and have as little impact as possible on other areas. Ideally, no adaptations to existing code are necessary, for example, when you want to extend the functionality.

The open-closed principle described by Bertrand Meyer was originally based on the concept of class inheritance. New functionalities or attributes are included, as shown in Figure 2.11, in new classes derived from the base class so that the existing code doesn't require adaptation. In our example, the `OriginalClass` class is extended by an inheritance hierarchy with the `attributeB` attribute and the `doExtendedSomething` method.

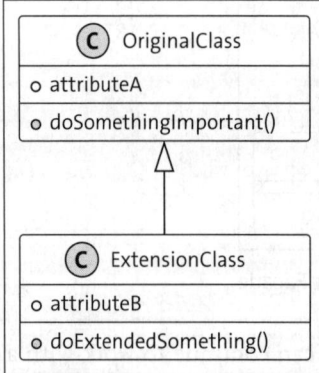

Figure 2.11 The Open-Closed Principle Using Inheritance

Over time, the implementation of the open-closed principle has shifted from a primary emphasis on inheritance to the use of polymorphism. The reason for this was, among others, Robert C. Martin and Joshua Bloch, the author of the book *Effective Java*, who repeatedly pointed out how the tight coupling introduced by the class inheritance has a negative impact on dependencies.

As shown in Figure 2.12, the polymorphic approach uses interfaces for which multiple implementations may exist. These interfaces are no longer changed, and functionality can be extended via additional concrete classes.

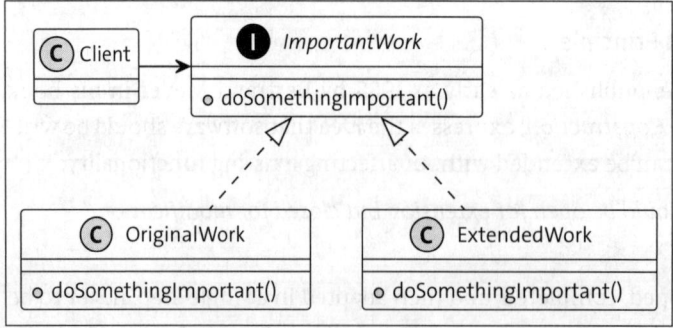

Figure 2.12 The Open-Closed Principle Through Polymorphism

The introduction of interfaces leads to a new level of abstraction in the application and thus enables a loose coupling of the individual independent components that implement the corresponding logic.

The Open-Closed Principle at the Architecture Level

Let's now turn to an extended example of developing some training management software to demonstrate the possible use of the open-closed principle in software architecture: We want this software to set up an internet presence with training courses. The required training data is stored in a database from which it should be read. In addition to displaying an HTML representation of a course in the browser, a user should be able to create a PDF document for each training course, the layout of which differs from that of the website.

Considering the single responsibility principle and its associated separation of responsibilities, a rough draft is created for our example, as shown in Figure 2.13. Even if the representation is not entirely UML-compliant, it should be made clear that the responsibility for loading from the database lies with the DBTraining component, and the responsibility for preparing the data for display lies with the PresentationView component. Two additional components handle the output as HTML and PDF (HTMLPresenter and PDFPresenter).

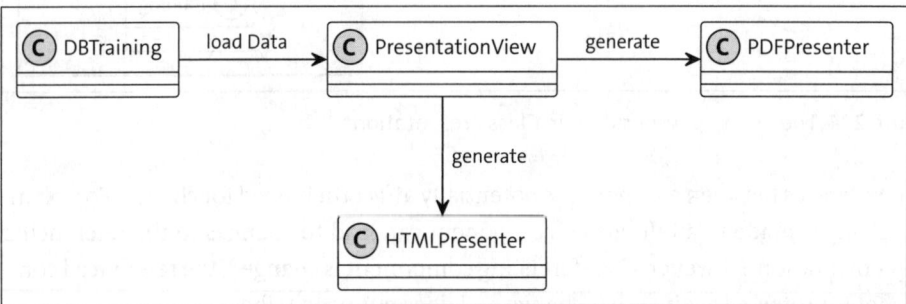

Figure 2.13 Basic Outline of the Training Application

Once the rough division of responsibilities has been made, the first classes and their dependencies can be defined.

Figure 2.14 shows the Controller package that can be created based on these considerations. Within it, the GeneratorController class and the PresentationGenerator and TrainingRepository interfaces are defined. The GeneratorController controls the entire generation process and uses a TrainingRepository and a PresentationGenerator implementation for this purpose.

This class diagram shows the dependencies that exist between the components and the division of these components into *higher and lower application layers*. A lower application layer has dependencies on a higher layer, but higher layers are never dependent on one of the lower layers. Each dependency in the diagram is visualized as a directional arrow and points exclusively in the direction of the dependency used. The principle of *dependency inversion* used in this context is discussed in more detail in Section 2.3.5.

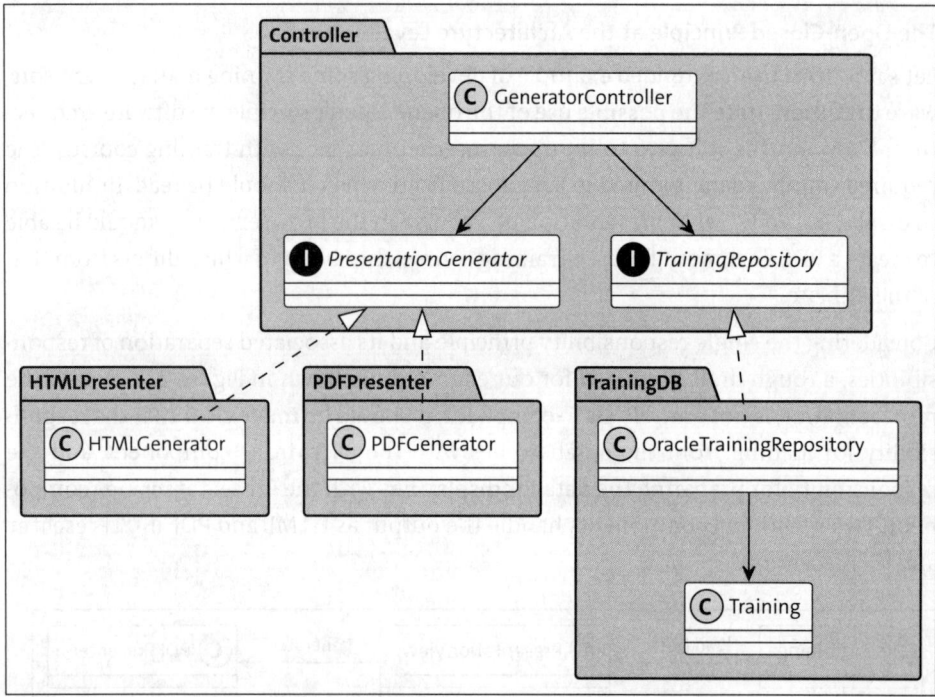

Figure 2.14 The Training Application in Class Presentation

Dependencies between components potentially affect their need for change. For example, changes made to a referenced component can lead to changes to the referencing implementation. However, if a referencing component is changed, the referenced component is protected against that change and does not need to be adjusted.

> *"If a component A should be protected from changes in component B, then component B should depend on component A."* —Robert C. Martin

For our example, this means that the controller package is located in a higher application layer because it has no further dependencies on other packages. The HTMLPresenter, PDFPresenter, and TrainingDB packages are located in a lower layer because they each have dependencies in the form of interface definitions in the higher layer of the controller package. If, for example, the implementation of the HTML output changes, this change will have no effect on the controller package. The same applies to the implementation of the PDF output. Changes within the control logic of the GeneratorController are considered unlikely in the example.

The division of a system and its components into dependency layers implements the open-closed principle at the architecture level. Higher layers are protected from changes in lower layers and cannot be changed by them. Extensions can be made by using polymorphism or, in other words, by implementing interfaces of a higher layer. Our example could also be extended with a `WordPresenter` without having to adapt the logic within the `controller` package.

The Options Pattern in Go: An Example of the Open-Closed Principle

The *options pattern* is a design practice in Go for creating functions with a variable number of parameters or with default values, without compromising the readability and extensibility of the code. In Go, this technique is often used to implement *constructors* or *factory methods* of structs with optional parameters.

> **The Options Pattern in Other Programming Languages**
>
> The *options pattern* can also be found in other programming languages. The *builder pattern*, which we describe in Chapter 4, Section 4.2, is also an implementation of the open-closed principle.

The following example shows the motivation for this pattern: New instances of the `Config` type can be created, as shown in Listing 2.58, using the `NewServerConfig` function. A list of `string` values is passed as a parameter for the generation.

```go
func NewServerConfig(param1, param2, param3, param3 string) Config {
    //doSth
}
```

Listing 2.58 Factory Method for Creating Objects Without the Open-Closed Principle (Go)

Without suitable parameter names, recognizing the semantics of the individual parameters can be difficult or almost impossible, as I have described in Section 2.2.2 on suitable parameters. Second, no additional parameters can be added without what's called a *breaking change* (i.e., an incompatible change) to the interface. Calling code must be adapted whenever changes are made and the open-closed principle is violated.

> **Beware of Too General and Generic Data Types as Parameters**
>
> Explicit data types should be used in an interface wherever possible. If, for the sake of simplicity, you only use general and generic data types (such as `string`, `int`, or `float`) as parameter types and pass a wide variety of data within them, the interface will also become difficult to understand: A `string` value can contain decimal numbers, XML strings, or IP addresses, for example, which makes an interface difficult to use, read, and maintain.

> Consider the following example:
>
> ```
> //Difficult to understand
> func NewServerConfig(param1 string, param2 string) {…}
> //Better readable due to the IpAddress and Port data types
> Func NewServerConfig(addr IpAddress, port Port) {…}
> ```

Using the option pattern, you can create a customized interface of the factory method from our earlier one, as shown in Listing 2.59.

```
func NewServerConfig (opts ...Option) ServerConfig {
    //doSth
}
```

Listing 2.59 Factory Method with Option Pattern (Go)

In contrast to the first variant, with its many parameters, the specific `Option` type is expected as a parameter in this case. Further, since the definition of the parameter is a *variadic parameter*, the parameter can be specified multiple times and as often as needed in a call.

The adaptation of the function results in various `Option` type implementations can be used as parameters and the list of possible options can be extended at a later date by new implementations. Figure 2.15 shows a corresponding UML diagram.

`Option` types are usually implemented in Go as *function types*, whereby the signature includes the type to be configured as a parameter. In the example shown in Listing 2.60, a `ServerConfig` object is to be filled with values accordingly.

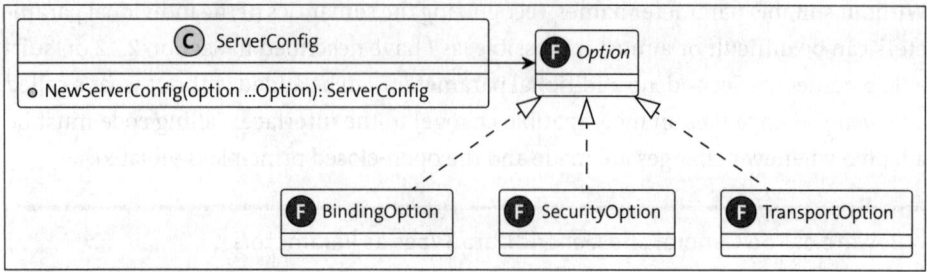

Figure 2.15 The Go Option Pattern

```
type Option func(config *ServerConfig)
```

Listing 2.60 Function Type for the Option Pattern (Go)

As shown in Listing 2.61, a concrete `Option` implementation can be implemented as a *closure*, which is returned by a function as a return value.

```go
func WithBinding(binding string) Option {
    return func(config *ServerConfig) {
        config.Addr = binding
    }
}
```

Listing 2.61 Concrete Option Implementation (Go)

Within the `NewServerConfig` factory method, regardless of the specific implementation of the `Option` type, all passed instances can be iterated to create a new configuration object. If configuration values are not passed, they can be filled in with default values when an object is initialized, as shown in Listing 2.62.

```go
func NewServerConfig(opts ...Option) ServerConfig {
    config := &ServerConfig{}
    for _, opt := range opts {
        opt(config)
    }
    return *config
}
```

Listing 2.62 Evaluating the Option Instances (Go)

The resulting function call is then easy to understand and expandable, as shown in Listing 2.63.

```go
config := NewServerConfig(WithBinding(":8080 "))
```

Listing 2.63 Using the Option Pattern (Go)

2.3.3 The Liskov Substitution Principle

Barbara Liskov—an American computer scientist who in 2008 became the second woman to receive the Turing Award for her work in computer science—presented her substitution principle for the first time at the OOPSLA conference in 1987 under the title "Data Abstraction and Hierarchy." She described that a subclass must fulfill specific conditions in relation to its base class so that the subclass can replace the base class without errors.

In other words, the principle states that an application that works correctly with objects of a base class must also continue to work correctly with classes derived from it (i.e., without having to adapt the code).

Computer scientist Barbara Liskov formally describes her idea of a correct class hierarchy as follows:

> "What is wanted here is something like the following **substitution property**: *If for each object o1 of type S there is an object o2 of type T such that for all programs P*

defined in terms of T, the behavior of P is unchanged when o1 is substituted for o2, then S is a subtype of T." —Barbara Liskov, OOPSLA, 1987

Such a correct class hierarchy can be achieved by complying with the following conditions:

▸ Derived classes may not restrict or tighten any preconditions of the base class.
▸ The postconditions of an application of the derived class may only be reinforced.
▸ Conditions that are true in the base type must also remain true in the derived type.

The terms *precondition* and *postcondition* indicate which conditions must be fulfilled before or after a call of the type or method so that the call can be considered error free. The object's interface may need to be called in a specific order, or certain correct parameters may need to be passed.

This requirement is also referred to as a *contract* that is concluded between the caller and the called and may not be directly visible in the programming interface. If a derived class deviates from this contract, this contradicts the *Liskov substitution principle (LSP)*, and the client cannot be used without errors if no adjustments are made.

You can check the correct behavior of different classes with a unit test, for example, which must return the same result for the base class as for a derived class.

Possible Issues with the Inheritance of Classes

The example shown in Figure 2.16 illustrates once again the problem of a Liskov substitution principle that is not adhered to.

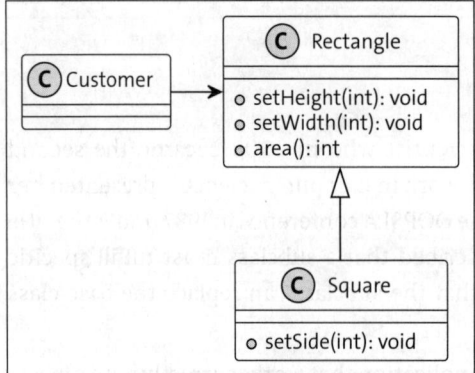

Figure 2.16 A Violation of the Liskov Substitution Principle

The Rectangle class represents a rectangle in which the height and width can be set using setter methods. The area of the rectangle can then be determined by calling the getArea method. Listing 2.64 shows how the class is implemented in Java.

```
public class Rectangle {
    protected int height;
```

```java
    protected int width;

    public void setHeight(int height) {
        this.height = height;
    }
    public void setWidth(int width) {
        this.width = width;
    }
    public int getArea() {
        return height * width;
    }
}
```

Listing 2.64 Rectangle Class for Calculating an Area (Java)

A caller can instantiate the class and, as shown in Listing 2.65, have the area calculated.

```java
Rectangle rectangle = new Rectangle();
rectangle.setWidth(10);
rectangle.setHeight(2);
System.out.println(rectangle.getArea()); // Output: 20
```

Listing 2.65 Using the Rectangle Class

Since a square is a special type of rectangle, the Square class is now derived from the Rectangle class, as shown in Listing 2.66, and adapted accordingly. When setting the height or length, the side lengths are always set to the same value using the new setSide method.

```java
public class Square extends Rectangle {
    @Override
    public void setHeight(int height) {
        setSide(height);
    }
    @Override
    public void setWidth(int height) {
        setSide(height);
    }
    public void setSide(int side) {
        this.width = side;
        this.height = side;
    }
}
```

Listing 2.66 Implementation of a Derived Class

However, the result in the client is now somewhat confusing, as shown in Listing 2.67.

```
Rectangle rectangle = new Rectangle();
rectangle.setWidth(10);
rectangle.setHeight(2);
System.out.println(rectangle.getArea()); //Output: 20

Rectangle square = new Square();
square.setWidth(10);
square.setHeight(2);
System.out.println(square.getArea()); //Output: 4
```

Listing 2.67 Confusing Result from Not Following the LSP

Although these are two variables of the `rectangle` type and are called in the same way, the result shown in Listing 2.67 is unexpectedly different. A unit test that tests the functionality of the base class could not be executed without any errors using an instance of the derived class. The result would deviate, and the Liskov substitution principle would be violated.

> **Must the Liskov Substitution Principle Always Be Adhered To?**
>
> The strict use of the LSP is sometimes considered problematic in practice. A derived class cannot always behave in exactly the same way as the base class.
>
> The following short example in Java shows the contradiction:
>
> `System.out.println(rectangle.toString().equals(square.toString()));`
>
> In this case, the `toString` method is called for both objects, which returns a text representation of the object. The output of the comparison is `false` because the two objects are not identical, and the potentially existing polymorphic behavior is desired. The LSP is violated.
>
> Despite this example, compliance with the LSP still makes sense in most use cases. For real-life use, the scope or context in which the principle should apply to the classes should be defined before use.

The Liskov Substitution Principle in Software Architectures

Not only can the principle of error-free substitution be used at the code level, as originally described by Barbara Liskov, but it can also be extended to software architectures.

One example is that some open-source projects provide interface-compatible server implementations for certain HTTP/REST interfaces of commercial products. Clients can then decide whether they want to connect to a commercial server instance or to a server based on the open-source solution. The goal is to facilitate the migration between the products and give users the option of switching.

The *MinIO* server, a published, open-source *object store*, provides an Amazon Simple Storage Service (S3)-compatible interface for managing files, for example. This server can therefore be used either as a replacement for Amazon S3 or for locally testing your own implementations.

If the implementations were to behave differently, the client code would have to be adapted, and a corresponding conditional statement would be necessary. The Liskov substitution principle would be violated in this case.

> **Caution with Conditional Statements When Switching Implementations!**
> An indication and *code smell*, which shows whether the Liskov substitution principle is violated is the presence of conditional statements in the code that distinguish between different implementations and make corresponding adjustments.

2.3.4 The Interface Segregation Principle

As I have already mentioned several times, one aim of good software architecture is to make software maintainable and expandable. Excessive coupling due to dependencies between the software components can make this more difficult or, in the worst case, prevent it entirely.

For this reason, the *interface segregation principle* focuses on the definition of interfaces in relation to the clients that use them and minimizes the resulting coupling of components. The principle states that no code should depend on other code that it does not need:

> "Clients should not be forced to depend upon interfaces that they do not use."
> —Robert C. Martin

An extensive interface with many defined methods often means that not all methods are implemented in new concrete implementations of the interface or that not all methods are used in the client code on the other side. This creates unnecessary dependencies that have to be maintained when extensions are made.

Example of the Interface Segregation Principle at the Code Level

I want to explain the abstract description of the principle using an example: A printer manufacturer develops software for its products that can be used to manage the devices. As only laser printers have been produced to date, the `LaserPrinter` class was implemented with the corresponding logic. As shown in Figure 2.17, the class contains three methods: `print`, `suppliesSummary`, and `supportsColor`.

The `suppliesSummary` method can be used to query the status of the consumables, the `supportsColor` method provides information on whether the printer is a color printer, and the `print` method can be used to trigger a print process.

2   Principles of Good Software Design

As the manufacturer plans to launch other products as well, the developers decide to create an abstract description for printers as an interface and define the three methods there. The previous `LaserPrinter` class now implements the new `Printer` interface, which is shown in Figure 2.17.

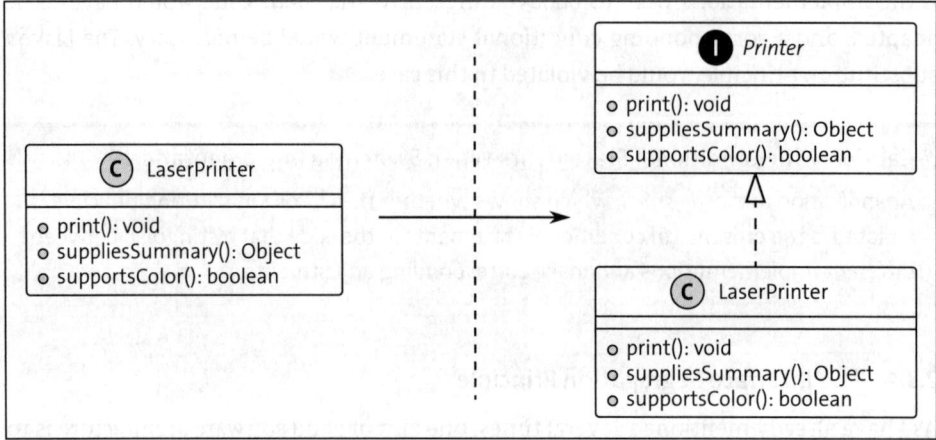

Figure 2.17 Unhooking the Printer Interface

In parallel to the printers, work is being carried out on the production of pure copiers, and the software for the copiers will provide similar functionality to the existing products. For this reason, a new `Copier` class is created, as shown in Figure 2.18, which also implements the `Printer` interface. This means that the new copiers can also be used polymorphically by the existing client software that uses the `Printer` interface.

To make sure that the entire functionality of the `Copier` class can be provided in the client software, the `Printer` interface is extended by the `startCopy` method, as the `Copier` class has a corresponding method.

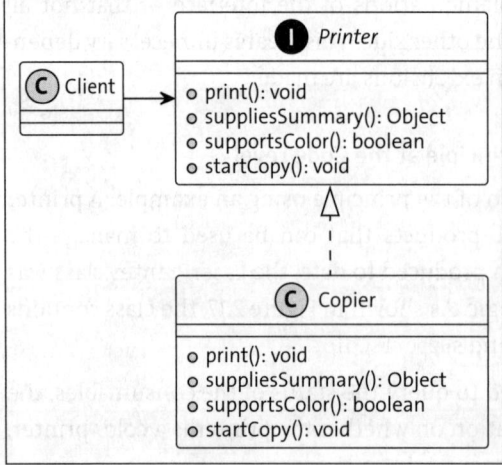

Figure 2.18 Copier Class Implementing the Printer Interface (Bad)

Because the `Copier` class has no print functionality, the `print` method is implemented empty or, as shown in Listing 2.68, immediately returns an error when called.

```java
//Unnecessary empty implementation is imposed!
public class Copier implements Printer {
    @Override
    public void print() {
        throw new UnsupportedOperationException("Not supported yet.");
    }
    ...
}
```

Listing 2.68 Implementation of the print Method in the Copier Class (Java)

The `LaserPrinter` class must also be adapted accordingly. It also returns an exception if the unnecessary `startCopy` method is called, as shown in Listing 2.69.

```java
//Unnecessary empty implementation is imposed!
public class LaserPrinter implements Printer {
    @Override
    public void startCopy() {
        throw new UnsupportedOperationException("Not supported yet.");
    }
    ...
}
```

Listing 2.69 Empty Implementation of the startCopy Method in LaserPrinter (Java)

The sample implementation makes it clear that the `Printer` interface is not suitable for both classes. In each case, an empty method had to be inserted into the corresponding concrete implementation.

What is even worse than the empty implementations in the classes is the dependency introduced between the `LaserPrinter` and `Copier` classes. If, for example, the functionality of the copier changes and the signature of the `startCopy` method in the `Printer` interface must be adapted, this change will affect the `LaserPrinter` class, although the class is not directly affected by the adaptation.

This problem can be solved by splitting the `Printer` interface into individual, independent interfaces for printers and copiers.

Each client receives an interface specifically adapted to its requirements. Common functionality, such as the `suppliesSummary` and `supportsColor` methods, are still defined in a common `PageDuplicator` interface. Figure 2.19 shows a corresponding class diagram.

2   Principles of Good Software Design

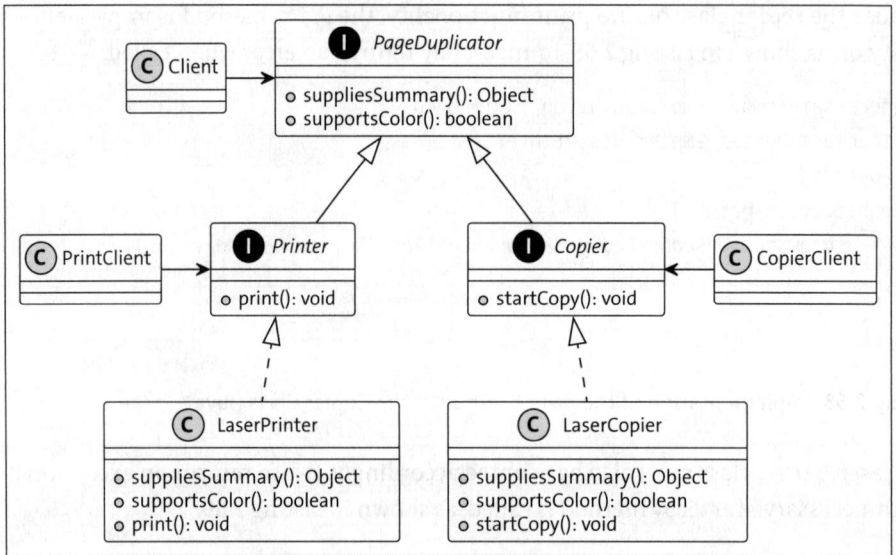

Figure 2.19 Separation of the Interfaces to Implement the Interface Segregation Principle

The introduction of the additional abstraction layer as the PageDuplicator interface enables the general client application to continue to work with both classes via this interface and to clearly delimit common behavior. Dependencies between the Printer and Copier interfaces are eliminated, and a potential change to the functionality of one interface has no effect on the implementation of the other interface. The close link between them has been removed.

Another positive side effect of the change is the following: If the printer manufacturer one day wants to offer new multifunction printers that can print and copy, a Multifunc-tionPrinter class can be created, for example, that implements both interfaces, as shown in Figure 2.20.

By abstracting interfaces, couplings and dependencies can often be minimized or completely eliminated. Rob Pike, one of the initiators of Go, for example, warns against too many methods in an interface:

"The bigger the interface, the weaker the abstraction." —Rob Pike

Abstraction and the creation of lean, client-related interfaces, as envisaged by the interface segregation principle, help to make code more maintainable and expandable.

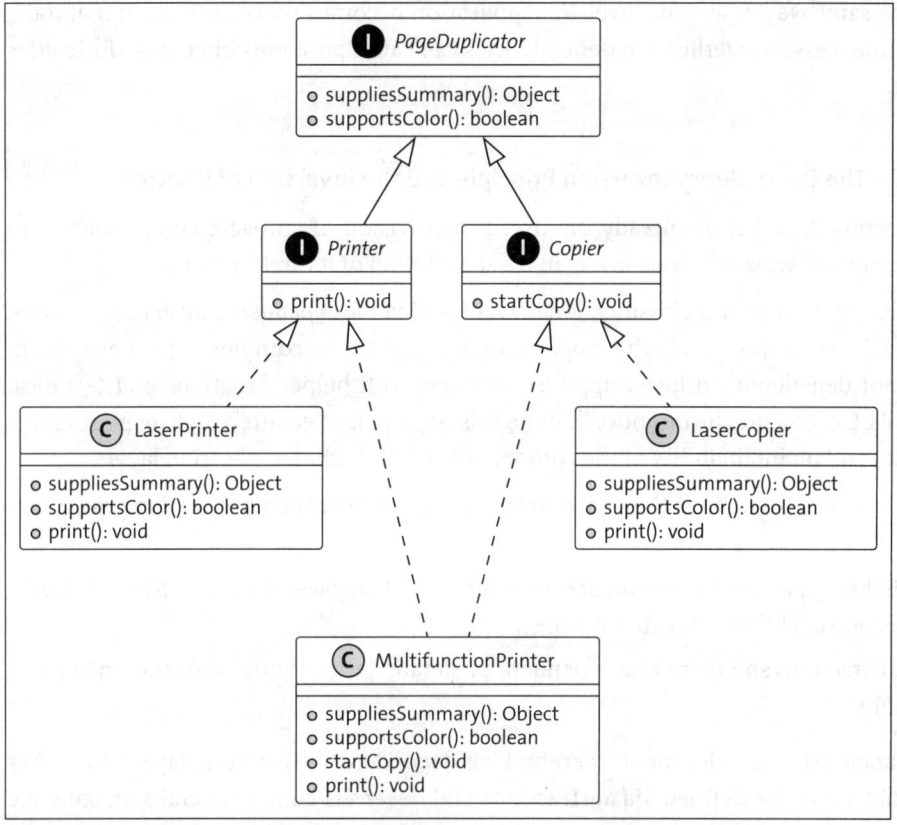

Figure 2.20 Multifunction Printers Implementing Two Interfaces

The Interface Segregation Principle at Software Architecture Level

The creation and use of lean, client-related interfaces can also be useful at the architecture level and reduce unnecessary dependencies.

If, for example, a library always loads configuration data from a database via a mandatory loading mechanism before it is used, but a calling application does not use a function that requires a corresponding configuration, the interface segregation principle is also violated in this case, as shown in Listing 2.70.

```
Library lib = new Library(oracle); // Parameter must be specified!
                                   // Specification of null not possible.
                                   // Configuration is loaded.
Result r = lib.addInteger(1,2);    // No configuration is required here!
```

Listing 2.70 Library Violating the Interface Segregation Principle: Unnecessary Database (Java)

The calling application therefore contains a dependency on a database that is not used at all, as shown in Listing 2.70.

In the same way as at code level, the application becomes more difficult to maintain, and unnecessary overhead or dependencies can cause problems or at least difficulties later.

2.3.5 The Dependency Inversion Principle and the Inversion of Control

In Section 2.3.2, I have already discussed the division of software components into *higher and lower application layers* in the description of its architecture.

The background to this classification is what's called the *dependency inversion principle*. Its goal is to ensure that higher application layers with the complex logic they contain are not dependent on lower application layers with helper functions and technical details. Once again, the intention behind this concept is to ensure a high degree of reusability and maintainability of the components of the higher application layers.

Robert C. Martin defines the goal of introducing the *dependency inversion principle* as follows:

- Higher application layers must not be dependent on lower layers. Both layers should depend exclusively on abstractions.
- Abstractions should not be dependent on details. Details should depend on abstractions.

For application development, therefore, all dependencies from one layer to another should always be defined via abstractions and never via concrete details. In concrete terms, you should always implement dependencies between layers via interfaces and not via classes.

The reason for this suggestion (as I have already explained with the open-closed principle) is that the dependencies potentially create a compulsion to change, which should be limited. Changes to a component can result in adjustments to the referencing component. As a rule, concrete implementations are subject to more frequent changes than their abstract interface.

Good software and interface design is therefore characterized, among other things, by the fact that you keep interfaces stable and, if possible, only make or require changes in specific implementations of these interfaces. Only referencing stable interfaces helps protect against changes.

Consider the following rules to set up a software design in which changes are not propagated to other components:

- Do not reference specific classes that change frequently.
- Do not build inheritance hierarchies on classes that change frequently.
- Do not overwrite any specific implementations.

> **Dependencies on Stable Interfaces**
>
> In practice, you could not define all component dependencies using abstract interfaces. If, for example, you use concrete classes or types from the standard library of a programming language, hiding them behind an abstract interface doesn't make much sense.
>
> The decisive factor for introducing an abstract interface is the probability of change or the frequency with which changes are made.
>
> Standard libraries of the various programming languages and the classes or types they contain can generally be regarded as stable and reliable for practical use. This is because *breaking changes*—incompatible modifications of the interfaces of these libraries—are avoided as much as possible. Otherwise, all existing programs in this programming language would have to be modified after an update to the new, adapted version. Fortunately, all programming language teams work hard to prevent these kind of disruptions!

The use of abstract interfaces between components results in the inversion of the direction of dependency (*inversion of control*) between them and thus also to a protection of the referencing component against changes.

The following example illustrates the dependency inversion principle: Figure 2.21 shows a section of an application in a class diagram in which data from a database is displayed via a user interface. We have three classes in three different application layers, whose dependencies are shown as arrows.

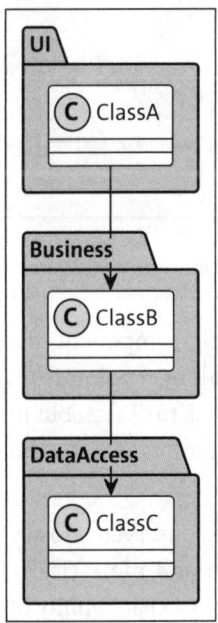

Figure 2.21 Dependencies Between Components

If you follow the dependency arrows of the application, notice how the display component in the example depends on the data access component (the UI component depends on the Business component and the Business component itself depends on the DataAccess component). Changes to specific classes in this component could therefore lead to adjustments in the presentation layer. This connection may still make sense for data objects that are extended by new fields intended for display, but certainly not if, for example, the database product is changed and the corresponding classes must be adapted.

According to the dependency inversion principle, the dependencies between the components should always take place via abstractions. In the example shown in Figure 2.22, following this principle leads to the introduction of the InterfaceC interface, which defines an abstract interface for storing and loading the data.

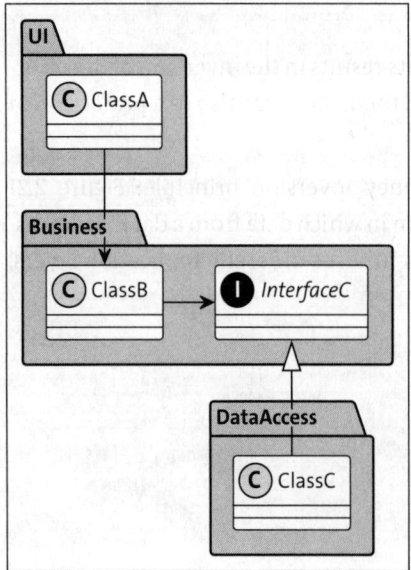

Figure 2.22 Dependency Inversion Principle for Component Dependencies

With the introduction of the abstract interface within the business component, the direction of the dependency has changed. The dependency arrow no longer points from the Business component to the DataAccess component or from ClassB to ClassC, but in exactly the opposite direction. Changes to the technical implementation of ClassC no longer affect the display component.

To summarize briefly, when the dependency inversion principle is observed, loosely coupled applications or components are created that are simpler and more clearly structured. This flexibility leads to better overall testability and easier expandability of the components. During the execution of unit tests, for example, a concrete implementation can also be replaced by a mock implementation.

2.4 Information Hiding

The principle of *information hiding* is a cornerstone of good software design, and I have already described it indirectly in a few places without mentioning it explicitly. According to this principle, data and functions within a component are hidden from the outside as far as possible, and only the important or relevant information is visible outside the component.

> *"Less is more, but not less than necessary."* —Wisdom of information hiding

The key principles of *information hiding* are *encapsulation* and *abstraction*. Both are intended to minimize the dependencies of the components in order to create a more flexible, robust, and maintainable software design.

The lower the dependencies between the components, the better the components are protected against external changes to a referenced component. Some overall advantages of information hiding include the following:

- Reduced dependencies between components
- Reduced complexity of the application
- More robust software
- Better maintainability and expandability
- Better reusability of components

Many of the principles I have presented in previous sections implement *information hiding* or support you in its implementation:

- **Encapsulation**
  Internal details are hidden by the encapsulation. The client only has access to an exported interface.
- **Abstraction**
  Abstraction is used to create interfaces that hide the details of a specific implementation.
- **Clear boundaries between components**
  Clear boundaries between components are created by a clear definition of the interface and the compartmentalization of details and information.
- **Interface segregation principle**
  The use of this principle makes it possible to define lean interfaces for clients and thus hide information and minimize dependencies.
- **Dependency inversion principle**
  The dependency inversion principle reverses the direction of a dependency by using abstractions. Abstraction ensures that dependencies are minimized.

The principle of *information hiding* can and should be used in many levels of software development, and you should always make sure to export as little information as possible, regardless of the current abstraction level. Always pay attention to the transferred parameters or return values of an interface! They determine the amount of information transferred and thus directly influence information hiding.

You can apply the principle to the definition of, among other things:

- Classes
- Interfaces
- Packages
- Modules
- Libraries
- Applications
- External interfaces (e.g., the REST interface)

2.5 Inversion of Control and Dependency Injection

Every program has a program flow or *flow of control* that defines the sequence of commands to be executed. For example, a component calls a reusable library to perform general tasks. In this case, the control over the program flow lies in the component itself, as it contains the code for the call.

With the *inversion of control (IoC)* principle, however, the flow of control of a component is no longer defined in the component itself, but by an external component or framework.

This approach is often humorously compared to the Hollywood movie industry: The actors don't call the production company all the time to check on the status of the bid, but they hand over the responsibility to an agent who does the placement for them and calls them when there's something new.

> "Don't call us, we'll call you." —*The Hollywood principle*

In software development, the IoC principle enables the implementation of generic code within libraries or frameworks that can be flexibly adapted to different requirements. At the same time, dependencies within the called component are minimized.

In many libraries and frameworks (e.g., in the Spring Framework in Java), IoC is used for dependency management and the associated *dependency injection* principle. Dependencies are no longer resolved by the component itself, but by a *container* or a *runtime environment*, which sets the dependency by calling methods from outside to or for the component.

If, for example, a database connection is required in a component, the component does not have to establish a connection itself but defines an interface via which it can receive a connection. Listing 2.71 shows an example without dependency injection.

2.5 Inversion of Control and Dependency Injection

```java
public class ShipService {
    public void storeShip(Ship ship) throws NamingException, SQLException {
        Context context = new InitialContext();
        DataSource dataSource =
            (DataSource)context.lookup("database_connection");

        Connection connection = dataSource.getConnection();
        //...
    }
}
```

Listing 2.71 ShipService Without Dependency Injection (Java)

By using *dependency injection*, the example changes as shown in Listing 2.72. The dependency of a `DataSource` instance is no longer resolved within the component using the *Java Naming and Directory Interface (JNDI)*, but is taken over by an external component (the container). The container uses the `setDataSource` method to set a corresponding `DataSource` implementation from outside:

```java
public class ShipService {
    private DataSource dataSource;
    public void setDataSource(DataSource dataSource) {
        this.dataSource = dataSource;
    }
    public void storeShip(Ship ship) {
        Connection connection = dataSource.getConnection();
        //...
    }
}
```

Listing 2.72 ShipService with DataSource Dependency as Dependency Injection (Java)

As a result of the adaptation, the resulting component contains fewer dependencies and becomes more maintainable and easier to test. In contrast to the customized version, the variant without dependency injection was dependent on an existing JNDI environment, for example. How and where the container in the customized version obtains the reference to the `DataSource` does not play a role in the component code.

In summary, it can be said that the principles of *inversion of control* and *dependency inversion* can be used to create cleaner, more modular, and more flexible software and that most frameworks promote this approach.

2.6 Separation of Concerns and Aspect Orientation

The *separation of concerns (SoC) principle* is another fundamental design principle in which software is subdivided into different sub-areas, each of which performs a specific task. Applications should not be developed as a complex, opaque block, but in smaller, more maintainable blocks. Among other things, the principle forms the basis for the two SOLID principles already presented: the *interface segregation principle* and the *single responsibility principle*.

A separation of concerns can be found in various areas and levels of software. In web development, for example, the textual content of a page is stored within an HTML file, while the appearance is created in separate files using CSS and the dynamic behavior is also created separately in JavaScript files.

The advantages of splitting into smaller blocks are obvious—as described in the corresponding SOLID principles:

- Better readability
- Increased maintainability
- Increased reusability

However, a rigid and unreflected separation of the application into different functional blocks is problematic in software development since topics or requirements often extend across multiple application parts or blocks and the separation leads to code duplication (as shown in Figure 2.23 within components A, B, and C). In 1997, Kent Beck coined the term *cross-cutting concerns* for such overarching topics and requirements. These requirements include, for example, the following areas:

- Security: Authentication, authorization, and encryption must be considered in virtually all applications.
- Logging: Every relevant function should log information.
- Transactions: Business logic that spans multiple calls and involves multiple database operations usually needs to be treated reliably as a unit.

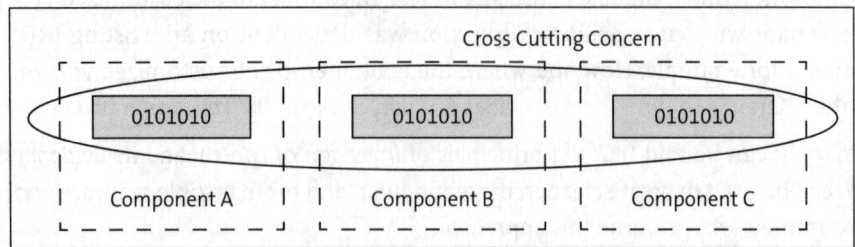

Figure 2.23 A Cross-Cutting Concern

Aspect-oriented programming (AOP) is one approach to tackling the code duplication that can arise from *cross-cutting concerns* and the associated problems of readability and maintenance. AOP combines generic functionality from the individually divided blocks into independent components and executes them at specified points via a configuration.

Gregor Kiczales and his colleagues at Xerox PARC presented their ideas on aspect-oriented programming at the European Conference on Object-Oriented Programming in 1997 and subsequently introduced a concrete implementation of their ideas with *AspectJ*. In the Java world, the AOP framework has now established itself as the default standard.

> **AOP in the Java Enterprise Environment**
>
> Aspect-oriented programming has now become a fixed and fundamental part of Java Enterprise development. AOP functionalities can be found both in *Java EE* and, for example, in the widely used *Spring Framework*. In both cases, shared functionality is extracted to central components and does not always have to be reimplemented. This approach makes both environments quite powerful.

In each AOP framework, a *join point model* is defined, which enables the dynamic definition of cross-cutting concerns by precisely determining their join points in the program code. AspectJ uses the following terms for this purpose:

- **Join point**
  A *join point* is an execution point at which the extracted functionality can be introduced. This functionality might be, for example, a method call, an object initialization, or a field access.
- **Pointcut**
  A *pointcut* marks a specific join point in the application. For example, the pointcut represents a call to a specific method of a concrete class.
- **Advice**
  An *advice* contains the actual extracted functionality as well as the assignment to a special pointcut.
- **Weaving**
  An advice is inserted into the existing code by means of *weaving*. With AspectJ, weaving can be achieved either at compile time using the what's called the *compile-time weaving* method or later at runtime using *load-time weaving*.

AspectJ can be used in various ways. Probably, the most common variant for defining a *join point model* for your own software is to specify annotations in the source code.

Listing 2.73 shows an example of an advice with annotations: The `@Pointcut` annotation of the `updatePositionPointcut` method defines a pointcut for all calls to the `updatePosition` method of the `com.example.Ship` class. This pointcut can subsequently be referenced via its method name, in this case, `updatePositionPointcut`.

> **Advice Types in AspectJ**
>
> In AspectJ, you can specify the execution time of an advice in three different ways. For each join point, the woven-in functionality can be included either before, after, or around it. With the enclosing variant, you can decide in the code, as shown in Listing 2.73, when the enclosed code should be called by calling the proceed method.
>
> You can use the following advice types and associated annotations:
>
> - @Before before the called join point
> - @Around around the called join point
> - @After after the called join point

The advice (i.e., the combination of pointcut and functionality) is defined in the example using the around method, as shown in Listing 2.73. This method has the @Around annotation with the specification of the corresponding updatePositionPointcut pointcut and contains the functionality to be woven in.

```java
@Aspect
public class ShipAdvice {

    @Pointcut("call(void com.example.Ship.updatePosition(*))")
    public void updatePositionPointcut() {
    }

    @Around("updatePositionPointcut()")
    public Object around(ProceedingJoinPoint pjp) throws Throwable {
        System.out.println("Calling " + pjp.getSignature().getName());
        Object methodResult = pjp.proceed();
        System.out.println("After " + pjp.getSignature().getName());     }
}
```

Listing 2.73 AspectJ Advice for the Ship Class (Java)

If the updatePosition method is called on an object of the com.example.Ship class, the woven-in functionality is called accordingly, as shown in Listing 2.74 and Listing 2.75.

```java
public class Ship {
    ...
    public void updatePosition(Position position) {
        System.out.println("Updating position: " + position);
        this.position = position;
    }
    ...
```

Listing 2.74 Excerpt from the Ship Class and updatePosition Method (Java)

```java
public static void main(String[] args) {
    System.out.println("Starting");
    Position pos = new Position();
    Ship ship = new Ship("Titanic", pos);
    ship.updatePosition(pos);
}
```

Listing 2.75 Sample Application with AspectJ (Java)

This example displays the lines shown in Listing 2.76 accordingly.

```
Starting
Calling updatePosition
Updating position: com.example.Position@5d41d522
After updatePosition
```

Listing 2.76 Output of the Sample Application

2.7 Quality Assurance with Unit Tests

Regardless of the software design or the programming principles used, software must work correctly and implement the defined requirements, for example, in terms of functionality, performance, or security. To ensure these requirements are fulfilled, software must be tested.

Testing can be performed using test cases in which an expected behavior for previously defined parameters is specified. A successful verification of the expected behavior leads to a test success. If the software delivers a different result or behaves differently, the test has failed.

At least one library or framework for creating tests is available for every modern programming language, and an execution is possible directly in most development environments. For example, you can use the *unit test* framework for Python, or *JUnit* for Java or the test framework already included in the standard library of Go. In addition to the functional requirements, other requirements can often also be tested in these environments, including performance requirements using benchmarks.

The definition of test cases is similar in all variants. Listing 2.77 shows a Python test that defines two test cases in which the string methods upper and isupper are checked. The first test case uses the assertEquals method to check values for equality. The first value passed to the method is the actual, calculated value; the second value represents the expected result. If the two parameters differ, the test fails.

The second test case works in the same way, but with assertTrue or assertFalse, in which actual values are checked for a Boolean value.

```python
import unittest

class TestStringMethods(unittest.TestCase):

    def test_upper(self):
        self.assertEqual('foo'.upper(), 'FOO')

    def test_isupper(self):
        self.assertTrue('FOO'.isupper())
        self.assertFalse('Foo'.isupper())

if __name__ == '__main__':
    unittest.main()
```
Listing 2.77 Unit Test Using unittest (Python)

If the implementation of the tested isupper or upper methods changes, the tests can be reexecuted to check whether the test cases are still fulfilled, and the software continues to function without errors.

The tests can be executed either manually or automatically, for example as part of a build pipeline. The second variant of automated execution is widespread in current software development and can be regarded as standard.

The automated and regular execution of the tests continuously ensures that all software requirements are still met even after changes have been made. The aim of these *regression tests* is to uncover unknown and problematic relationships between components and to ensure that components not directly affected by the changes continue to work without errors.

Tests are often differentiated according to their dependencies or their isolation. For example, if tests only affect a single unit or component, they are referred to as *unit tests* or *component tests*. If multiple units or components are tested together in a test, you're conducting *integration tests*. No clear line exists between these terms, and you'll often see other phrases like *service tests* or *UI tests* thrown in. Regardless of the terminology, when creating a test case, defining a clear boundary between the tests and applying the *single responsibility principle* to tests, for example, are important.

Mike Cohn, one of the cofounders of the *Agile Alliance* and the *Scrum Alliance*, has developed what's called the *test pyramid*, a concept that organizes tests hierarchically in a pyramid according to their execution speed and the costs they incur. The concept states that unit tests should form the basis of a test concept, as they can be used to test software quickly and in isolation from other components and thus provide rapid feedback on the status of the software. Tests with a higher level of integration with other components are executed more slowly and should therefore not be carried out as frequently since they are more expensive. Figure 2.24 shows the test pyramid with its corresponding classifications.

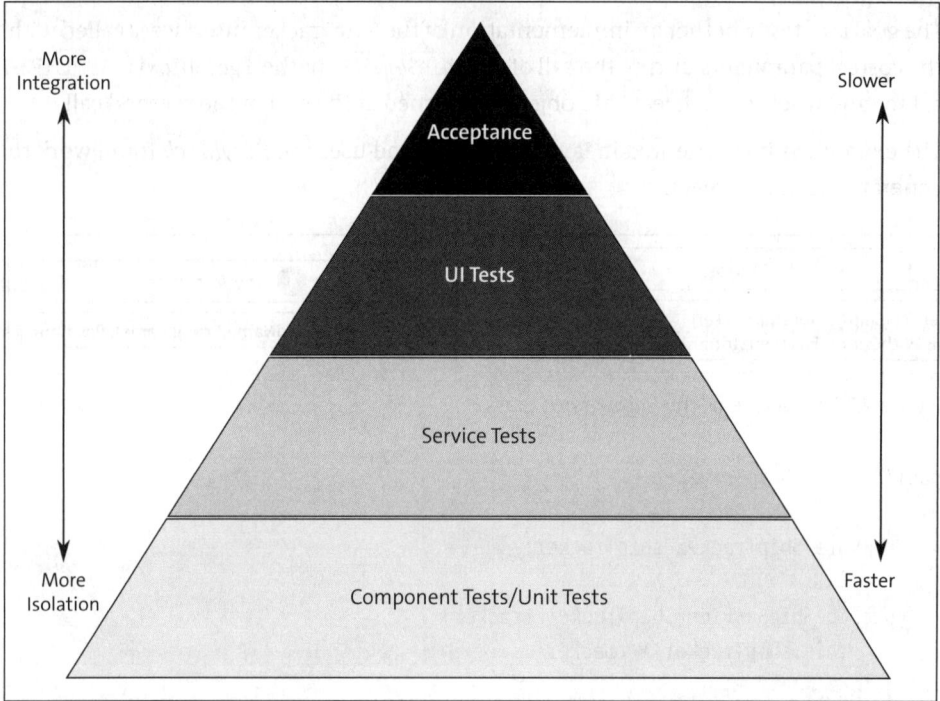

Figure 2.24 Testing Pyramid

For effective testing, software components must therefore be designed in such a way that they can be tested individually and in isolation as far as possible. This separation requires a software design that modularizes components and clearly defines their boundaries and dependencies. The SOLID principles and their application play an important role in this context since they support such a design. Well-structured components are much easier to test than unstructured components with many dependencies.

In particular, the concept of *dependency injection* in components can simplify the test process, as the required dependencies can be replaced by alternative implementations at test time. In this way, specific scenarios can be mapped and undesirable side effects excluded.

Using *mock frameworks*, which are available for the various programming languages, you have the option of creating substitute or *mock objects* for existing interfaces or classes at test time. They have the same interface, can be used analogously and provide the advantage that their behavior can be configured for the respective test case during or prior to the test.

Listing 2.78 and Figure 2.25 show the `ShipService` class whose `getShipWithActualPosition` method is being tested. The service has a dependency on the `ShipTracker` interface, which provides the `getPositionForShipName` method to determine the position of a ship. The goal is to test whether an implementation of the `ShipTracker` interface is called with the correct parameters during the call of the `ShipService` method `getShipWithActualPosition` and whether a correct `Ship` object is returned at the end of the method call.

The example is implemented in Java using *JUnit* and uses the *EasyMock* framework to generate the mock object.

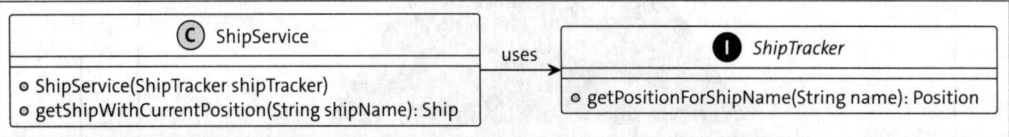

Figure 2.25 Structure of the ShipService Class

```java
public class ShipService {

    private ShipTracker shipTracker;

    public ShipService(ShipTracker tracker) {
        this.shipTracker = tracker;
    }

    public Ship getShipWithCurrentPosition(String shipName) {

        Position shipPosition =
            this.shipTracker.getPositionForShipName(shipName);
        return new Ship(shipName, shipPosition);

    }

}
```

Listing 2.78 Example Being Tested Using EasyMock (Java)

The `ShipTracker` interface only contains one method, shown in Listing 2.79.

```java
public interface ShipTracker {
    public Position getPositionForShipName(String name);
}
```

Listing 2.79 The ShipTracker Interface (Java)

2.7 Quality Assurance with Unit Tests

Within the `testGetShipWithCurrentPosition` test method, as shown in Listing 2.80, no "real" `ShipTracker` implementation is used. Instead, a *mock object* is created by Easy-Mock. The instance is created via the JUnit 5 extension of EasyMock and the `@Mock` annotation, which is specified for use in the context of an attribute definition.

Prior to each test run, a new `ShipService` instance is created using the `setUp` method, which receives a reference to the `ShipTracker` mock object via dependency injection.

At the start of the actual test, the mock object is configured in such a way that it returns a `position` object when the `getPositionForShipName` method is called.

```
expect(shipTrackerMock.getPositionForShipName(shipName))
    .andReturn(new Position());
```

At the end of the test, the `assertEquals` method checks whether the generated `Ship` object has the appropriate name and whether the methods of the mock object were actually called. This check is performed by the `verify` method from the EasyMock library.

```
@ExtendWith(EasyMockExtension.class)
public class ShipServiceTest {

    ShipService shipService;

    @Mock
    ShipTracker shipTrackerMock;

    @BeforeEach
    public void setUp() {
        shipService = new ShipService(this.shipTrackerMock);
    }

    @Test
    public void testGetShipWithCurrentPosition() {

        //given
        String shipName = "Titanic";
        expect(shipTrackerMock.getPositionForShipName(shipName))
                .andReturn(new Position());
        replay(shipTrackerMock);

        //when
        Ship titanic = this.shipService.getShipWithCurrentPosition(shipName);
```

```
        //then
        assertEquals(shipName, titanic.getName());

        verify(shipTrackerMock);
    }

}
```

Listing 2.80 Unit Test for ShipService with EasyMock (Java)

An important part of professional software development, unit tests help to detect errors at an early stage and improve the quality of the software and reveal potential side effects that might only occur in production systems without the tests.

Chapter 3
Source Code and Documenting the Software Development

Software must be documented. Good documentation not only consists of detailed information about the source code but also covers the architecture of a software system. Documentation enables developers to understand, adapt, and maintain software efficiently and thus contributes to the success, traceability, and quality management of a system. This chapter describes how you can create comprehensive, high-quality documentation for your software.

Documentation often makes a decisive contribution to the success of a project. However, each target group has different interests in software documentation. For example, users and customers need user documentation that describes how to use the finished software. Developers, project managers, even perhaps your quality management team, are more interested in technical details, information on the structure of the software, or even fundamental decisions made during development.

The goal of creating software documentation is to provide clear insights into the structure, functionality, and implementation of the software and thus create a better understanding among all team members. Detailed, meaningful information allows the code to be understood, expanded, and maintained more efficiently.

Many projects are not created, expanded, and maintained with a constant project team: Developers leave the project, while new developers join the project team. Documentation is therefore an indispensable part of the development process to ensure the long-term success of software and create a uniform understanding of the solution.

This chapter shows how software can be documented in the development process for various technically interested target groups.

3.1 Comments in the Source Code

Comments in the source code can be enormously helpful in making code that is difficult to understand more comprehensible for others or, after a certain time, for yourself. For example, they can provide further, additional details that enables the expansion or adaptation of the corresponding code. Poorly formulated or even outdated comments in the code often have the exact opposite effect and lead to confusion or incorrect assumptions.

For this reason, one goal in software development should be to write expressive, readable code that speaks for itself and that a developer can understand even without separate, explicit comments within the source code. Robert C. Martin, the author of the *Clean Code* book, even describes certain comments as errors, because these comments reveal that the developer couldn't express themselves via the source code alone and couldn't produce comprehensible code. However, Martin also makes clear that, in some cases, comments are necessary, but you should never be proud of them.

In Martin's opinion, every comment should be checked to see whether the code itself can be written more clearly—and the comment thus becomes superfluous.

Bad code cannot be improved by a comment. You'll often find comments at particularly confusing code positions about what is done or intended in the respective section. If you have the feeling during development that a code point should be documented for clarity, you should consider rewriting this code block so that no additional comments are needed.

Writing fewer comments in the code also has a major advantage: It reduces the effort involved in maintaining the software. In any case, comments—like the source code itself—must be well maintained and carefully considered. If fewer comments are needed to understand the code, less maintenance is required. In projects that have undergone several revisions, comments are often outdated, incorrect, or in the wrong place due to code shifts. Unfortunately, comments cannot be fully relied upon.

A major challenge is to keep the comments synchronized with the current code. However, developers are often fail to make this effort or overlook this task.

This problem can be mitigated with code that is self-documenting and self-explanatory. Comments then fade into the background as they are no longer needed, and revisions to the code do not directly lead to outdated comments. If code is moved to another location, for example, the corresponding comment does not have to be moved as well.

"Truth can only be found in one place: the code." —Robert C. Martin

Although you might be getting the impression that no comments are useful, you must not regard all comments or all documentation as mistaken, bad, or superfluous. In some situations, comments might be useful or even absolutely necessary due to various conditions (e.g., licensing terms). Public interfaces, such as the application programming interfaces (APIs) of libraries, must or should always be documented so that other developers can use them.

In the following sections, I will discuss how you can create comments in the source code and how you can minimize the scope of these comments or avoid them entirely by using more expressive code. I will show you examples where comments are useful or superfluous.

3.1 Comments in the Source Code

> **Formatting Comments in Your Source Code**
>
> In most programming languages, source code comments are introduced either by two slashes (//) or by the combination of a slash and an asterisk (/*). The latter is used to define comment blocks, which must then be closed using */. The two slashes, on the other hand, are used for single-line comments.

3.1.1 Documenting Interfaces Using Comments

In most languages, you can document a programming interface with the specific position and syntax of the comments. You can then use the appropriate tools to create a document that you can make available to other developers as documentation.

Documenting Java Source Code Using JavaDoc

In Java, for example, you can use the `javadoc` command-line tool to generate a clear collection of HTML pages from Java source code, as shown in Figure 3.1. Such documentation is available for all versions of the standard Java library.

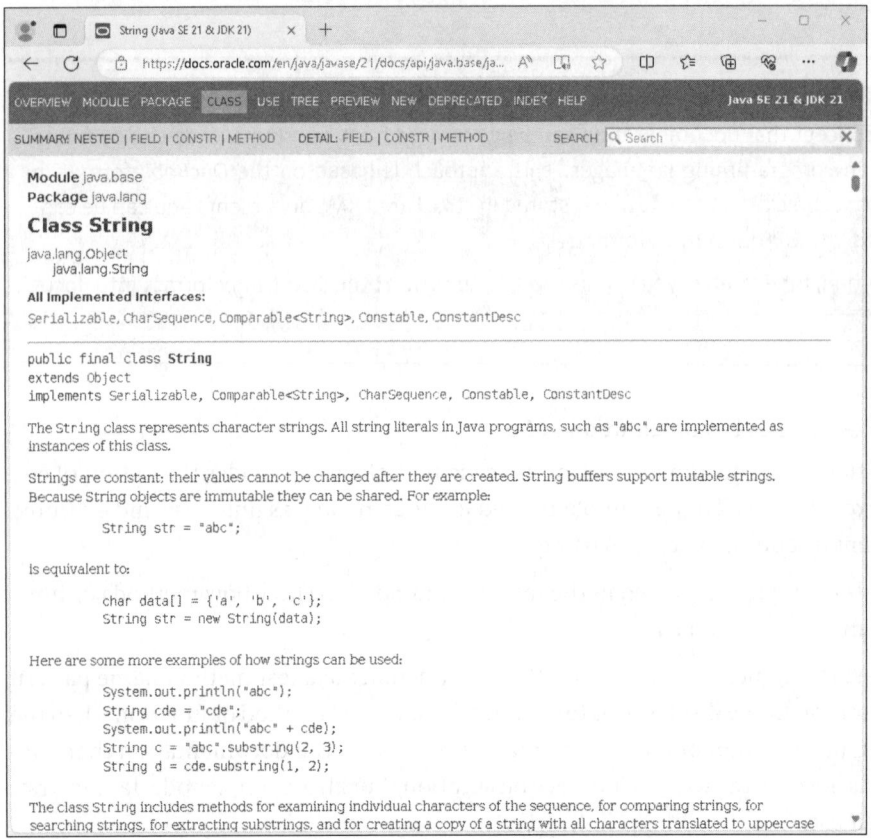

Figure 3.1 JavaDoc Documentation from Source Code Comments

145

The documentation of the API is included via comments directly above the implementation. A few keywords or *annotations* control the appearance and meaning of the specified values. Using @param, for example, you document a parameter, while @return or @throws describe return values or indicate that exceptions may be thrown, as shown in Listing 3.1.

```
/**
* Allocates a new {@code String} so that it represents the sequence of
* characters currently contained in the character array argument. The
* contents of the character array are copied; subsequent modification of
* the character array does not affect the newly created string.
*
* @param  value
*         The initial value of the string
*/
public String(char[] value) {
this(value, 0, value.length, null);
}
```

Listing 3.1 JavaDoc Example for the java.lang.String Class

> **Generating Documentation from Source Code**
>
> The concept that documentation can be generated from the source code can be found in many programming languages. This approach is based on the *DocBook* format in which technical documentation is stored in structured XML documents and can be converted into various output formats.
>
> In Python, for example, you can use *Sphinx* to convert the *DocStringx* format into documentation.

Code Examples as Documentation in Go

In addition to the documentation options via text, Go also provides the option of creating *example tests*. They are implemented in the same way as unit tests and executed like a unit test during a test run using go test.

The code examples contained in the tests are intended to show how methods or functions can be used correctly.

Each test is assigned to a specific method or function via a test method name pattern and is automatically displayed as text under the specified method or function when the HTML representation of the documentation is created. The documentation therefore contains not only the written interface descriptions but also example code demonstrating their use.

> **Creating Go HTML Documentation Using Godoc**
>
> To create an HTML representation of your Go documentation, you can install and use the godoc tool. The installation is carried out via a go install command, which downloads the required files and installs them locally:
>
> go install golang.org/x/tools/cmd/godoc@latest
>
> If the Go environment is configured correctly and the GOBIN directory is contained in the operating system environment variable PATH, you can generate the documentation via the following command and it will be accessible at *http://localhost:6060* on your local machine:
>
> godoc -http :6060

Because the code examples are implemented as tests and checked during each test run, the examples are always executable, and the documentation remains up to date. A user can rely on the examples given.

The checks and test conditions for the sample code are written using a special syntax. At the end, a check can be initiated using the //Output: string. The character string following this expression must have been output within the test via an fmt.Print output.

Our next example, shown in Listing 3.2, is an example test in which a Car object is created, and the Color attribute, which is pre-initialized to the value Blue, is changed to the value Red by a corresponding method call. This step is followed by an output and a check using the //Output: string.

```go
func ExampleCar_PaintRed() {

    a := &Car{Color: "Blue"}
    a.PaintRed()
    fmt.Println(a.Color)
    //Output: Red
}
```

Listing 3.2 Example Test with a Color Check (Go)

Many use cases exist in the standard Go library. Our next example, shown in Listing 3.3, illustrates a Get function from the http package, which is implemented without verification. It should "only" show the use of the function.

```go
func ExampleGet() {
    res, err := http.Get("http://www.google.com/robots.txt")
    if err != nil {
        log.Fatal(err)
    }
```

```
        body, err := io.ReadAll(res.Body)
        res.Body.Close()
        if res.StatusCode > 299 {
            log.Fatalf("Response ..: %d and\nbody: %s\n", res.StatusCode, body)
        }
        if err != nil {
            log.Fatal(err)
        }
        fmt.Printf("%s", body)
}
```

Listing 3.3 Example Test from the "net/http" Package for the Get Function (Go)

Figure 3.2 shows the representation for this code shown, which is an example of the Get function from the net/http package.

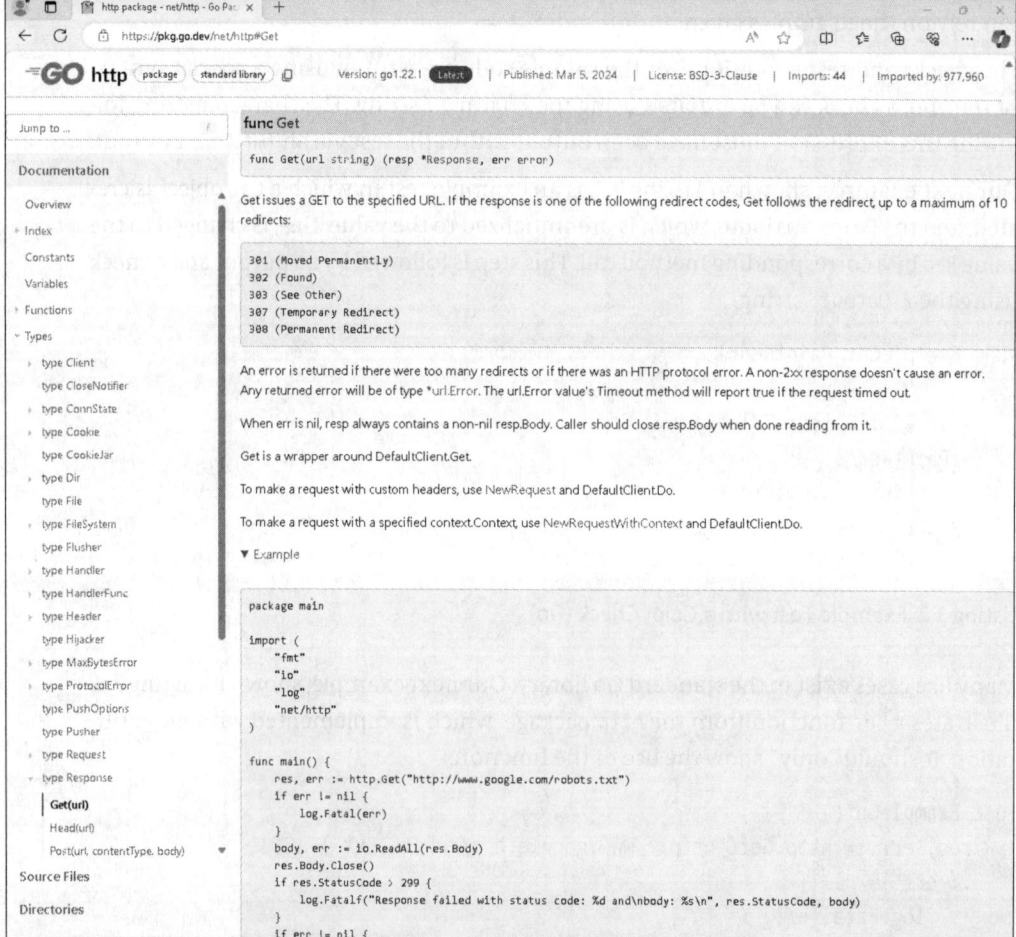

Figure 3.2 Presentation of the Test in the Documentation

> **The doctest Package in Python**
>
> In Python, you can use the *doctest* package to enter code within DocStrings and have the code executed at test time.
>
> The following example shows the syntax in the documentation of the `factorial` function:
>
> ```
> """
> This is the "example" module.
>
> The example module supplies one function, factorial().  For example,
>
> >>> factorial(5)
> 120
> """
>
> def factorial(n):
>     ...//Implementation was omitted
> ```

3.1.2 Creating Expressive Code

I already mentioned it in the introduction to this chapter: The more readable and self-explanatory a code section is, the fewer comments are needed to explain the meaning and purpose of the section to other developers. The basic principles of clean code or the use of software patterns can make an important contribution to clarity and readability.

Simple refactoring techniques, such as the *extract method* or *decompose conditional*, can be used, for example, to make details in the code that are difficult to understand clearer and thus avoid comments.

In the code example shown in Listing 3.4, a comment is helpful to understand quickly the nature of the check.

```
// Check if date is in summer
if (!aDate.isBefore(plan.summerStart) &&
    !aDate.isAfter(plan.summerEnd)){
...
}
```
Listing 3.4 Complex If Condition (Java)

However, extracting a method for the condition check makes the code clearer and easier to read, as shown in Listing 3.5. The comment can then be removed since the code is self-explanatory in this case.

```
if (isSummer(aDate)) {
...
}
```
Listing 3.5 Decompose Conditional Refactoring for Greater Clarity (Java)

3.1.3 Necessary and Meaningful Comments

As mentioned earlier, one of the goals in software development should be to create source code that is readable and expressive and for these reasons does not require explicit comments to explain code passages. However, in some situations, comments are important and helpful or are mandatory due to various framework conditions, such as licensing terms.

The following sections contain examples of source code comments to illustrate when the use of a comment in the code can be helpful or even necessary.

License Terms

Licensing information at the beginning of each source code file provides clarity about the conditions for the use and redistribution of the corresponding software.

Some companies and many open-source licenses, for example, require that the license text be provided together with the source code in order to provide transparent information about the applicable license terms and to ensure that the code is used accordingly.

In addition, licensing information ensures transparency with regard to company or developer copyright and helps comply with these rights.

In development environments, this kind of standard header can be configured for source code files so that the same comments can be automatically included in every file.

When commenting on license terms, you should ensure that the comment does not contain the complete license terms, only a reference to where the specific version can be found.

In the *Apache Commons* library for Java, for example, you can find the information shown in Listing 3.6 in each file.

```
/*
 * Licensed to the Apache Software Foundation (ASF) under one or more
 * contributor license agreements.  See the NOTICE file distributed with
 * this work for additional information regarding copyright ownership.
 * The ASF licenses this file to You under the Apache License, Version 2.0
 * (the "License"); you may not use this file except in compliance with
 * the License.  You may obtain a copy of the License at
```

```
 *
 *      http://www.apache.org/licenses/LICENSE-2.0
 *
 * Unless required by applicable law or agreed to in writing, software
 * distributed under the License is distributed on an "AS IS" BASIS,
 * WITHOUT WARRANTIES OR CONDITIONS OF ANY KIND, either express or implied.
 * See the License for the specific language governing permissions and
 * limitations under the License.
 */
```

Listing 3.6 Example License Terms for an Apache Commons File

Useful Information

Short and clear information for a specific code passage can help you to understand it quickly. The example shown in Listing 3.7 demonstrates how a dense regular expression can be made easier to understand with a comment. Without the comment, someone might ask why only certain characters, and why precisely those specified, are permitted.

```
// regularExpression validates a "Vehicle Identification Number" (VIN)
// that it has a valid length and no illegal characters.
var regularExpression = regexp.MustCompile("^[A-HJ-NPR-Z0-9]{17}$")
```

Listing 3.7 A Regular Expression with a Comment (Go)

As an alternative to an explicit comment, renaming the variables in this example could further improve readability and make the comment superfluous again, as shown in Listing 3.8.

In this case, when reading the calling code, you shouldn't need to read the comment or jump to the corresponding entry in the documentation.

```
var vinLengthAndValidCharacterExpression =
    regexp.MustCompile("^[A-HJ-NPR-Z0-9]{17}$")
```

Listing 3.8 Variable Renamed for Easier Understanding (Go)

Renaming variables is certainly not possible in all situations; a variable name can rarely replace a longer, more complete, or more informative comment.

We could also combine both aspects and use a more expressive variable name and a short informative note, as shown in Listing 3.9, possibly with a reference to an external resource. In this case, the code is more readable, since what the regular expression checks is clear, and details of the expression can be looked up in the documentation.

```go
// regularExpression validates a VIN that it has a valid length
// and no illegal characters.
var validVinExpression =
    regexp.MustCompile("^[A-HJ-NPR-Z0-9]{17}$")
```

Listing 3.9 Combining an Informative Comment and a Variable Name (Go)

Explanatory Comments

In some situations, code passages appear incorrect or at strange when reading. A short comment on why you have implemented this passage in this way and a brief explanation of your decision could help others to understand the block. Regardless of whether the passage could have been solved differently, the intention will become clearer.

The example shown in Listing 3.10 contains an excerpt from the standard Go library with a brief explanation of why the loop was written in the way it was.

```go
// The loop condition is < instead of <= so that the last byte does not
// have a zero distance to itself. Finding this byte out of place implies
// that it is not in the last position.
for i := 0; i < last; i++ {
    f.badCharSkip[pattern[i]] = last - i
}
```

Listing 3.10 Example from the Standard Library with Explanatory Comment in Go

Clarifications

If code cannot be formulated more clearly, for example, due to confusing return values or other parameters, short, expressive comments can make your code more readable or easier to understand.

First and foremost, you should look for a solution with clearer code so that no comments are necessary. However, clearly not every code can be adapted. If you use an external library, for example, or if data is mapped into a data object, as shown in Listing 3.11, you still might need a comment for clarity.

```go
type Subscription struct {
    // true/1 = switched on (no restriction)
    // false/0 = switched off (everything affected)
    PrivacyEnabledByUser    bool
}
```

Listing 3.11 Clarifying Values via Comments

Like any comment, explanatory comments can be wrong and should be replaced by meaningful code.

Warnings and Requirements

You can use comments to point out side effects that are caused by a change as shown in Listing 3.12.

```go
if err != nil {
    //Attention: this string/message is used for alerting in splunk
    return fmt.Errorf("failed to doSth: %v", err)
}
```

Listing 3.12 A Warning with a Comment on Possible Consequences (Go)

Even seemingly awkward design decisions or implementations can benefit from warnings or notes on their consequences so that overly diligent, refactoring-obsessed developers are slowed down and don't "make the code worse." Listing 3.13 shows an example.

```java
//SimpleDateFormat is not thread safe!
//We need to create each instance independently.
SimpleDateFormat df =
    new SimpleDateFormat("MMM dd yyyy HH:mm:ss");
```

Listing 3.13 A Warning about the Consequences of Multithreading (Java)

Side effects or requirements for calling code must be fully and clearly formulated in comments or documentation. The example shown in Listing 3.14 therefore clearly indicates the responsibility of the client.

```go
// Body is the request's body.
//
// For client requests, a nil body means the request has no
// body, such as a GET request. The HTTP Client's Transport
// is responsible for calling the Close method.
// ...
Body io.ReadCloser
```

Listing 3.14 Documenting Requirements for User Code (Go)

Still to Do: TODO Comments

Many developers leave hints in the source code that indicate changes are still necessary in certain places, which they have not yet achieved for various reasons, as shown in Listing 3.15.

Such //TODO comments only make sense if the change cannot actually be made at the moment. The example shown in Listing 3.15 refers to a bug that must first be solved in a third-party system.

```
if testv.Builder() == "darwin-amd64-10_14" {
    // TODO(#23011): When the 10.14 builders are gone, remove this skip.
    t.Skip("skipping due to platform bug on macOS 10.14;
           see https://golang.org/issue/43926")
}
```

Listing 3.15 Notes on Dependencies (Go)

You should not use TODO comments to mark code blocks with notes indicating that no ideal solution has yet been found for this block! Some developers also consider every TODO in the code as a built-in bug because it will probably never be fixed.

For this purpose, most integrated development environments (IDEs) provide an automatic check or display of TODO comments prior to the check-in into a version control system (VCS). This check-in/check-out mechanism prevents unfinished code from being accepted by mistake.

3.1.4 Comments Not Needed

In addition to the comments that can and should be used in a reasonable manner, some comments in the source code are just superfluous. Ideally, well-structured and clearly structured code does not need any comments to explain itself.

Listing 3.16 shows an example of a superfluous comment that provides no added value.

```
public interface Validated {
    //Empty
}
```

Listing 3.16 Is the Interface Still Empty? (Java)

If you write a comment, it should definitely make sense and provide the reader with additional information. Every comment should be well considered and make things clear. Don't leave the reader with new questions!

In the example shown in Listing 3.17, the comment still leaves questions unanswered. Why is an attempt being made to disable the feature, or why can an error be ignored? Is the feature not important after all? What are the consequences if the feature cannot be deactivated?

For this reason, in this example, a helpful step would be to add at least one further comment with the reason why the feature should be deactivated.

```
@Override
protected DocumentBuilderFactory createDocumentBuilderFactory(int
    validationMode, boolean namespaceAware)
    throws ParserConfigurationException {
    DocumentBuilderFactory factory =
```

```
            super.createDocumentBuilderFactory(validationMode, namespaceAware);
        try {
            factory.setFeature("../features/..", false);
        } catch (Throwable e) {
            // we can get all kinds of exceptions from this
            // due to old copies of Xerces and whatnot.
        }
        return factory;
    }
```

Listing 3.17 Comment in Java Apache CXF Source Code (Feature Name Shortened)

Avoiding Duplicate Statements via Comments

Each comment should provide additional information about the code. If the code is listed again in the comment in different words, the comment is simply superfluous.

If code seems to need "structuring" via comments, as shown in Listing 3.18, a better approach is to revise the code instead.

```
//do the calculation
result := thing.calculate()

//check the result if it is valid
isValid := result.IsValid()

//return error if not valid
if !isValid {
....return errors.New("result is not valid")
}
```

Listing 3.18 Duplicate Statement with Comments

In some projects, getter and setter methods are documented via comments due to project or company rules. What added value is the documentation shown in Listing 3.19 supposed to provide? This duplication can be omitted since it does not provide any additional information.

```
/**
 * Sets the bus
 * @param bus the bus
 */
public void setBus(Bus bus) {
    this.bus = bus;
}
/**
 * Gets the bus
```

```
 * @return the bus
 */
public Bus getBus() {
    return bus;
}
```

Listing 3.19 Sample Documentation on Getter and Setter Methods in Apache CXF

Likewise, the statement shown in Listing 3.20 applies to the documentation of the fields of a class or for things that can be clearly read from the code.

```
//the license
private String license;
```

Listing 3.20 Documentation of a Field

> **History of a File**
>
> Back when version management was not as widespread as today, and source code was passed on via file repositories, developers often included the history of a file as a comment in the code. Fortunately, these times are over, and version management systems such as Git have become established practice. History comments have thus become unnecessary and should therefore no longer be used.

Development environments often automatically generate documentation or comment templates. You should therefore evaluate whether each automatically introduced comment is necessary and what additional information it provides.

In turn, you must also check whether the code is comprehensible and clearly formulated. You may need to modify it or add useful comments so that the code is better understood and can be used correctly.

Commented-Out Code

Even though commented-out code is not a comment in the traditional sense, I would like to discuss this topic here.

Commented-out code often remains in the code base, even if it is no longer needed. The reasons for commenting out code are usually unclear, which can confuse other developers reading the code: You cannot judge whether the commented-out code is still relevant or not. The code may have been replaced by a better alternative and simply forgotten at this point. Such ambiguities make it difficult to understand code.

If code is commented out, the reason for this should either be noted as a comment or the code should be deleted completely. Commented-out code contributes to what's called *technical debt*, which describes the additional costs and effort caused by an unstructured, outdated, or insufficiently maintained code base.

3.2 Documenting the Software Architecture

The documentation of a software and especially its architecture fulfills several purposes.

Robert C. Martin's quote that the truth lies in the code is true, but the truth usually involves more than just looking at a section. The documentation of a software architecture has a greater focus; it represents and documents more complex systems.

"The code doesn't tell the whole story." —Simon Brown

In their book *Documenting Software Architectures: Views and Beyond*, Paul Clements and his coauthors have identified three task areas for the documentation of software architectures:

- Development
- Communication
- Basis for analysis and further development

> **Documenting Software Architectures: Views and Beyond**
>
> In 2003, the first edition of *Documenting Software Architectures: Views and Beyond* by Paul Clements and colleagues on the subject of architectural documentation was published. The second edition, published in 2010, was revised and extended and is often described as a groundbreaking, indispensable reference work.
>
> The book not only presents the importance of architecture documentation for communication but also considers it instrumental for supporting decisions, minimizing risks, and ensuring the long-term maintainability of your software.

If new developers or software architects join an existing team, documentation about the architecture helps onboard new members to the existing solution and its structure. With a corresponding presentation, new members and even interested external parties can obtain a structured overview and expand or deepen their knowledge of the software.

In addition to this type of knowledge transfer about the software, the documentation serves to ensure more efficient communication between the members of a development team or across its boundaries.

One of the main tasks of a software architect is to communicate solutions that have been developed or are yet to be developed. The software architecture documentation supports this communication and provides a clear and precise way of exchanging information. In addition to the specific solutions, it also documents the decisions made and the associated justifications so that they can be retraced later, for a deeper understanding.

In many cases, the architect himself is one of the main consumers of the documentation. In most cases, the decisions made and their justifications are interrelated in a complex network, and remembering every decision and its justification is impossible without informative documentation.

Decisions that have been documented can form the basis for future developments and influence them. Cumbersome design decisions can be revised; good decisions can be retained and reapplied during implementation.

In addition to successful decisions, negative experiences can also be documented. In this way, you can learn from your mistakes, and transparent documentation means you can develop more successfully in the future. Mistakes can and should also have an influence on future development and the associated decisions.

Fundamental decisions made early on about the software design within a project should be documented and communicated as soon as possible. This approach prevents similar considerations from being made more than once and possibly leading to slightly different approaches at different points in the application.

Good documentation leads to long-term maintainability of software. Knowledge is passed on and stored and can be used in future considerations.

3.2.1 Documenting Quality Features

In addition to the technical requirements (for example, that some code performs calculations correctly), most software must fulfill further quality features. Software that can generate the right result but can't deliver it correctly or at the right time has little or no added value for a user. Non-technical quality characteristics of a software solution may include the following:

- Performance
- Stability
- Security
- Maintainability
- Capability for incremental updates

High performance can be achieved, for example, by parallelizing tasks or with optimized remote calls that generate a smaller volume of data or are executed less frequently. The possible solutions are varied and depend on the area of application.

If software is to remain maintainable, it can be subdivided into individual components, each of which fulfills its specific task and can be developed independently of one another. Applying the single responsibility principle can help.

All of these quality features must be considered and addressed by the software architecture. Accordingly, these features must be included as requirements in the software architecture documentation, and the selected solution must be documented.

Three questions should be answered for each quality feature:

- What quality feature was addressed with the architecture?
- What solution was chosen to solve the challenge?
- Why does the solution ensure the quality feature?

3.2.2 Rules for Good Software Architecture Documentation

In *Documenting Software Architectures: Views and Beyond*, Paul Clements and his coauthors set out seven rules for good software architecture documentation. Let's now take a closer look at these rules.

Write from the Reader's Point of View

All documentation should be created with a certain target group in mind, and it is precisely for this target group that the documentation should be written.

The authors of the book refer to Edsger Wybe Dijkstra who claimed to spend two hours thinking about how he could make a single sentence in the documentation clearer. He was of the opinion that comments are read by many people and that these two hours quickly pay off if each person can save one to two minutes of confusion.

Dijkstra's principle can be applied to source code, which is usually written once, but read several times.

Target group-oriented documentation is generally not only read and understood but can serve as a reference later. If the reader does not receive the information they need, the documentation will play no further role for them.

Some tips to bear in mind when writing documentation include the following:

- Know the readers: Uninformed assumptions about the readership should be avoided. Know what is expected of the documentation. A brief exchange, if possible, can be helpful.
- Structured documentation: Information should be prepared in a structured manner and appear in appropriate places.
- Use clear terminology: The documentation is read by a wide range of people, sometimes even by people not familiar with the specialist terminology or jargon. When special terms are used, they should therefore be clearly defined, for example, in a glossary.
- No overloading with acronyms: Acronyms (i.e., short words made up of the first letters of their components, such as "API") sometimes make communication within a

project more efficient. However, in documentation, these abbreviations should be used with caution and should always be listed in a glossary.

Avoid Unnecessary Repetition

The same documentation should not be repeated in multiple places. You run the risk that, due to adjustments or different formulations, the information can vary one day. Which part is decisive and actually represents the correct information may not be clear.

Repetition may be useful if the same information is presented in several places with a slightly different perspective or in a slightly different focus.

Avoid Ambiguity

As soon as documentation is not formulated clearly enough and can be interpreted differently, ambiguities will arise. This problem leads to differing opinions and false conclusions.

This kind of problem can be avoided or mitigated with the help of standardized formulations or presentations.

> **Requests for Comments: Indicate Requirement Level**
>
> The *requirement levels* defined for RFCs are an example of standardized text formulations. They are used, for example, to define the meaning of the expressions "must," "must not," "should," "should not," and "may" within the RFC documents.
>
> For more information, see *https://datatracker.ietf.org/doc/html/rfc2119*.

Regardless of which format is used, the corresponding meaning must be clearly defined. Classic *box-and-line diagrams*, such as those often found on whiteboards, can easily cause misunderstandings.

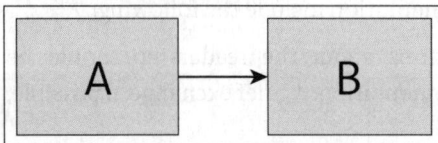

Figure 3.3 Box-and-Line Diagram

As shown in Figure 3.3, for example, the arrow can mean several different things: Does A call B synchronously? Does A instantiate type B, or is a message sent asynchronously from A to B? Is class A possibly a derivative of class B?

In many cases, the best approach is to use widespread, standardized representation formats, for example, Unified Modeling Language (UML) or the C4 model, to create clarity about the representation. Detailed descriptions of both formats will be provided later in this book.

Use a Standardized Structure

The documentation should be created using a predefined structuring template—i.e., using a template, which already specifies sections that are filled with the actual documentation.

This is an advantage for the reader, as the clear structure makes it easier to find information. The template can also be used in more than one project. This means that you can familiarize yourself with new projects and find your way around them more quickly.

In addition, the predefined structures also provide many advantages at the time of creation. Information can be added directly to the respective section as soon as it is created—regardless of whether upstream areas have already been filled or not. In this way, the template serves as a suitable repository for the snippets of information, which together make more sense.

At all times, there is transparency about how much documentation is available and whether sub-areas still need to be filled. If sections are not filled, information is missing; if all sections are complete, the documentation is accordingly complete.

For example, the *arc42* template has become widespread as a standardized structure.

Keep Track of Decisions and the Underlying Reasons

In software development, many individual decisions together lead to a software architecture. Every decision should be made consciously and ultimately result in a well thought-out, suitable software architecture.

Unconsciously made decisions often lead to unclear structures within the application or to unintended dependencies in the source code or in external systems. Changes can be more difficult to implement, and the stability or security of the software may suffer.

Overall, unconscious decisions in a software architecture can lead to a variety of problems and should therefore be avoided at all costs.

Some of the most important decisions that need to be made are, for example:

- Programming language or technology stack
- Chosen architectural style
- Identification of components and interfaces
- Data modeling
- Integration and interoperability
- Security guidelines, mechanisms, protocols, etc.

Every important decision made must be recorded in the documentation with the corresponding justification. It is also best to list the alternatives considered and the arguments in favor of the chosen solution. This means that the decision can be retraced at a later date and the reasons why an alternative is not used can be reviewed.

Keep the Documentation Up to Date, But Not Too Up to Date

For documentation to be useful, it must be up to date and correct. Outdated documentation that does not represent the actual situation causes confusion and will be avoided by the target group as soon as they discover the issue. Questions that could be answered by the documentation then require research elsewhere and generate additional work for the readership.

All documentation should be up to date. But in return, not every innovation requires immediate documentation. If alternative approaches are evaluated or new approaches are quickly discarded, the documentation can be delayed to minimize unnecessary effort. The first thing to consider in this context is how sustainable an innovation is.

Updating the documentation should be a fixed point in the development process. For *Scrum*, for example, the term *definition of done* has become popular. This definition serves as a checklist of items that must be completed before a task can be considered finished. One item on the list should be updating the documentation.

Check the Documentation for Suitability for Use

Every documentation has its own specific target group. Only this group of people should give feedback on the content and decide whether the information their members are looking for is included in an appropriate way.

For this reason, the documentation should be read and reviewed regularly by the relevant target group.

3.2.3 arc42 for the Complete Documentation

In German-speaking countries, *arc42* has spread as a structured architecture description. Initially developed by Gernot Starke and Peter Hruschka, this document structure often serves as the basis for a separate architecture documentation or is sometimes as a process model for an architecture design. These open-source templates can be used directly in your own projects in various formats. For each section, the template also provides a brief description of what should be included in the corresponding section.

Although *arc42* is open source and available in several languages, it is not yet widely used in English-speaking countries. However, the adoption of the iSAQB (International Software Architecture Qualification Board), an international association that offers a standardized training program and certifications for software architects, is certainly contributing to its further spread. The C4 model by Simon Brown, which is often seen as an alternative, will also be presented later.

arc42 provides a clear, simple, and efficient structure for the documentation and communication of software architectures and comprises twelve sections that cover the most important aspects of an architecture or application. Figure 3.4 shows the structure of the entire document template.

Not all structure points are always required, and some chapters can be omitted in documentation based on arc42.

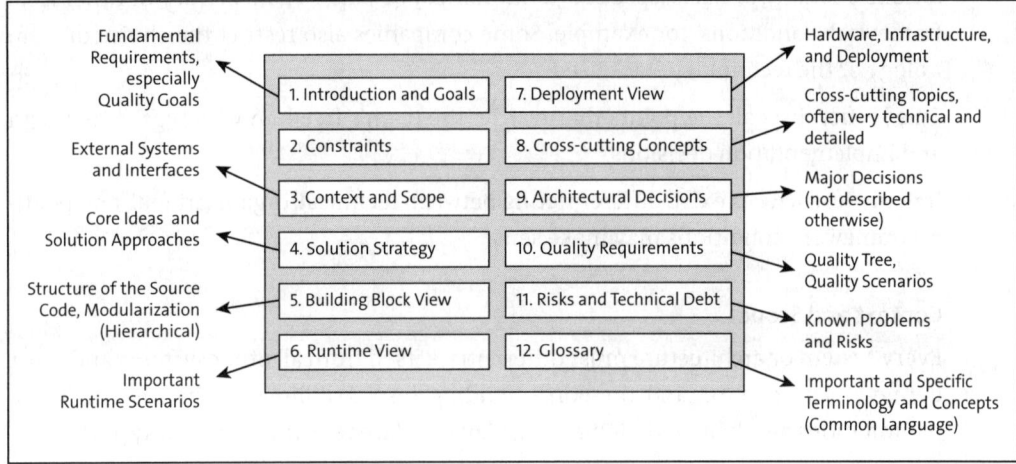

Figure 3.4 Overview of arc42 (Source: https://arc42.org/overview)

> **arc42 Documentation and Examples**
>
> The arc42 templates can be found at *https://arc42.org*. Either blank or prefilled templates are available for download; the prefilled documents themselves represent an extended documentation of arc42 itself.
>
> In addition, you will find detailed examples of various projects at *https://arc42.org/examples*.
>
> We'll use the HtmlSanityCheck sample application by Gernot Starke is used for the presentation. You can find it at *https://hsc.aim42.org*.
>
> The graphics used are licensed as follows: *https://creativecommons.org/licenses/by-sa/4.0/deed.en*.

The following sections walk through the structure points of the arc42 template.

Introduction and Objectives

The introductory section describes the basic requirements and objectives of the system as well as the intention and motivation behind the development of the software solution.

In addition to the technical requirements and the basic business objectives, the most important quality requirements for the architecture are described, as well as all important stakeholders and their expectations about the software.

Constraints

Most software architectures are not created without specifications or framework conditions. In many projects, special attention must be paid to data privacy or other legal framework conditions, for example. Some companies also restrict the choice of technology or the technology stack used.

This section lists all the points that restrict the team's freedom with regard to design and implementation decisions.

If many restrictions exist, differentiating between technical, organizational, and political framework conditions may make sense.

Context and Scope

Every system or architecture must differentiate itself from all other systems and must define a clear context and the corresponding responsibilities of the system. These delimitations also help to define and display interfaces to other external systems.

A context should always be documented from a business- or domain-related perspective. In the event that infrastructure or hardware play an important role, a technical perspective can also be presented.

The contexts and their system dependencies can be presented as text or with the help of diagrams, for example with UML diagrams, as shown in Figure 3.5.

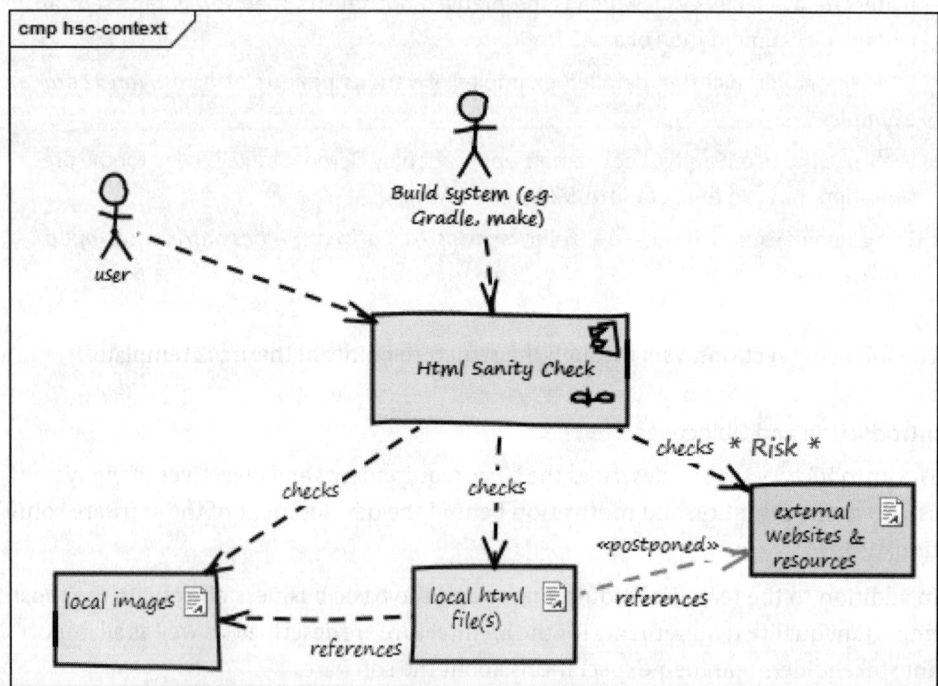

Figure 3.5 HtmlSanityCheck Sample Application: Context

Solution Strategy

This chapter provides an overview of all the fundamental decisions and approaches that determine the design and implementation of the system. They form the basic framework of the application and influence many other decisions and implementation rules.

Let's say, for example, you've selected a Go-based technology stack in the solution strategy. This decision also influences later decisions on the selection of external libraries.

The list of solutions and decisions usually contains the following items:

- Technology decisions
- Representation of the modularization at the highest level of abstraction
- Use of formative design or architectural patterns
- Decisions on the most important quality features
- Relevant organizational decisions

Building Block View

The building block view can serve as an initial overview and shows the static breakdown of a system into its components and their relationships with each other. The included modules, components, subsystems, classes, interfaces, packages, libraries, frameworks, layers, partitions, and more are best listed using diagrams. This view refines the context view in which the system is displayed as a large box.

The abstract representation of the structure of the system can help you obtain a better overview of the source code and can serve as a basis for communications at an abstract level. Implementation details do not play a role in this context.

A distinction can be made in the presentation between *white boxes* and *black boxes* as well as between multiple levels with different degrees of detail.

In the first level, the most important subsystems, components, or parts of the system are listed, and in the second level, if necessary, these elements can be described in more detail in the white boxes.

Not every component from level 1 must be shown in the second level.

The diagram shown in is a representation of level 1 of the *HtmlSanityCheck* (HSC in the diagram) example from *https://hsc.aim42.org/*. The individual boxes and the corresponding interfaces with their tasks are described in table format in the documentation for the respective level.

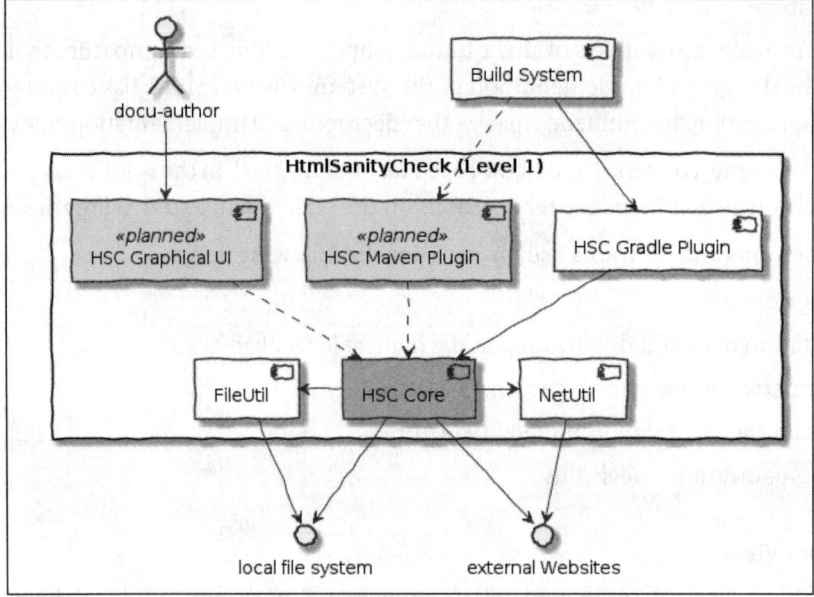

Figure 3.6 HtmlSanityCheck Sample Application: Level 1 (Source: https://github.com/aim42/htmlSanityCheck)

In level 2, the HSC core components are shown in detail using a white box, as shown in Figure 3.7. Here too, the individual components are explained in the documentation in a table with their corresponding tasks.

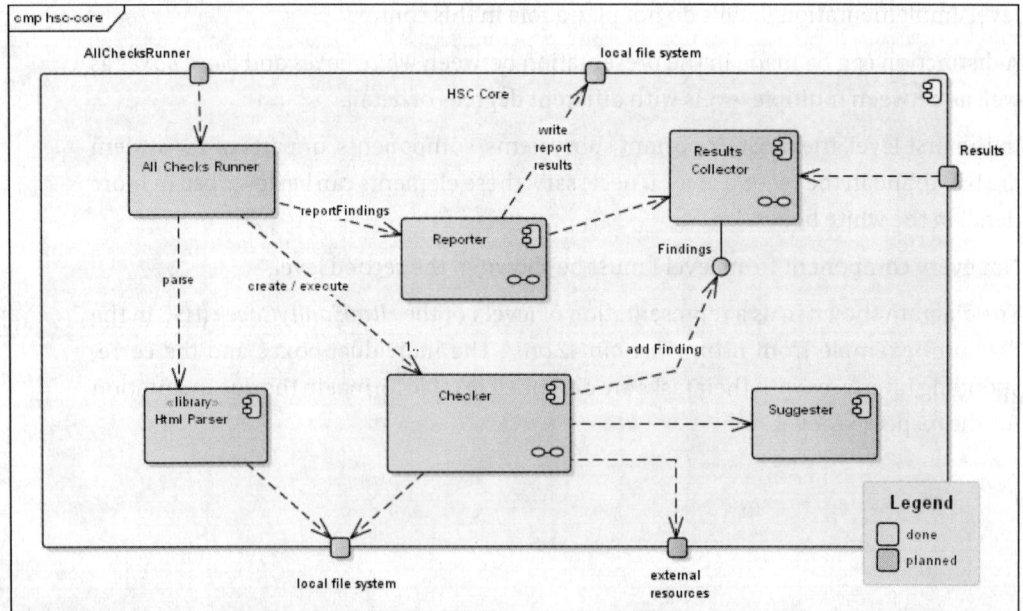

Figure 3.7 HtmlSanityCheck Sample Application: Core Display on Level 2 as a White Box

Runtime View

The runtime view shows important or critical concrete processes, tasks, and relationships between the building blocks of the architecture. These building blocks can be internal or external components.

Scenarios are used to deepen the knowledge of the individual components and their interactions with other components. Individual scenarios can be described using text or diagrams.

Deployment View

Software is executed on hardware. The distribution view shows the technical infrastructure with all environments, computers, processors, networks, and network topologies and assigns the software modules accordingly, as shown in Figure 3.8.

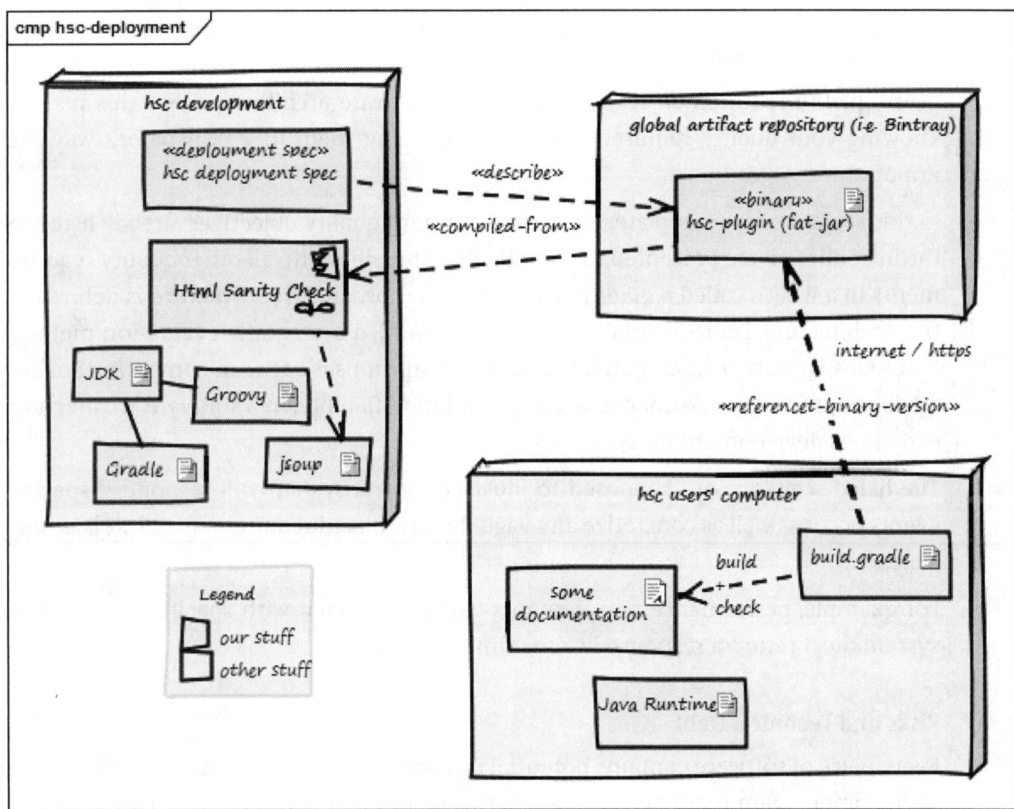

Figure 3.8 HtmlSanityCheck Sample Application: Distribution View of the Application

CrossCutting Concepts

This section documents key regulations, solutions, and concepts that affect multiple building blocks. These elements can involve widely different topics, such as the following:

- Models, especially business models
- The architecture style used
- Architecture or design patterns used
- Rules for the specific use of technologies or libraries
- Implementation rules

Architectural Decisions

Architecture decisions should always be clearly thought through, consciously made, and documented with sound reasons.

In this section, the important architecture decisions are documented with a corresponding justification, and the documentation of rejected alternatives should also be included here.

Quality Requirements

Every quality requirement can influence the software architecture. For this reason, knowing your quality requirements and presenting them in a transparent way are important.

In this section of the template, the most important quality objectives already listed in the introduction are presented as completely as possible with all other quality requirements in a what's called a *quality tree* with scenarios. This tree structure is defined in the Architecture Tradeoff Analysis Method (ATAM), a systematic evaluation method that helps identify risks and understand the compromises between competing quality attributes, such as performance, security, and modifiability, in a software architecture early in its development lifecycle.

The listed scenarios are supposed to illustrate how a system will respond if specific events occur as well as concretize any vaguely formulated requirements as well as possible.

For example, performance requirements could be defined with specific values: "The system must return a response in a maximum of 100 ms."

Risks and Technical Debt

Every piece of software contains potential problems or immature areas with which the development team is not yet satisfied. In this section, these problems are presented in a prioritized list and are therefore transparent.

The items in the list can already be provided with suggested solutions so that they can be scheduled directly.

Glossary

The glossary explains all important domain and technical terms used for communication with individual stakeholders. Generally known acronyms such as HTTPS or REST do not need to be listed here.

A glossary can also be helpful as a reference for translations in international environments.

3.3 Representing Software in Unified Modeling Language

The *Unified Modeling Language (UML)* is a modeling language that was developed, among other things, to improve communications between software developers and make it easier for them to understand the software. The graphical language elements of UML can be used to specify, design, document, and visualize software systems.

The first version of UML was developed in the 1990s, after several modeling languages and methods for object-oriented software development had already been created. Grady Booch, Ivar Jacobson, and James Rumbaugh were instrumental in defining the language during their time at Rational Software and are therefore considered the "fathers" of UML.

In 1997, UML was handed over to the *Object Management Group (OMG)* and accepted as standard. Since then, UML has been continuously developed.

In today's software systems, UML is a widely used tool for modeling and for standardized documentation, even if the full range of functions of the language is rarely used. The familiar graphical notation is only one aspect of the representation of the models described by UML.

The UML 2.3 specification defines 14 different diagrams, which can be roughly divided into two main categories:

- **Structure diagrams**
    - Class diagram
    - Object diagram
    - Package diagram
    - Component diagram
    - Deployment diagram
    - Profile diagram
    - Composite structure diagram
- **Behavior diagrams**
    - Activity diagram
    - Use case diagram

- Interaction overview diagram
- Communication diagram
- Sequence diagram
- Timing diagram
- State diagram

UML diagrams are often used in modern software development. For this reason, the most well-known and widely used diagrams are presented in the following sections. This discussion will enable you to create your own diagrams; however, for reasons of space, I cannot offer a complete introduction to UML introduction in this book. For the diagrams in this book, I also use the UML notation for visualization. The widely used, easy-to-use, and (unlike many commercial tools) lightweight open-source modeling tool *PlantUML* allows you to create graphical UML diagrams from text files, usually directly from the development environment. I will present this tool in more detail in Section 3.5.2.

> **PlantUML**
>
> *PlantUML* is an open-source component for creating diagrams from text files. You can find PlantUML and its documentation at *https://plantuml.com/*.

3.3.1 Use Case Diagram

A use case diagram can illustrate the functions or actions of a system and their interactions with users or other systems. The objective is to show the functional requirements of the system and visualize how a user can interact with the system to achieve their goals.

Three elements are defined, as shown in Figure 3.9:

▶ Actor: An *actor* is a user who interacts with a system. Actors are depicted as stick figures. They can be people, but also external systems.

▶ Use case: This is a single action that can be performed by a user. It is displayed as an ellipse and can be subdivided by corresponding relationships or extended by means of inheritance.

▶ Context: The system context is drawn as a rectangle and encloses the use cases defined in it.

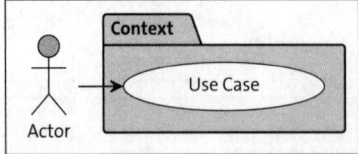

Figure 3.9 Elements of a Use Case Diagram

Use case diagrams do not contain any further details, such as information on the order in which the individual interactions must be carried out. They are primarily intended to provide an overview of the system and its context and are usually supplemented by a textual description.

Figure 3.10 shows examples of use cases for a training company in which a customer can book training courses, either online or by telephone. Once a booking has been received, the reservation is confirmed by a member of the training management team, who must first book a corresponding room. The appointment confirmation includes dispatching a confirmation message to the participant.

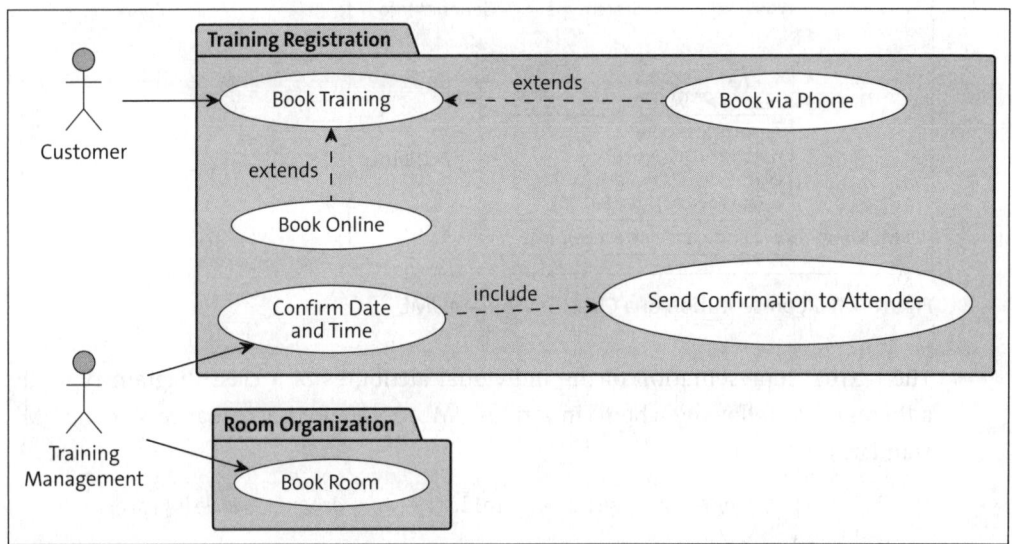

Figure 3.10 A UML Use Case Diagram

3.3.2 Class Diagram

Class diagrams are by far the most frequently used diagrams in object-oriented programming. They are used for the graphical representation of classes and interfaces and their relationships with each other.

Each class and each interface are displayed as a rectangle that contains the attributes and methods of the class or interface in addition to the name.

Figure 3.11 shows the PlantUML modeling tool for classes as it is used in this book. The following information can be displayed for a class or an interface:

- Name: The name of the class or interface.
- Type: Whether it is a class, an abstract class, or an interface. In PlantUML, classes are labeled with "C"; interfaces, with "I"; and abstract classes, with "A."
- Stereotype: The purpose of a class can be represented with *stereotypes*. Some types, such as the entity type used in our example, have already been defined in UML 2.1.

Classes marked with entity, for example, are classes for implementing business logic.

- Generics information: Some programming languages allow classes to be parameterized. In Java or Go, for example, these classes are called *generics*. With this addition, called *template arguments*, the corresponding information can be specified.
- Attributes: A list of all attributes with their corresponding visibilities.
- Methods: Like attributes, all methods are listed with their visibility, parameters, and return types.

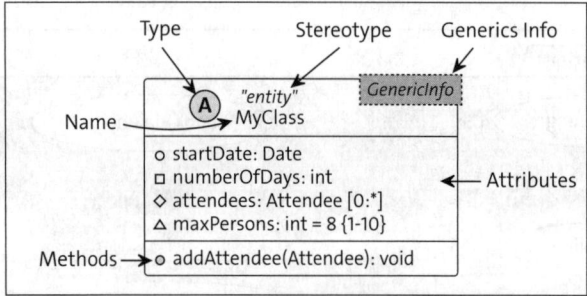

Figure 3.11 Representation of a Class Using PlantUML

The textual representation of the individual attributes of a class diagram typically adheres to the following schema in various UML tools (and in accordance with the UML standard):

[Visibility] [/] name [: Type] [Multiplicity] [= default value] [{property value*}]

In PlantUML, the visibility of the elements is defined accordingly within the text files via individual characters and visualized via symbols in the graphical representation. The PlantUML syntax uses the following characters and symbols:

- A circle corresponds to public or "+"
- A square corresponds to private or "-"
- A diamond corresponds to protected or "#"
- A triangle corresponds to package or "~"

Method signatures are described and visualized in the same way. An important point in this context is that the return value is included at the end of the signature, not at the beginning as in many programming languages:

[Visibility] name [({Parameter})] [: Return type] [{property value*}]

The example shown in Figure 3.12 is a class diagram for the administration of a training session, in which relationships between the classes are also shown.

3.3 Representing Software in Unified Modeling Language

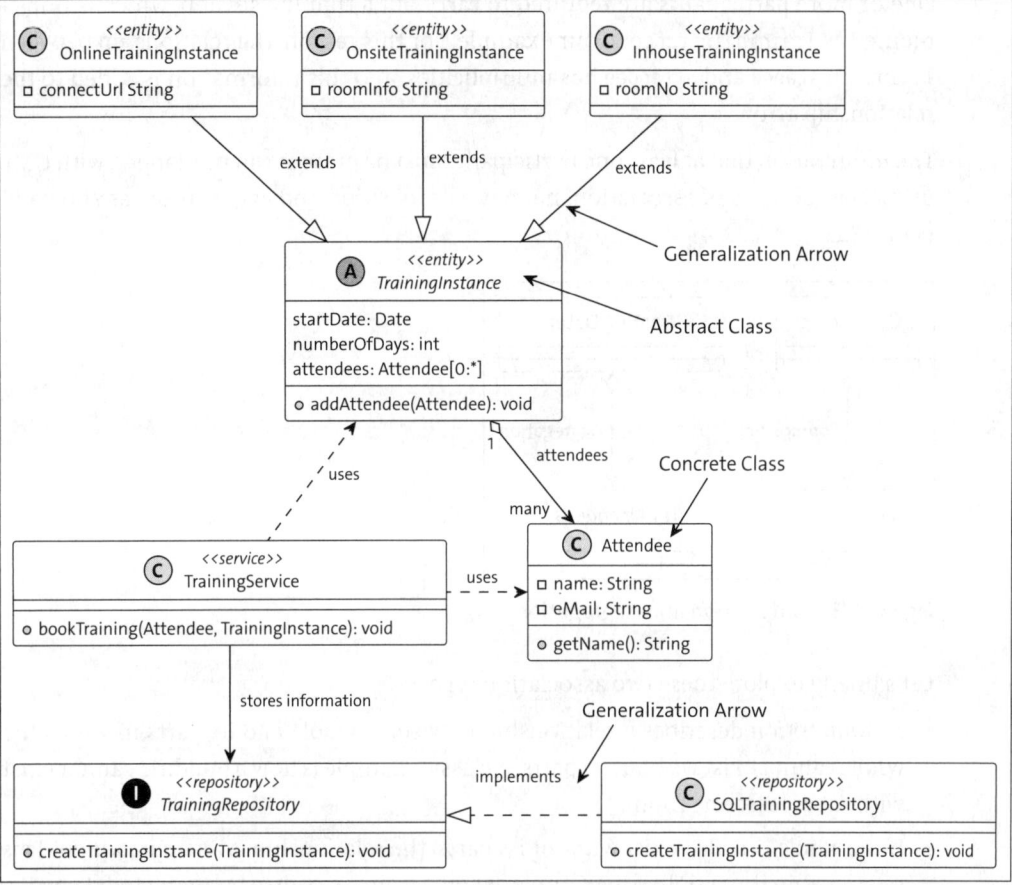

Figure 3.12 Example Class Diagram

This example contains the abstract TrainingInstance class, which is extended by the OnlineTrainingInstance, OnsiteTrainingInstance, and InhouseTrainingInstance classes. This generalization is represented by a straight, solid arrow that leads from the specific to the general implementation and has an unfilled arrowhead.

Interface implementations are also displayed using a generalization arrow, but the line of the arrow is dashed. In this example, the SQLTrainingRepository class implements the TrainingRepository interface.

Relationships between classes, *associations*, are represented by simple arrows. For example, TrainingService uses TrainingRepository to store data. Multiplicities are often used to specify how many of the referenced objects are related to the other objects. This information is then written to the arrow as additional information. In our example, TrainingService has exactly one TrainingRepository instance and therefore is not specified here.

One or more participants are required to carry out a training session, which is implemented as `TrainingInstance` in our example. For this reason, the relationship between `TrainingInstance` and `Attendee` has multiplicities, and this information is added to the relationship arrow.

The information that at least one participant must be present can be mapped with UML using special types of association, namely, composition and aggregation, as shown in Figure 3.13.

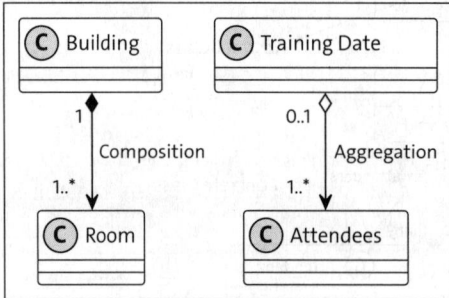

Figure 3.13 Composition and Aggregation

Let's briefly explore these two association types:

▶ A *composition* describes a relationship between a whole and its parts in which the whole cannot exist without its parts. A classic example is how a building cannot exist without at least one room.

▶ If the whole can also exist without its parts, the relationship is an *aggregation*. This relationship is possible, for example, when a training course is conducted: Even if a `TrainingInstance` without an `attendee` makes little sense, it must be possible to create a `TrainingInstance` without an `attendee` during planning.

An aggregation is shown with an unfilled diamond, which is located at the end of the relationship on the whole. In a composition, a filled diamond is used instead, as shown in Figure 3.13.

3.3.3 Sequence Diagram

Sequence diagrams represent interactions in a system and model the exchange of messages between different objects.

Each object has what's called a *lifeline* in the diagram, which serves as the starting point for sending or receiving a message and is shown as a dashed line below the object. The exchange of messages between the objects is represented by arrows.

A branching or decision syntax is not provided for in sequence diagrams, and a variant of a sequence is always shown. If multiple variants of a process are required, multiple diagrams must be created accordingly.

The example shown in Figure 3.14 is a sequence diagram for booking a training. A customer, whose lifeline is not shown for better readability, first triggers a new booking in the booking system.

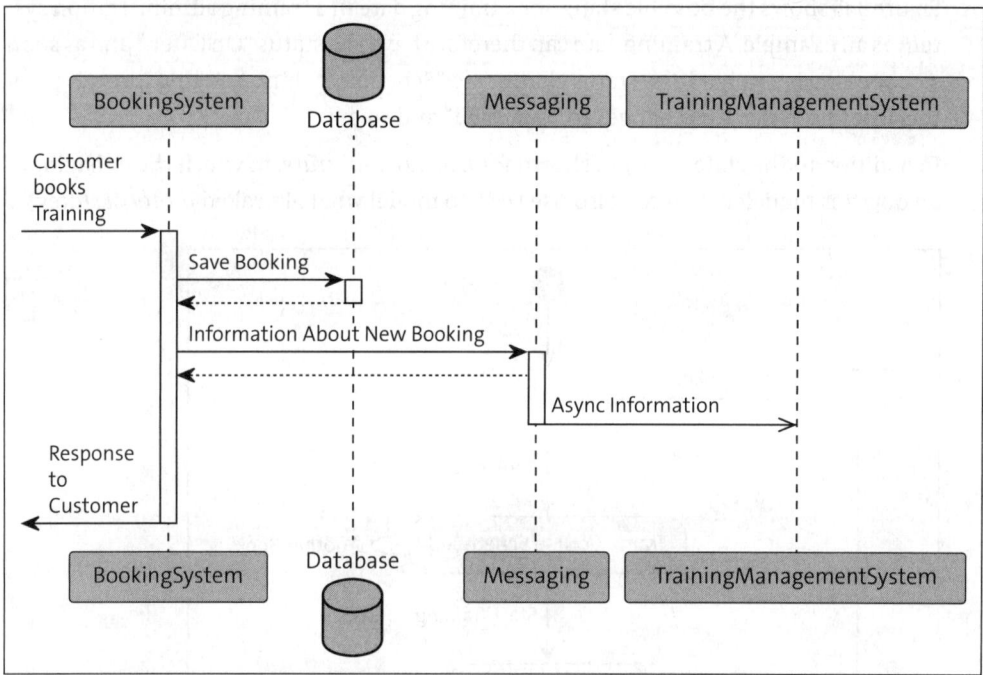

Figure 3.14 Sequence Diagram Example

The system then saves the data in a database and generates a message in the messaging system that a new booking has been received. The process is completed for the customer for the time being. However, the message with the information that a new booking has been created is transmitted asynchronously to the TrainingManagementSystem.

The distinction between synchronous and asynchronous messages is made via the tip of the arrow during message transmission: If the arrowhead is filled in, it is a synchronous message. If the arrowheads are not filled in, this message is an asynchronous message, as shown in our example, on the right, when a message is redirected to the TrainingManagementSystem.

3.3.4 State Diagram

Each object in a system can assume different combinations of internal information and thus different states during its life cycle. The possible states and their transitions can be visualized using a UML state diagram.

Each state diagram has a start point and an end point, which are also displayed as such. Unlike the start point, the end point has a border.

An object's states and their possible transitions are listed between "Start" and "End." Transitions are represented by arrows with an optional label; states are visualized as rectangles with rounded corners.

Figure 3.15 shows the possible states for a training date in a training administration system as an example. A training date can therefore have the status "On Offer," and as soon as a booking is received, the status changes to "Scheduled." If no booking is received by the deadline, the status changes to "Canceled" and so on.

In addition to the state diagram shown for a *behavioral state*, in which the behavior of an object is modeled, you can also use UML to model what are called *protocol states*.

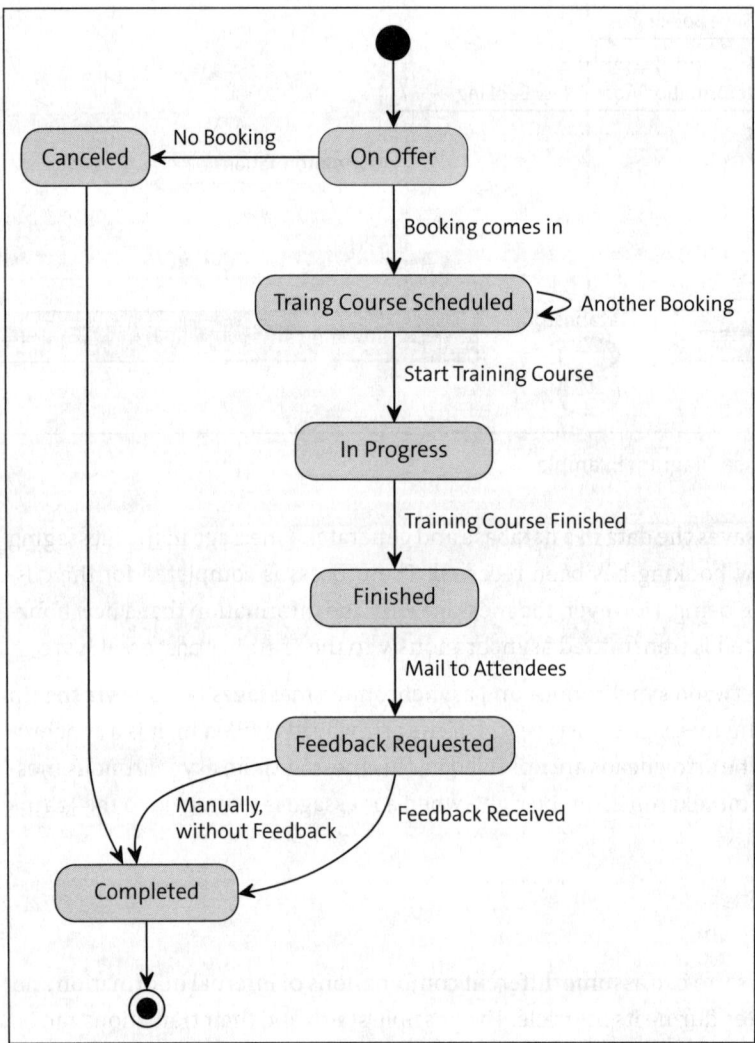

Figure 3.15 TrainingInstance State Diagram (Behavioral State)

Protocol states describe the permissible use of an object or its behavior, as in our example shown in Figure 3.16. The diagram specifies the possible methods that can be called in addition to the states.

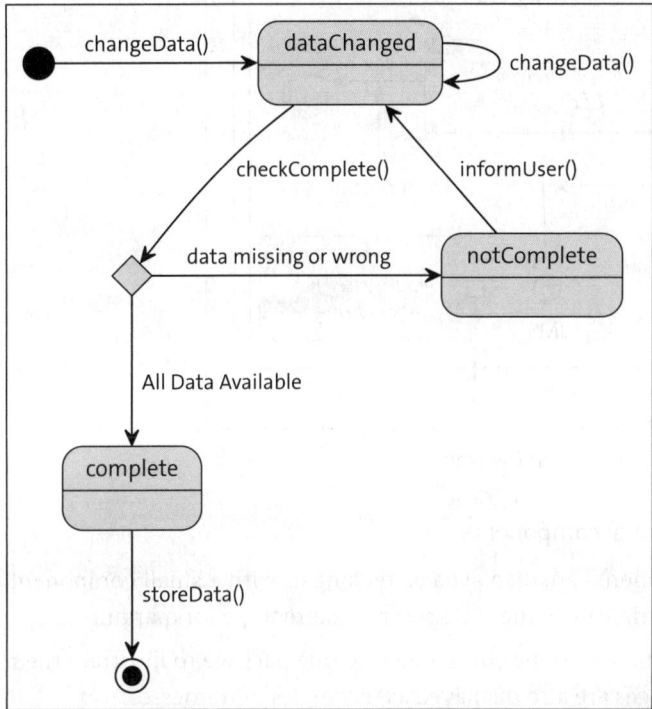

Figure 3.16 Protocol States with State Diagram

3.3.5 Component Diagram

A *component diagram*, another structural diagram in UML, represents components and their relationships to other components in the system.

In UML, the term *component* refers to a module that consists of several classes and can be regarded as an independent system or subsystem. Interfaces define the connections to other components of the rest of the system.

Component diagrams can be used to visualize software systems at a high level of abstraction in order to provide the best possible overview of all components involved. This representation is particularly useful in cloud or microservice-based applications, where multiple self-contained components interact with each other.

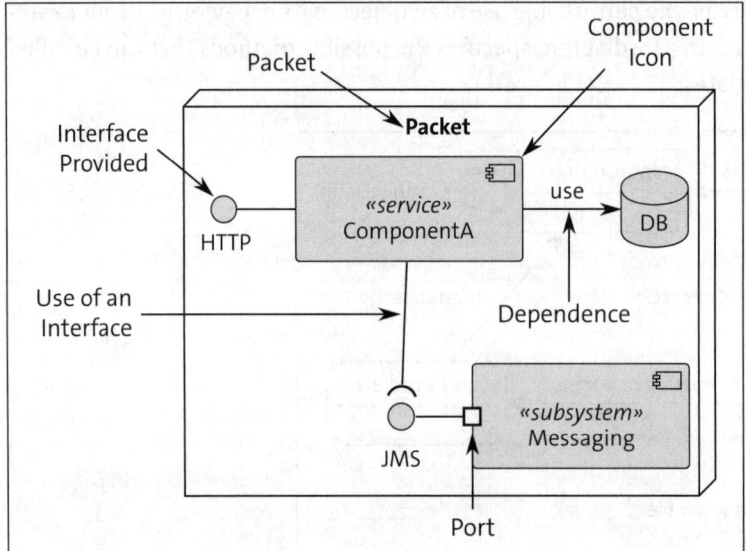

Figure 3.17 Components of a Component Diagram

The diagrams consist of several components:

- Component: The components are displayed as rectangles with a small component symbol. Some display variants use the <<component>> stereotype for marking.
- Package: Several components can be combined into one package to illustrate their close relationship. Packages are also displayed as rectangles or frames.
- Interface: Each component can provide one or more interfaces for the communication with other components. These interfaces can be used as interaction points. Available interfaces are represented by a circle with a connecting line. Used interfaces are documented with a semicircle and a corresponding line.
- Port: Sometimes, calls to a provided interface are delegated to internal classes, which is visualized via ports represented as squares.
- Dependency/relationship: As in other UML diagrams, dependencies are drawn as a line between the components. As before, the arrow indicates the direction of the dependency.

Figure 3.18 shows a component diagram for a training administration software. The diagram shows three independent application components and a cloud-based messaging solution that are connected to each other.

A booking can be triggered via the Booking component, which is redirected to the Administration component as an event using the Messaging solution. The Print component is called via an HTTP endpoint and produces documents for the training course, which are saved in a file repository.

3.3 Representing Software in Unified Modeling Language

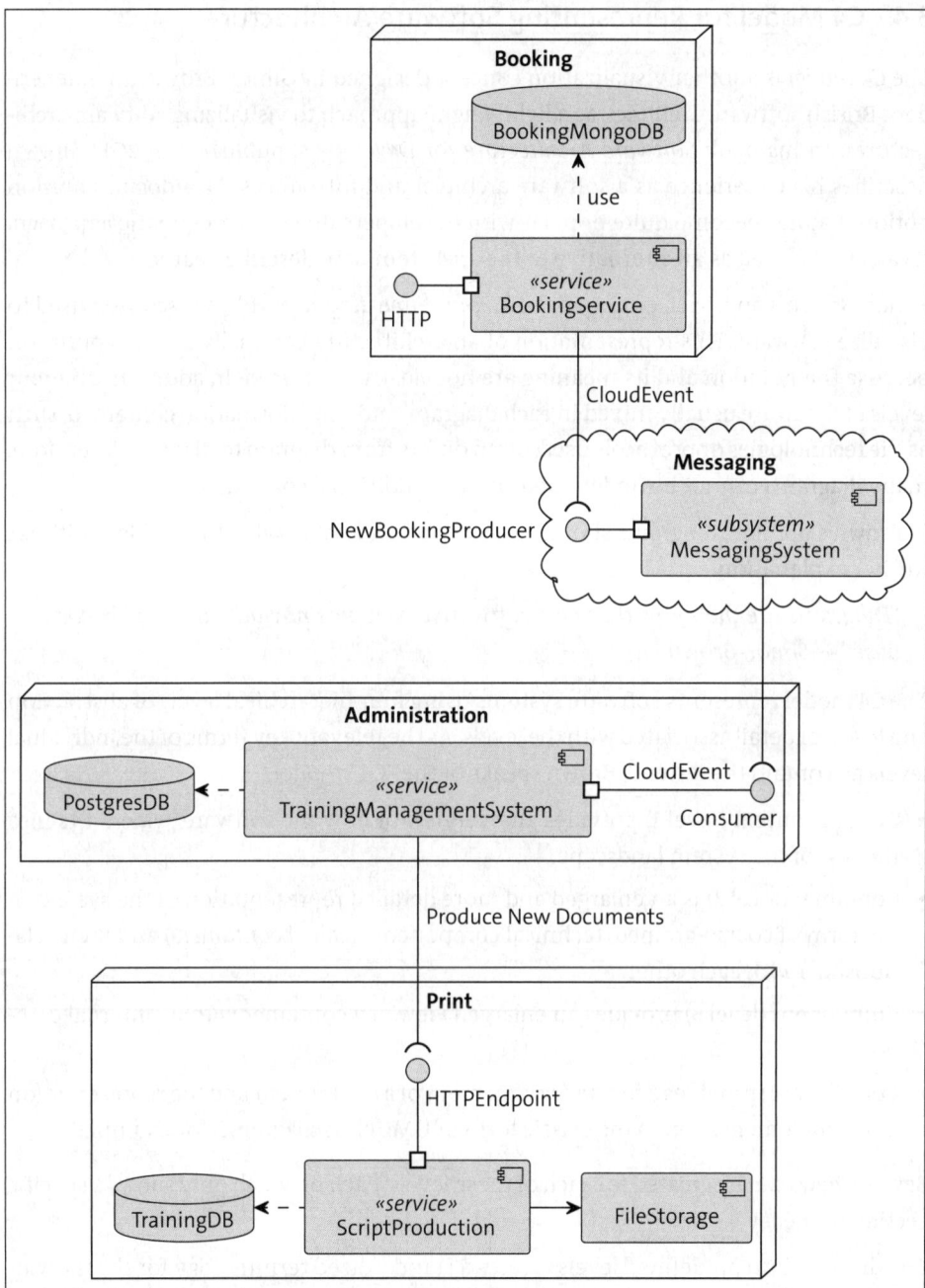

Figure 3.18 Component Diagram for the Training Application

3.4 C4 Model for Representing Software Architecture

The *C4 model* is another visualization concept designed by Simon Brown, an independent British software architect, as a lightweight approach to visualizing software architectures. In his book *Software Architecture for Developers*, published in 2012, Brown describes his experience as a software architect and introduces C4, a documentation option that has become quite popular with developers due to its pragmatic approach. It can also be used as an alternative to the arc42 template described earlier.

Simon Brown's approach critiqued the *box-and-line diagrams* that are so often used to visualize software. This representation of an architecture can easily lead to confusion because the notation and its meaning are not clearly expressed. In addition, different levels of detail are usually mixed in each diagram, and the information contained, such as the technologies or protocols used, often differs from diagram to diagram. Therefore, many diagrams cannot be understood without additional context.

In Brown's opinion, diagrams should be clear and understandable to outsiders without further explanation:

> *"Diagrams are the maps that help software developers navigate a complex codebase."* —Simon Brown

The C4 model represents software systems using four hierarchical levels of abstraction and levels of detail associated with the levels. As the relevant key terms of the individual levels all contain the letter C, Brown speaks of the "C4" model:

- System context (level 1) provides an overview of how the software system fits into the rest of the system landscape.
- Container (level 2) is an enlarged and more detailed representation of the system in the form of coarse-grained, technical components (called *containers*) and their relationships with each other.
- Component (level 3) provides an enlarged view of a container with its internal components.
- Code (level 4) provides a further enlargement of a component and the representation of its implementation in the classic form of UML class diagrams, for example.

Several *views* can be created for each of these levels, each of which only shows a specific section of the level.

The division into predefined levels creates a standardized terminology for the individual building blocks of an architecture and describes corresponding levels of detail for a representation. Figure 3.19, Figure 3.20, Figure 3.21, Figure 3.22, and Figure 3.23 provide an overview of C4 representations and their levels of detail and scope. More precise details cannot yet be seen in this overview. I will describe them in the following sections.

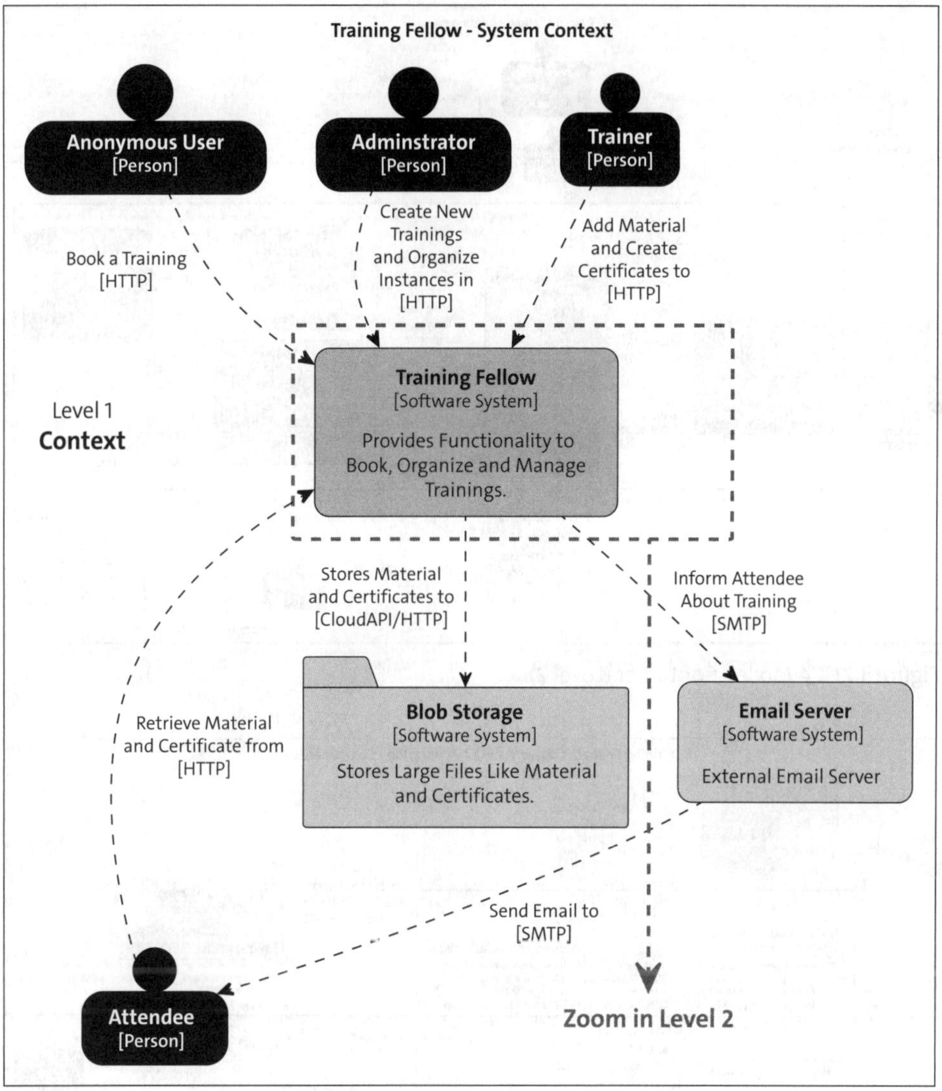

3.4 C4 Model for Representing Software Architecture

Figure 3.19 C4 Model: System Context (Level 1)

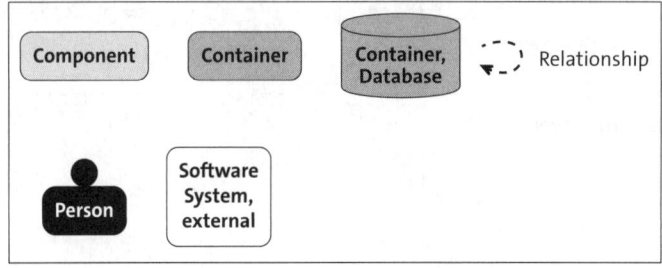

Figure 3.20 C4 Model: System Context (Level 1), Legend

3 Source Code and Documenting the Software Development

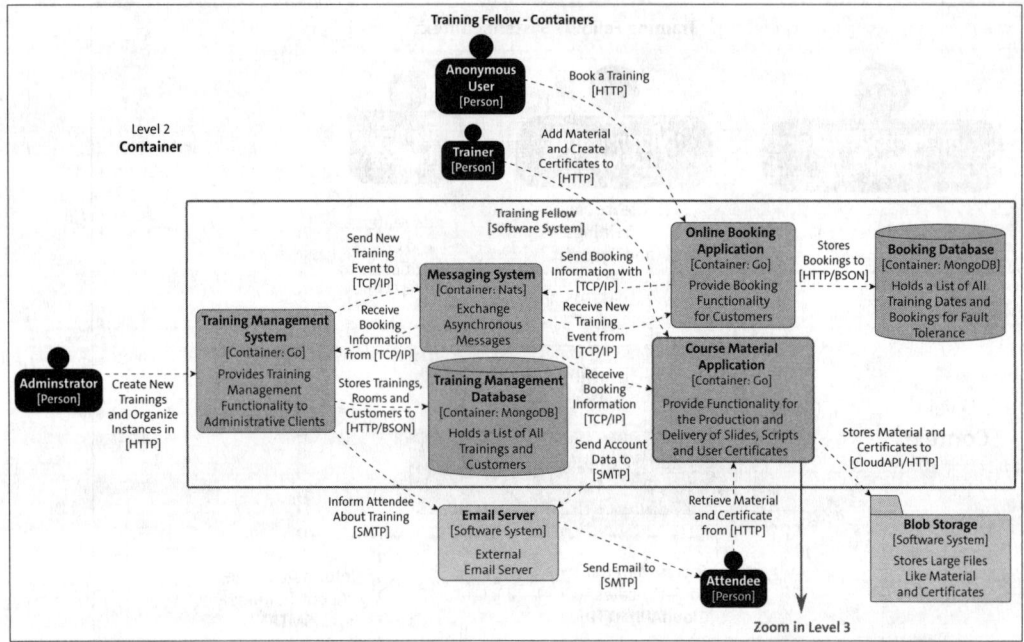

Figure 3.21 C4 Model: Container (Level 2)

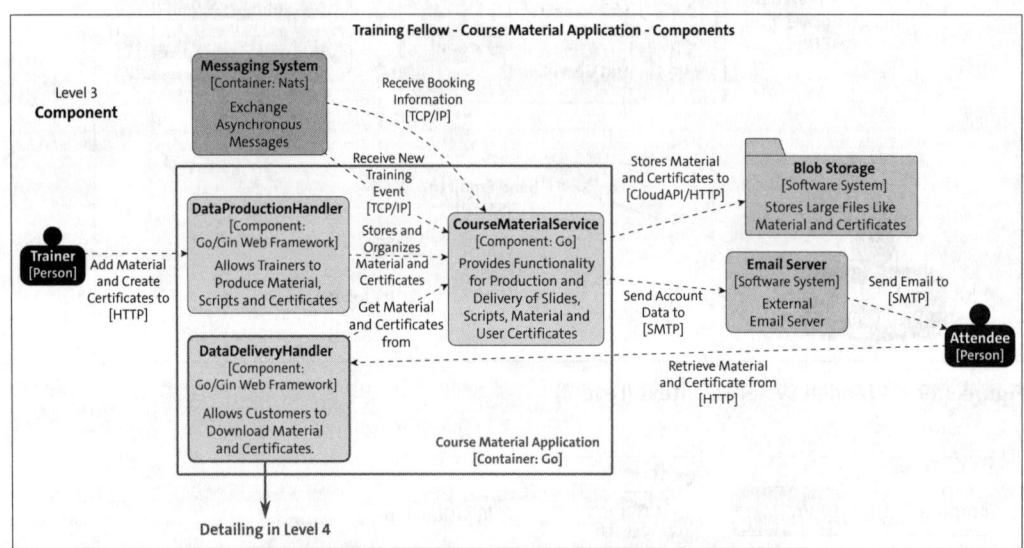

Figure 3.22 C4 Model: Component (Level 3)

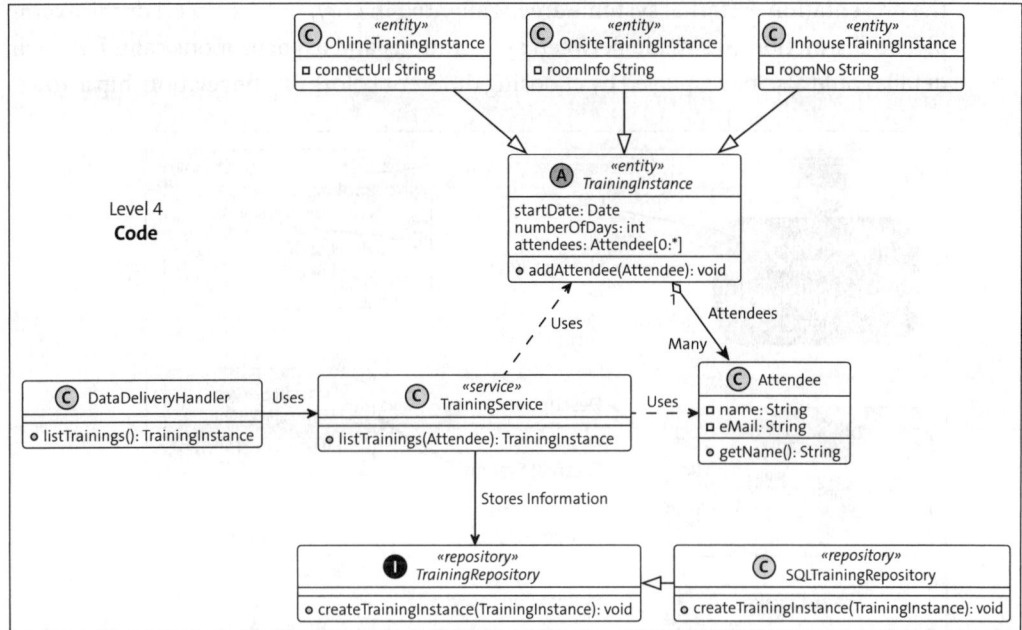

Figure 3.23 C4 Model: Code (Level 4)

With various tools, you have the option of displaying the models in such a way that you can switch to the corresponding refinement level during display, thus allowing the viewer to navigate interactively through the model.

C4 diagrams are often compared with maps, for instance, Google Maps. In these cases as well, various information is displayed at different levels of detail, and special views provide selected details. For example, a railway map does not contain any information about the highway network, which is not relevant in this context. You can zoom in for more information and zoom out again for a better overview.

In contrast to UML, since no fixed notation or design language is prescribed for C4 diagrams, you can define and use your own custom elements or designs.

However, as recommended by Brown, if you create your own elements, you should create a legend either directly in the diagram or in a separate explanation. Doing so allows you to clearly indicate the meaning of shapes, line type, color, borders, and abbreviations and thus increase comprehensibility. Figure 3.20 shows a legend for the diagram shown in Figure 3.19.

A notation of the C4 diagrams, as generated for the examples using *Structurizr*, is widely used. Structurizr is a modeling tool that enables you to generate different views from a common textually described model. I present the tool in more detail in Section 3.5.3.

The common forms of presentation are summarized in Figure 3.24. Each element has a name, a specified element type, and a description. Technical details can also be included, which can be displayed depending on the view's level of detail. Colors support the presentation. External systems are highlighted in gray, for example. For the example, no distinction was made between synchronous and asynchronous calls, but such details could also be displayed by choosing different colors for the relationship arrows.

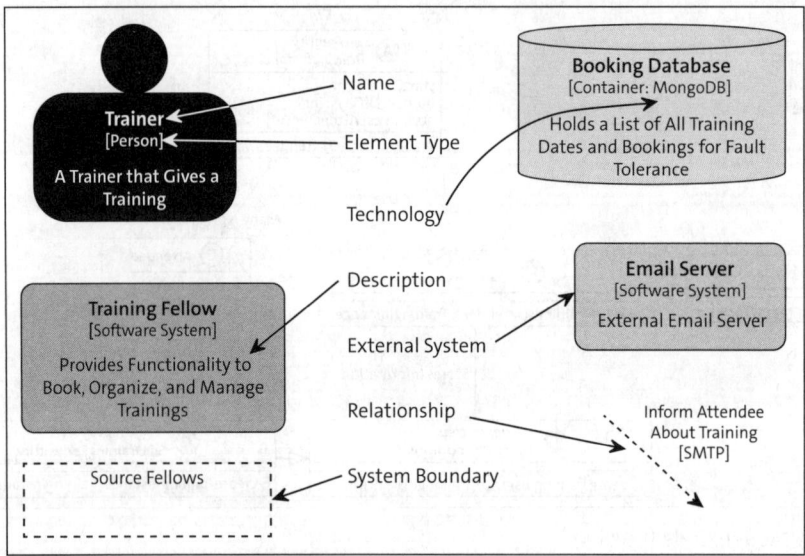

Figure 3.24 Legend for the Notation Used

3.4.1 System Context (Level 1 or C1)

The system context diagram is intended to give both technically and non-technically interested parties an overview of the system and answer the following questions:

- What is the software system that has been or will be created?
- Who uses the system?
- How does it fit into the system landscape?

At this level, only a few details are displayed, as shown in Figure 3.25 for the Training Fellow sample application. The goal is to provide a rough overview of the system and its dependencies, the *big picture*.

The system is represented as a central box around which the other participants are arranged. All dependencies to external systems and users are drawn in the diagram, and ideally, each relationship is provided with brief information on why and how communication with the external system or user takes place.

3.4 C4 Model for Representing Software Architecture

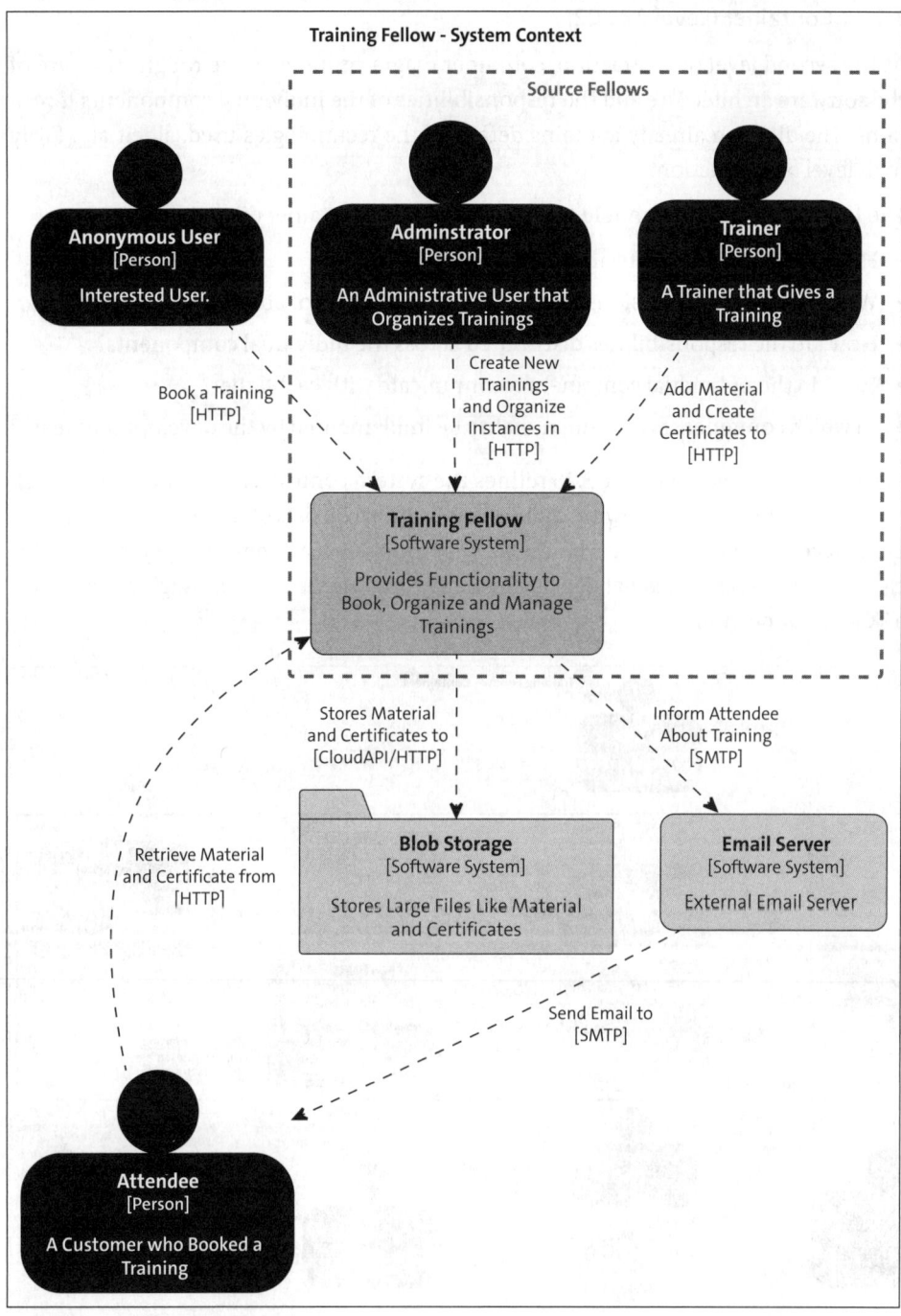

Figure 3.25 C4 System Context Example

3.4.2 Container (Level 2 or C2)

At the second level of abstraction, container diagrams indicate the rough structure of the software architecture and the responsibilities of the individual components it contains. The diagram already contains details of the technologies used, albeit at a fairly high level of abstraction.

The following questions should be answered by the container diagrams:

▶ How is the system's software architecture structured?

▶ Which technology decisions were made at a high level of abstraction?

▶ How are the responsibilities distributed across the individual components?

▶ How do the individual containers communicate with each other?

▶ In which container should functionality be implemented by the development team?

The example shown in Figure 3.26 refines the system context view of the Training Fellow sample application, showing that the application consists of three separate containers that communicate with each other asynchronously via a messaging system. In this case, the messages are exchanged between the container and the messaging system via a TCP/IP connection.

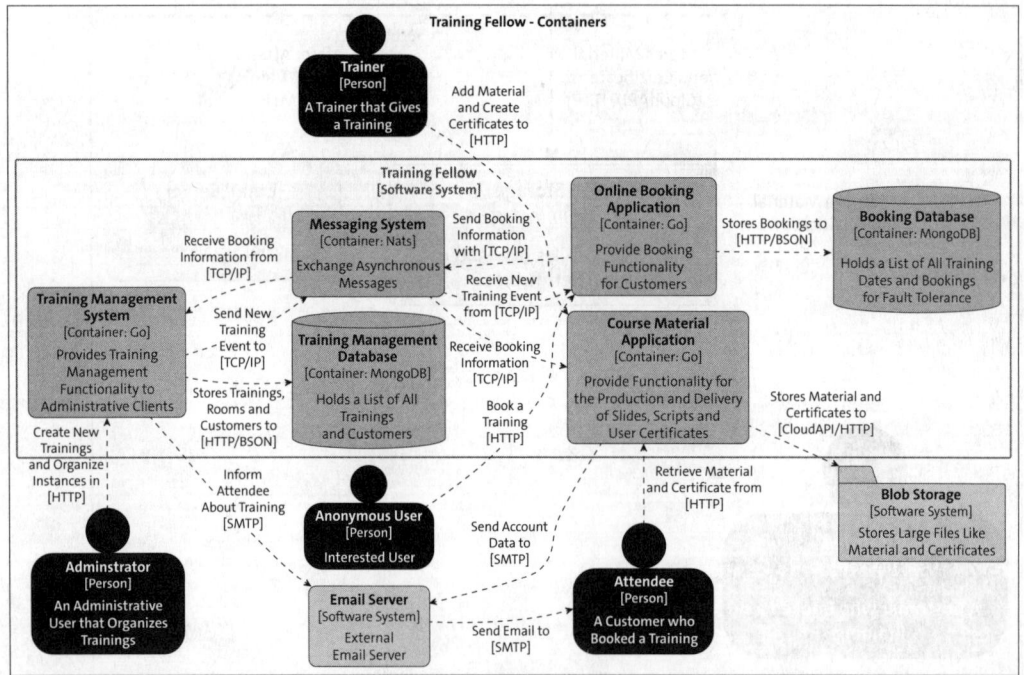

Figure 3.26 Example of a C4 Container Diagram

The responsibilities of each component are noted in each case so that extensions can be made at the appropriate points.

The diagram captures how the training management system (TMS) container is responsible for managing training courses, for example, and makes this functionality available to an administrative user via an HTTP connection. The container is implemented in Go and uses its own MongoDB database for its tasks, in which all training courses and customer data are stored. The diagram also shows that the TMS container sends status changes of training bookings to the participants via email and exchanges messages with the other containers via the messaging system. The TMS is informed about newly scheduled training dates and receives messages when new bookings have been made, which is indicated by the arrows pointing toward the container.

The level of detail of a container diagram is not explicitly specified and can be determined by the user. No technical details need to be provided, but certain decisions can be better understood by presenting more information.

3.4.3 Component (Level 3 or C3)

The third level of the C4 diagrams shows individual components with additional details on their technologies and internal structures. Figure 3.27 shows the Course Material application, which is responsible for creating and managing training materials and certificates.

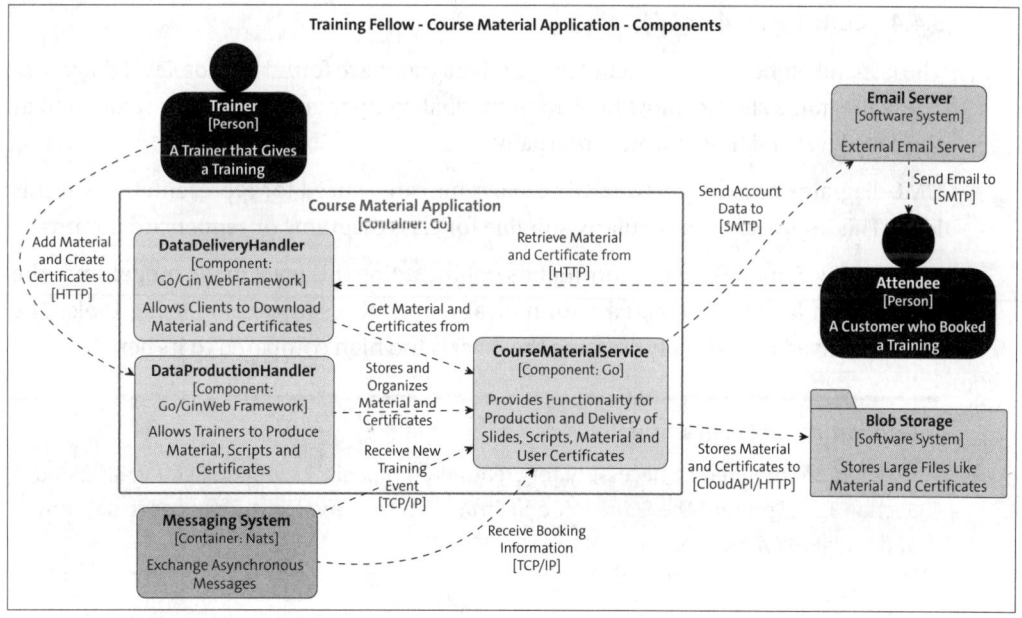

Figure 3.27 The Course Material Component in the C4 Model

The diagram always shows just one container and is intended to answer the following questions:

▶ What components does a container consist of?

- Are all components assigned to a container?
- Is it clear on an abstract level how the software works?

Three internal components are listed for the Course Material application in our example: two handler implementations, which are responsible for HTTP-based communication with users, and a service component, which contains the business logic. Each component has its own responsibilities as well as additional information on the internal technical details. The container diagram shows that our app is a Go-based application, and from the additional details, we can clearly see that the *Gin* web framework is used for its handler implementations.

No infrastructure components or *cross-cutting concerns*, such as those required for logging, are listed in this sample diagram because these components are not business-critical for the use case. If this information is important or critical, the required components can be included, or their use can be indicated by an additional comment. You can also display the affected components separately by color-coding and describing the coding in a legend.

The same rule applies to shared functionalities. If a functionality is business critical, the information should be included; if not, a reference is usually sufficient.

3.4.4 Code (Level 4 or C4)

The last and highest level of detail in the C4 diagrams are found in code-level diagrams. These diagrams should show how an individual component is set up or structured at the code level and how it works internally.

UML diagrams or excerpts from UML diagrams can be used for representations at this level. This approach is particularly suitable for class diagrams or sequence diagrams.

Incidentally, Simon Brown recommends not including this level in the model since the content can largely be created automatically from the source code using tools: The effort involved in manually creating the level is too high compared to its benefits.

> **Generating C4 Diagrams**
>
> No tools are required or necessary for creating C4 models. Hower, in Section 3.5.2 and Section 3.5.3, I present the *PlantUML* and *Structurizr* tools, which you can use to describe and visualize C4 models.

3.5 Doc-as-Code

In the past, software documentation or architecture documentation would often be created and managed in parallel with development work using separate tools, such as a standalone UML editor, Microsoft PowerPoint, Microsoft Word, or Wiki systems like

Atlassian Confluence. These tools cannot be integrated into the development process, or only integrated poorly, and developers are often forced to leave their familiar environment (i.e., the development environment) and switch to the documentation tools for their work. The necessary tools may not be available for their development platform and must then require virtual machines or remote connections.

Tools that are difficult or even impossible to integrate into the development process have disadvantages: New or adapted features, for example, must be consistently documented or subsequently documented separately. Even if the development team has the necessary discipline, a great deal of effort is required to keep the documentation up to date and complete.

In addition, some tools prevent team members from collaborating in a meaningful way because their file formats cannot be edited by several people at the same time or because manual workflows are required in the release process. If, for example, UML diagrams are integrated as PNG images, they cannot be worked on collaboratively.

Doc-as-code, on the other hand, takes the approach of viewing documentation as source code and creating, editing, and managing it accordingly. The individual documents are created in lightweight text formats and versioned and managed in the source code repository used. Different versions can be compared with the usual development tools and restored to a specific state if necessary.

Changes to the documents, like source code changes, can be coordinated with team members and integrated via pull requests or review processes, for example.

Storing documents within a source code repository provides advantages, not only during creation and editing: Various output formats or scopes can be generated from the predominantly plain text formats by integrating them into existing build processes. A special output document with the appropriate content can be provided for each target group from the same text source.

You can also aggregate information from multiple such as text files and source code. Automatically created UML diagrams can be integrated directly into the documentation and don't require separate updating or redundant storage.

Instead, with each release of an application, up-to-date documentation is automatically created and delivered. The documentation itself becomes an artifact to be delivered and is integrated into the agile development process known as *continuous documentation*.

In the following sections, I will introduce you to formats and tools that you can use to pursue a doc-as-code approach.

3.5.1 AsciiDoc

AsciiDoc is a text format for creating plain text documents for structured documentation. A human-readable, platform-independent markup language, AsciiDoc is similar to *Markdown*, but with many additional functions and a clearly defined syntax.

AsciiDoc was developed to bridge the gap between plain text and complex markup languages such as HTML or LaTeX. Various output formats can be generated from the documents, such as HTML, PDF, Word, or EPUB.

> **AsciiDoc and Asciidoctor**
>
> *AsciiDoc* is a simple text markup language for creating technical content. You can find the language specification at *https://asciidoc.org/*. Plugins are available for common development environments that provide support for editing and allow documents to be previewed.
>
> The transformation of AsciiDoc documents into the various source formats is performed by *Asciidoctor* (*https://asciidoctor.org/*), which can be easily integrated into a separate build pipeline with its many extensions.

The AsciiDoc format is quite similar to other markup languages. The text file shown in Listing 3.21 can be generated as a PDF using Asciidoctor, for example, as shown in Figure 3.28.

```
= Hello, AsciiDoc!

The syntax description can be found at https://asciidoc.org[AsciiDoc].

== Section Title

* A list item
* Another list item

[source,go]
----
fmt.Println("Hello, World!")
----
```

Listing 3.21 AsciiDoc Example

The advantages of AsciiDoc when creating technical documents, such as architecture documentation, include the tooling, which can be easily integrated into your own build pipeline, and the standardized syntax. This syntax contains all known markups (such as tables, lists, and more), but also provides for the insertion of external content as well.

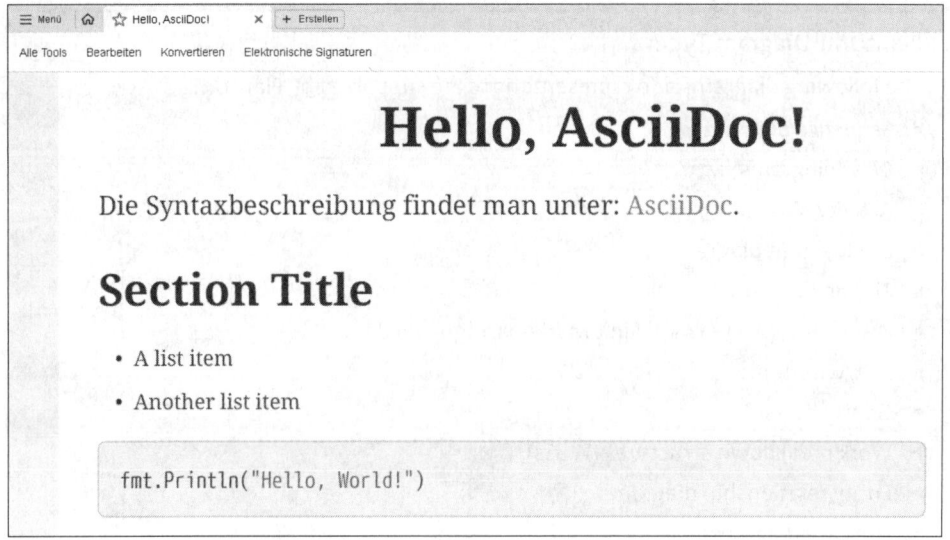

Figure 3.28 PDF Document Generated from the AsciiDoc File

The core of the tooling is the *Asciidoctor* text processor that converts AsciiDoc into various formats such as HTML and PDF.

Some of the advantages of AsciiDoc include the following:

- A defined syntax, not multiple grammars
- Generation of various output formats possible
- Extensive tooling for integration into the build pipeline
- Easy inclusion of content from external files (images, graphics, source code, AsciiDoc documents, and more)
- Possible use of variables in documents
- Creation of target group-oriented documentation through the customized aggregation of content

3.5.2 PlantUML

With the help of the *PlantUML*, an open-source project, you can create various graphical representations from simple text descriptions. Several UML diagrams are supported, as well as other diagram types such as *wireframes* or *mind maps*.

Strictly speaking, PlantUML is a drawing tool, as the models or drawings created are not validated or checked for consistency.

3   Source Code and Documenting the Software Development

> **PlantUML Diagram Types**
>
> The following diagrams and representations are supported by PlantUML:
>
> - Sequence diagrams
> - Class diagrams
> - Activity diagrams
> - Deployment diagrams
> - Timing diagrams
> - Displaying YAML (YAML Ain't Markup Language) data
> - Salt/wireframe
> - Gantt charts
> - Work breakdown structures (WBSs)
> - Entity relationship diagrams
> - Use case diagrams
> - Object diagrams
> - Component diagrams
> - State diagrams
> - Displaying JavaScript Object Notation (JSON) data
> - Network diagrams
> - ArchiMate diagrams
> - Mind maps
> - Math

The diagrams can either be created in a text editor or in a development environment or generated from source code using various tools.

As soon as the text descriptions are ready, they can be transformed into a graphical representation using the available command-line tool, for example, in PNG or SVG files. PlantUML takes over the entire layout of the diagrams unless you override its defaults with your own configuration.

Plugins with syntax highlighting and a direct preview of the diagrams are available for most development environments.

The syntax is quite similar for the different diagram types. Elements are defined directly or implicitly, and they are connected to each other via relationships.

The example shown in Listing 3.22 is a *sequence diagram* in which the `Alice` and `Bob` elements are defined implicitly and communicate using calls described as arrows (`-->`). Figure 3.29 shows the generated diagram.

```
@startuml
Alice -> Bob: Authentication Request
Bob --> Alice: Authentication Response
Alice -> Bob: Another authentication Request
Alice <-- Bob: Another authentication Response
@enduml
```

Listing 3.22 PlantUML Example of a Sequence Diagram

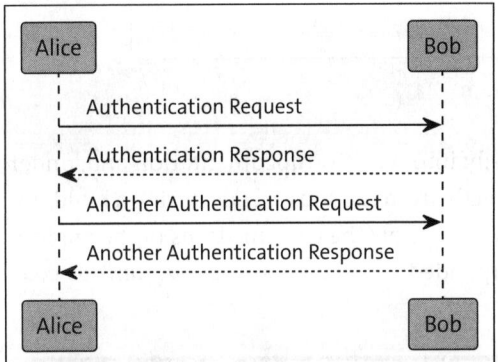

Figure 3.29 Generated PlantUML Sequence Diagram

An explicit definition of elements is useful for class diagrams, for example, as soon as the class has attributes and methods.

The example shown in Listing 3.23 is the explicit definition of class A with the counter attribute and the abstract start method. Class B is implicitly defined and extends class A. Figure 3.30 shows the generated result.

```
@startuml
class A {
{static} int counter
+void {abstract} start(int timeout)
}
note right of A::counter
This member is annotated
end note
note right of A::start
This method is now explained in a UML note
end note

A <|-- B

@enduml
```

Listing 3.23 Definition of a Class in PlantUML: Explicit and Implicit

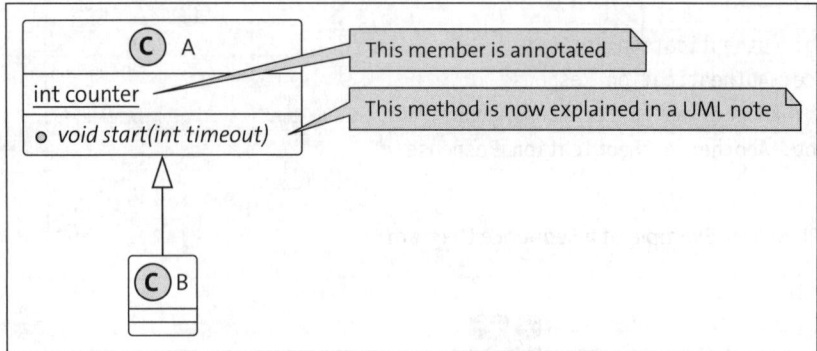

Figure 3.30 Generated PlantUML Class Diagram

PlantUML diagrams can be integrated easily into AsciiDoc documentation and, under certain circumstances, even generated directly from the source code during a build run using plugins. In this way, the diagrams always show the current status of the application, and the *doc-as-code* approach is expanded to become also a *diagrams-as-code* approach.

> **Mermaid as an Alternative to PlantUML**
>
> As an alternative to *PlantUML*, the open-source tool *Mermaid* is becoming increasingly popular. Mermaid is a JavaScript-based diagram and chart tool that also processes text files and displays them graphically. Although the syntax of both tools is similar, some users consider Mermaid more user-friendly and easier to learn, offering an even more intuitive syntax and additional graphical editors.
>
> You can find this tool at *https://mermaid.js.org/*.

3.5.3 Structurizr

Structurizr is a *DSL (Domain Specific Language)* for describing complete C4-based software architecture models as text files and managing them as code. Although you can also use other tools, such as PlantUML, Structurizr is a particularly simple method of modeling.

For each architecture model created, you can generate several views or diagrams with different levels of detail, as shown in Figure 3.31.

The starting point for the model is what's called a workspace, in which the component, container, or softwareSystem components known from C4 are defined in the area of a model. Various views and their content are then defined below the views element.

The example shown in Listing 3.24 is an architecture model that contains the User user and the software system named Software System. The software system consists of the Web Application and Database containers. The relationships between the components are described via arrows (->) and are included in the model. In our example, User calls the Web Application container.

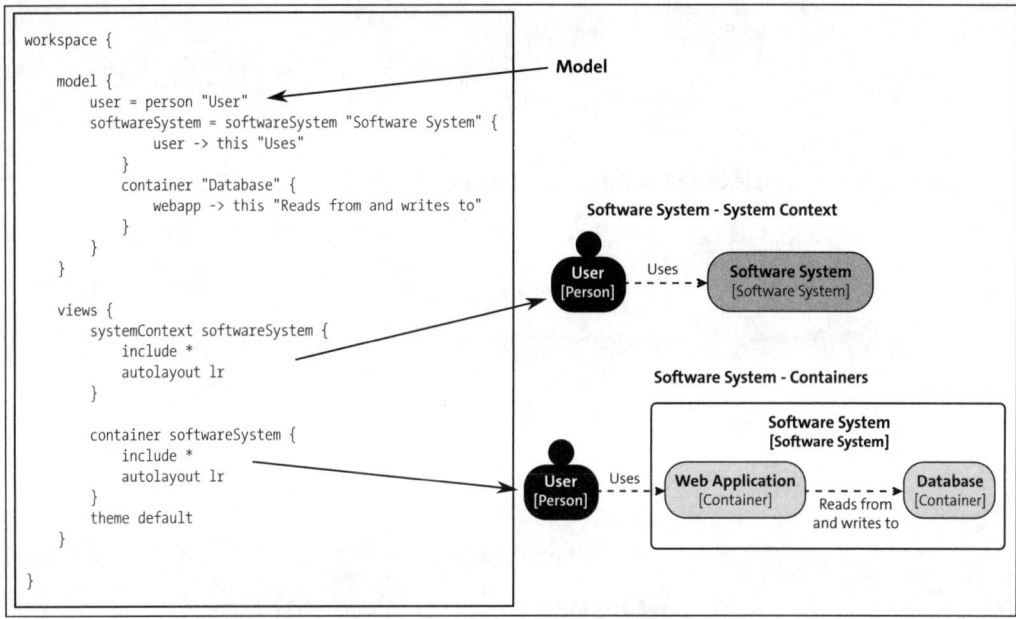

Figure 3.31 Structurizr Model and Views

Two views are defined in the views area, in each of which all components are to be displayed.

The diagrams are generated either via the *Structurizr* command-line tool or via a server that is either self-managed or used as a cloud service.

Structurizr diagrams can also be easily integrated into AsciiDoc documents and generated from code using tools.

```
workspace {
    model {
        user = person "User"
        softwareSystem = softwareSystem "Software System" {
            webapp = container "Web Application" {
                user -> this "Uses"
            }
            container "Database" {
                webapp -> this "Reads from and writes to"
            }
        }
    }
    views {
        systemContext softwareSystem {
```

```
            include *
        }
        container softwareSystem {
            include *
        }
    }
}
```

Listing 3.24 Example of a Structurizr Architecture Model

Chapter 4
Software Patterns

Tried and tested recipes in programming, design patterns provide solutions to recurring problems that developers encounter time and again. This chapter introduces common software patterns in detail, covering their structure, typical areas of application, and practical examples to make it easier for you to understand and apply them in real-life projects.

The objective of a design pattern is to describe and standardize consistent and proven solutions to recurring problems in software development. These patterns serve as blueprints to help developers create robust, efficient, and maintainable software and software architectures.

The use of software design patterns can improve the quality of your code and also save on time and costs. By using proven templates, developers can work more efficiently and don't need to invent a new solution for every challenge that arises.

In addition, the use of established names for design patterns can simplify the documentation process and make your software easier to understand, as developers can immediately recognize the function of the code from the name without having to study the source code in detail or read every detail in the documentation. In this chapter, I describe a selection of common design patterns and *Gang of Four (GoF) design patterns* as well as some commonly used *anti-patterns*. The selection of patterns is based on my experience of which patterns are most commonly used in practice.

I'll describe each pattern adhering a fixed structure based on the GoF's original book, including the following sections:

- Origin
- Problem and motivation
- Solution
- Sample solution
- When to use the pattern
- Consequences
- Concrete example

I have presented the sample solutions as Unified Modeling Language (UML) diagrams and created them in various programming languages. A real-world example is also provided for each of the patterns, showing how they are used in the "wild."

In this chapter, I'll cover these patterns in the following sections:

- Factory method (Section 4.1)
- Builder (Section 4.2)
- Strategy (Section 4.3)
- Chain of responsibility (Section 4.4)
- Command (Section 4.5)
- Observer (Section 4.6)
- Singleton (Section 4.7)
- Adapter/wrapper (Section 4.8)
- Iterator (Section 4.9)
- Composite (Section 4.10)

Chapter 5, Chapter 6, and Chapter 7 deal with further patterns at the software architecture level and at the system level.

4.1 Factory Method

Origin: Gang of Four—Creational Pattern

Belonging to the category of creational patterns, the *factory method pattern* can uniformly create objects that are derived from a common abstract base type. In this pattern, the abstract interface for creating the objects is defined in a basic type and is implemented accordingly in concrete versions of this type.

4.1.1 Problem and Motivation

Various training courses can be managed in our training management software. A room is required for each of these training courses, which must be reserved, booked, and confirmed.

The first version of our software is based on the assumption that all training sessions are carried out on site at the training provider's premises, as shown in Listing 4.1. The software uses the Training and Room classes internally for this purpose. Accordingly, instances of the Room or Training class are used directly in many places in the code. These are instantiated as needed and used accordingly.

As the number of customers increases, so does interest in corporate events. These training courses aren't always carried out at the training provider's site but sometimes also directly at an end customer's location. Rooms are also required in these cases, but they must be reserved and booked in different ways.

The reservation, booking, and confirmation of company rooms is usually conducted over emails by the responsible department. In contrast, the internal rooms owned by the training provider can be booked directly via a programming interface from other applications.

As the previous software works directly with the Room class and its logic for reservations and bookings, all places where this class is used must now be adapted or extended so that the newly introduced OnsiteLocation class can now also be used, as illustrated in the class diagram for internal and external training courses shown in Figure 4.1.

```java
if (training.isOnsite()) {
    OnsiteLocation onsite = new OnsiteLocation();
    //... additional code for reservation and confirmation
    //...
} else if (training.isInhouse()) {
    Room room = new Room();
    //... additional code for reservation and confirmation
}
```

Listing 4.1 If Conditions for Room Booking (Java)

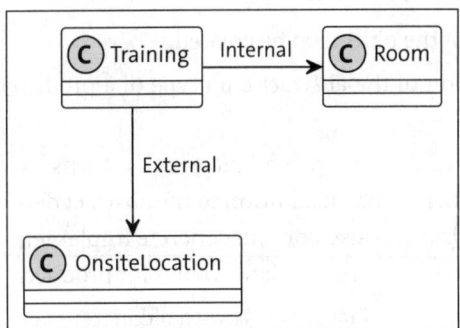

Figure 4.1 Class Diagram for Internal and External Training Courses

If a decision is made at a later date to provide additional online training, further work would also be required to adapt the code again in the same places, which runs the risk that these code sections become confusing and error prone due to a large number of if conditions and alternative code sections. Ultimately, special logic must be implemented for the reservation or booking for each training venue.

4.1.2 Solution

The *factory method pattern* solves this problem by no longer creating the objects, which are called *products*, directly via the class's own constructor, but by initiating the creation via a separate method, namely, the *factory method*.

As shown in Figure 4.2, this factory method is, in turn, defined as an abstract method of a factory class, for which a corresponding implementation exists for each of the specific product types.

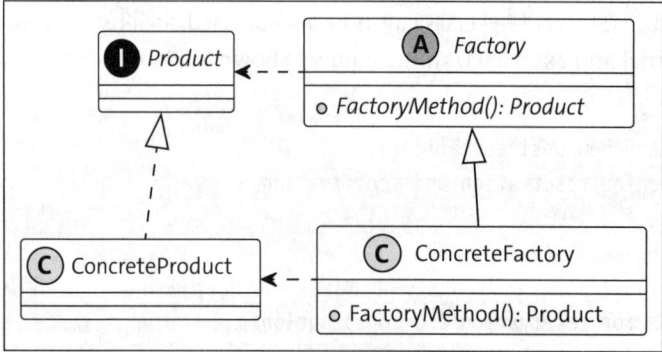

Figure 4.2 Structure of the Factory Method Pattern

The structure and participants in this pattern are as follows:

- Product: Describes the abstract data type of the objects to be created.
- Concrete product: Concrete implementation of the abstract data type of a product and therefore the objects to be created.
- Factory: Abstract class with the definition of the *factory method*, which returns the abstract data type of the product as the return value. In addition to the abstract definitions of the factory methods, a factory class can also contain concrete implementations of these factory methods and use them to return default values for products.
- Concrete factory: Implementation of the abstract factory class with a concrete version for the factory method. This is where the specific products are created. There is a separate concrete factory for each product type.

> **Abstract Versus Concrete Factory Methods in the Factory Class**
>
> The factory method in the factory (base) class can be implemented either as an abstract method or a non-abstract method. If defined as an abstract method, the concrete factories must contribute their own implementations. If, on the other hand, the method is not defined as an abstract method, a standard implementation can be implemented within the factory (base) class, which can be adapted by the specific factories.

4.1.3 Sample Solution

A Java-based implementation of the *factory method pattern* for our example training company is shown in Figure 4.3.

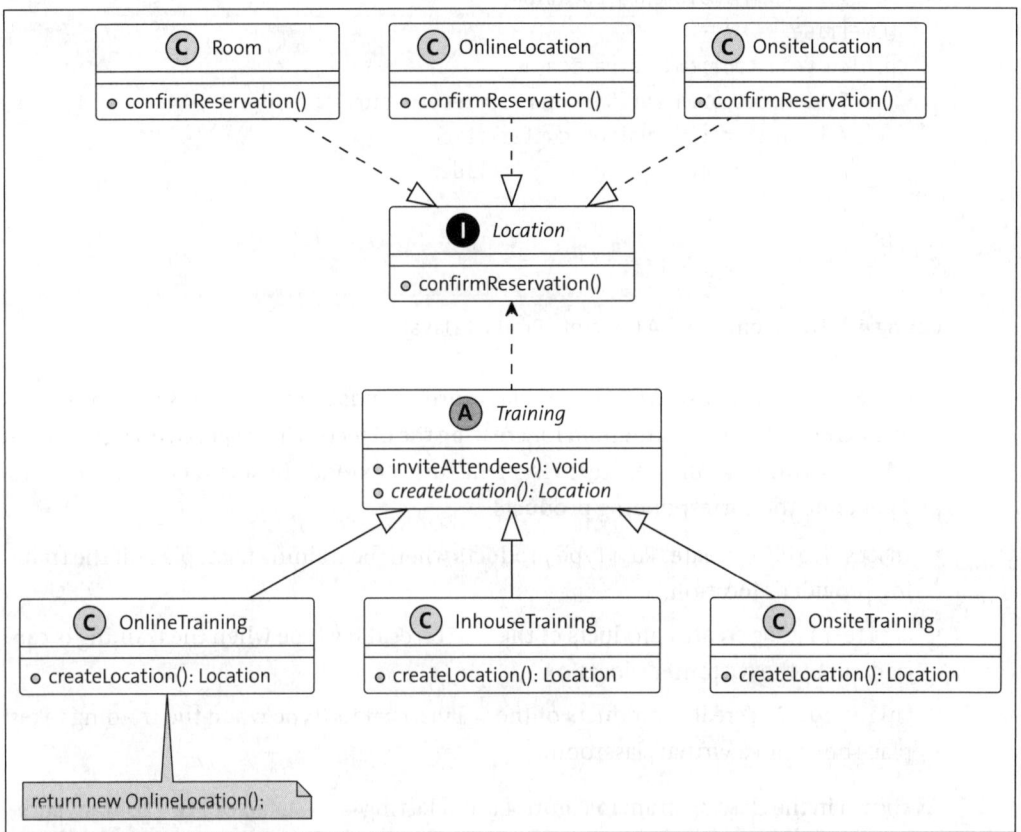

Figure 4.3 Factory Method Pattern: Training Example

The Location interface is introduced, as shown in Listing 4.2, which is implemented by the concrete types Room, OnsiteLocation, and OnlineLocation. This interface contains the confirmReservation method, in which the respective logic for confirming a reservation is implemented for the different location types. The Location objects therefore represent the concrete products of the *factory method pattern*.

```
package com.sourcefellows.patterns.factory;

public interface Location {
    void confirmReservation();
}
```

Listing 4.2 Location Interface in Java: The Abstract Product

The `Location` implementation `Room` is shown in Listing 4.3.

```
package com.sourcefellows.patterns.factory;

public class Room implements Location{
    @Override
    public void confirmReservation() {
        System.out.println("Room.confirmReservation");
        // Logic for reservation confirmation
        // of a room at the training provider
        ...
    }
}
```

Listing 4.3 The Room Class: A Concrete Product (Java)

In this example, the abstract `Training` class corresponds to the pattern's factory, which defines the `createLocation` method for creating the objects. The respective `OnlineTraining`, `InhouseTraining`, and `OnsiteTraining` factories extend the abstract `Training` class and generate the corresponding products:

- `InhouseTraining` creates `Room` type products when the training takes place at the training provider's location.
- `OnsiteTraining` creates products of the `OnsiteLocation` type when the training is carried out at the customer's location.
- `OnlineTraining` creates products of the `OnlineLocation` type when the training takes place online in a virtual classroom.

As shown in the class diagram in Figure 4.3 and Listing 4.4, the factory can contain additional logic, such as the sending of invitations to individual attendees (`inviteAttendees` method). This logic is implemented uniformly for all use cases. Invitations to attendees are always sent by email, for example, and contain further information about the location. No further distinction is made beyond this.

```
package com.sourcefellows.patterns.factory;

public abstract class Training {

    abstract Location createLocation();
    void inviteAttendees() {
        ...//consistent logic for all locations
    }
}
```

Listing 4.4 Abstract Training Class: The Abstract Factory (Java)

> **Tasks of a Factory**
>
> Factories do not always *merely* take on the task of creating instances. They can also contain overarching business logic. In our example, invitations to events are added.

The specific factories are responsible for creating the associated objects. In the example shown in Listing 4.5, `InhouseTraining` creates a `Room` instance.

```java
package com.sourcefellows.patterns.factory;

public class InhouseTraining extends Training {
    @Override
    public Location createLocation() {
        System.out.println("InhouseTraining.createLocation");
        // Logic for direct reservation via API ...
        return new Room();
    }
}
```

Listing 4.5 The InhouseTraining Class: A Concrete Factory (Java)

The calling code can then create new rooms via the specific factory and confirm the reservation uniformly. The associated logic for the booking is implemented in the corresponding implementation, and it is no longer necessary to differentiate calls via `if` conditions, as shown in Listing 4.6.

```java
Training training = new OnlineTraining();
Location location = training.createLocation();
location.confirmReservation();
```

Listing 4.6 Using the Factory Method Pattern (Java)

4.1.4 When To Use the Pattern

Some examples of when to use the pattern include the following:

- **The types to be generated are not (yet) known**
  The *factory method pattern* can always be used when new objects or products must be created within an application, but their exact types or dependencies are not yet fully known. Further product versions may be added at a later date, which can be seamlessly integrated into the existing software—and above all into its logic—without any adjustments.

- **Complexity or variability in the creation of objects**
  The pattern separates the instance creation of individual products from their actual use. If, for example, changes must be made to the creation code, these changes can

be implemented independently of the product use. New products can be added using new factories. In this case, too, no adjustment is required when using the products.

In addition, the separation of logic for the creation of the instance provides an advantage in that even complex processes for instantiation can be implemented that don't make the use of the products more difficult.

- **Possibility of extending a library**
  The *factory method pattern* can be used in libraries to make them extensible. Existing basic functionality can continue to be used for new products without adjustments. For example, users can create a new concrete factory through inheritance and thus use the existing internal logic. In our earlier example, the `inviteAttendees` method can be used independently of the actual creation of a room.

- **Catching or pooling of products**
  Since instances of a concrete product type are returned within the concrete factory methods, you can also replace the creation of the instance with a cache or pool so that new objects are not created with every call.

4.1.5 Consequences

Some consequences of using these patterns are as follows:

- **Decoupling object creation and use**
  The use of the pattern decouples the creation of objects from their use, which makes the code more flexible and easier to adapt, as changes in object production do not necessarily affect the rest of the code.

- **Extensibility**
  Applications or libraries can be extended by adding new product variants without changing the existing logic. The new variants can also make use of existing basic functionality (see also Chapter 2, Section 2.3.2).

- **Using abstraction and polymorphism**
  The structure of the pattern can be implemented with the concept of polymorphism and through the use of abstraction.

- **Improved testability**
  By using the *factory method pattern*, and the resulting separation between the creation and use of product objects, not only concrete products but also mock objects, for example, can be returned in test cases, which ultimately increases the testability of the software.

- **Additional code and complexity; inheritance hierarchies**
  Additional code is required to create products when using the pattern. Instead of directly calling the constructor of a class to create an object, the inheritance hierarchy of various factories can be used. In smaller use cases, this hierarchy of inheritance might seem excessive—especially if the logic of the individual object creations is simple.

Particularly in more complex systems with many different product classes and variants, the overall complexity can increase significantly when using the pattern and make the development process more time consuming.

4.1.6 Real-World Example in Open-Source Software

Various data compression formats can be used in Java via the *Apache Commons Compress* library. In addition to the "classic" formats such as TAR, ZIP, 7z, and BZIP2, more unknown formats or compression algorithms such as ar, Pack200, or LZMA can also be supported. A complete list of supported formats is available at *https://commons. apache.org/proper/commons-compress/index.html*.

The library distinguishes between compression and archive formats. *Compression formats* usually compress a file or a Java stream, while *archive formats* can manage multiple files in a structured form. A common compression format is Gzip, for example.

To use the different formats, the library uses the Compressor and Archiver implementations, which are each implemented as Java streams. In both cases, these instances are created via factories and can then be used.

Figure 4.4 shows the CompressorStreamFactory, which we'll use to create a CompressorInputStream, namely, GzipCompressorOutputStream.

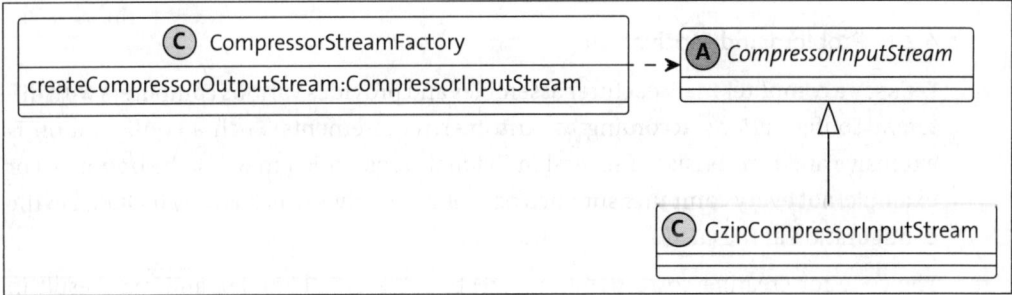

Figure 4.4 Factory in Apache Commons Compress

The *Apache Commons Compress* library does not differentiate between an abstract factory and a concrete factory when generating products. However, the documentation for the extension notes the following:

> *"To add other implementations you should extend CompressorStreamFactory and override the appropriate methods (and call their implementation from super, of course)."* —JavaDoc

Adding your own compression implementations to the library can be performed accordingly by deriving the existing factory.

The example shown in Listing 4.7 illustrates how to create a compressed file using the *Apache Commons Compress* library. The data to be compressed is written to a CompressorOutputStream, which is generated by the CompressorStreamFactory.

```
var myOutputStream = Files.newOutputStream(Paths.get("my.gz"));
CompressorOutputStream gzippedOut = new CompressorStreamFactory()
        .createCompressorOutputStream(CompressorStreamFactory.GZIP,
    myOutputStream);

gzippedOut.write("Hello World".getBytes(Charset.defaultCharset()));
```
Listing 4.7 Creation of a Compressor Using Apache Commons Compress (Java)

4.2 Builder

Origin: Gang of Four—Creational Pattern

Another creational pattern, the *builder pattern* enables complex objects to be created flexibly, step by step. The logic for object creation is separate from the actual object to be created, which enables both iterative creation and the use of the same code to generate several different objects.

4.2.1 Problem and Motivation

Let's say a computer manufacturer assembles and provides various computers with different configurations according to customer requirements. Such a configuration is extensive and can consist of several individual steps, which may also be optional. For example, not every computer supplied has a Blu-ray drive, which is only included in the configuration at the customer's request.

The code for creating corresponding Computer objects is complex and may result in rather extensive implementations within the constructor or to the distributed initialization code within the application. Code may require redundant maintenance, specifically at every point where the objects are created.

For creating objects that represent computers with standard specs, a constructor with two parameters might be sufficient. For complex computer configurations, additional constructors with a large and varying number of parameters would have to be provided, as shown in Listing 4.8.

This complexity means that either a corresponding constructor must be created for each potential configuration or that a constructor with a large number of parameters must be defined. In the latter case, not all parameters might be already set for the different configurations. In both cases, the code is confusing and error prone and cannot be maintained centrally.

```
//Constructor for standard office workstation PCs
public Computer(int ramSizeInGB, int harddiskSizeInGB) {
...
}
//Constructor with "all" details
public Computer(int cpuCoreCount,
                int ramSizeInGB,
                HardDisk harddisk,
                int usbConnectorCount,
                boolean silentConfig,
                boolean hasBlueray,
                boolean hasCardReader,
                ...) {
...
}
```

Listing 4.8 Computer Constructor with Many Parameters (Java)

Alternatively, as shown in Figure 4.5, a solution could also consist of implementing a derivation of the Computer type for each configuration type, with the individual subclasses each representing a specific configuration. The various Computer versions could thus be generated via a simple constructor call.

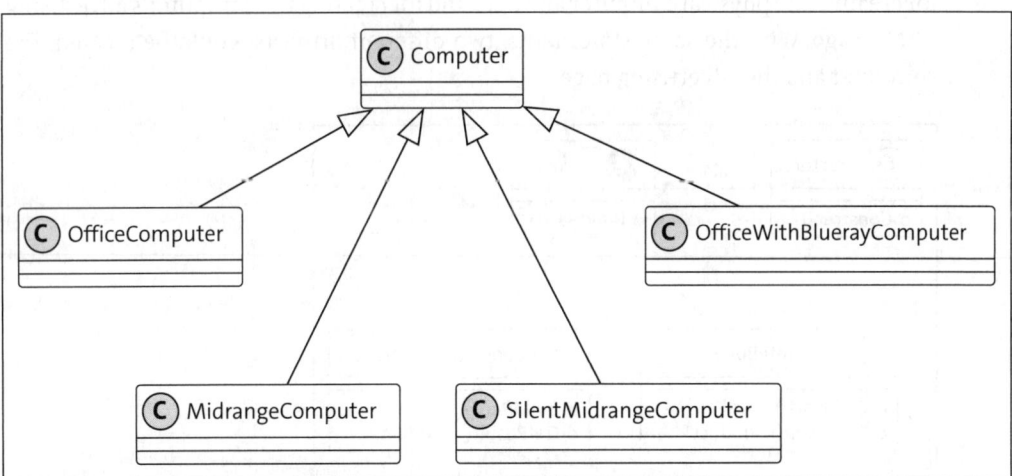

Figure 4.5 Computer Inheritance Hierarchy for Configurations

However, as shown in Figure 4.5, this variant potentially creates a large inheritance hierarchy. A new, additional class is required for each configuration permutation, which inevitably leads to a confusing and complex structure within the application.

In addition, the configuration logic implemented within the individual Computer constructors cannot be used for other use cases. If, for example, a configuration is to be created for the display of a computer on an HTML page, the same logic for creating a configuration may also have to be repeated for processing the data for a PDF file. The two objects, the HTML page and the PDF file, would be configured in the same way, but created differently.

4.2.2 Solution

The *builder pattern* is a creational pattern in which the logic for creating complex objects is extracted from its own class. Instead of direct object initializations, methods in a separate class (the *builder*) are used to create individual, complex objects.

Within the builder, the required individual steps or configurations for object creation are defined separately and do not require full execution to instantiate an object. In our computer assembly example, the configuration of the Blue-ray drive is optional and would therefore be defined as an optional single step in the builder.

The separate steps of a creation process are defined in an abstract implementation, the *builder* interface, and implemented using concrete implementations, as shown in Figure 4.6. This means that multiple different objects or products can be created with the same configuration logic. These products can be completely independent of each other and do not have to have a common base type, for example.

In the computer example, one concrete builder implementation could be responsible for creating the physical computer and a second for creating the computer's advertising HTML page. With the same statements, two different artifacts would be created: the computer and the advertising page.

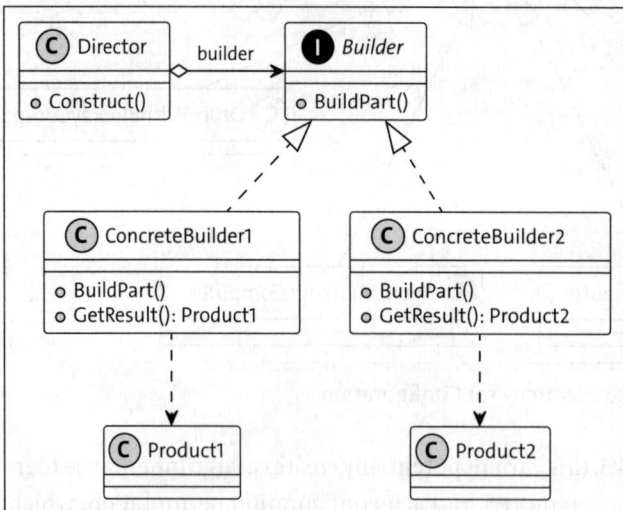

Figure 4.6 Structure of the Builder Pattern

The standardized interface for the different concrete builder implementations enables users of the interface to create different products with the same configuration logic and to make the code more reusable.

As an extension, the possibly more complex coordination of the individual steps can be taken over by a *director*. It is used to summarize more extensive processes and provide users with a simpler interface—without details on the creation of objects via the Builder.

For the computer example, a director implementation could provide methods to create a standard Office PC or a Max Power PC with a few options.

The structure and participants in this pattern are as follows:

- Builder: describes the abstract interface with the individual steps for creating the objects or products.
- Concrete builder: concrete implementation of the abstract description of the builder responsible for creating the products. The state of the object or product to be created is saved within the builder instance. The specific products created are returned by calling a separate method.
- Director: The director orchestrates a more complex process for creating objects via the abstract builder interface, providing users with a simpler and more usable interface.
- Product: the product or products to be produced. Each specific builder implementation can create its own objects or products.

4.2.3 Sample Solution

The configuration of the `Computer` objects described above can be implemented using the *builder pattern* as shown in Figure 4.7.

A corresponding `Builder` interface defines the individual steps and options required to configure a computer. For the example, these are methods for setting the hard disk size, the RAM size, and the CPU name. The interface also contains a `Reset` method that can be used to cancel or reset a configuration.

Two specific versions of the `Builder` interface take care of creating the respective objects: The `ComputerBuilder` creates physical `Computers`, the `ManualBuilder` creates a matching instruction `Manual`.

To keep the complexity of the example to a minimum, as you can see in Listing 4.9, explicit data types for the size specifications for hard disk space or RAM size have been omitted. The variables are specified as primitive data types. As a result, the `Builder` interface looks as follows:

```
package model

type Builder interface {
    SetHardDisksizeInGB(disksize int)
    SetRamInGB(ram int)
    SetCPU(cpu string)
    Reset()
}
```

Listing 4.9 Interface Definition: The Builder (Go)

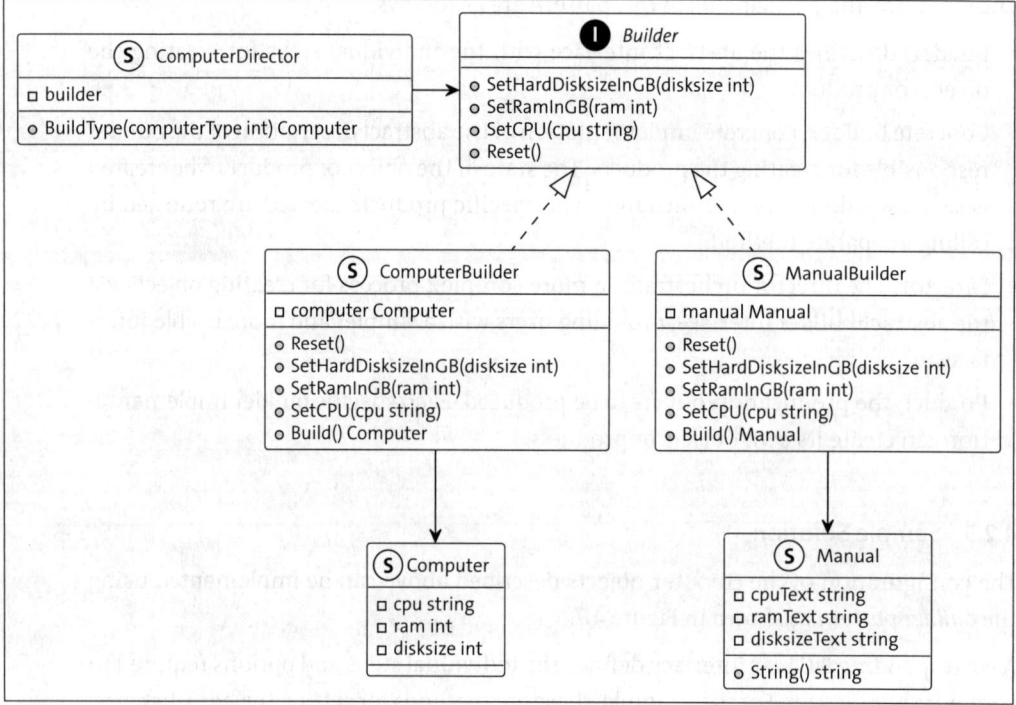

Figure 4.7 Sample Implementation of the Builder Pattern for Computers

The Computer type is defined as a product with three fields, as shown in Listing 4.10.

```
package model

type Computer struct {
    cpu      string
    ram      int
    disksize int
}
```

Listing 4.10 The Computer Implementation: The Product (Go)

The specific builder implementation—ComputerBuilder—is shown in Listing 4.11. As it is located in the same package as the Computer struct, it can access the individual, non-exported fields of the Computer type and fill them in.

A Computer instance is held within the concrete builder implementation, which is successively configured with values. Calling the Build() method returns the instance and resets the builder.

```go
package model

func NewComputerBuilder() *ComputerBuilder {
    return &ComputerBuilder{computer: Computer{}}
}

type ComputerBuilder struct {
    computer Computer
}

func (c *ComputerBuilder) Reset() {
    c.computer = Computer{}
}

func (c *ComputerBuilder) SetHardDisksizeInGB(disksize int) {
    c.computer.disksize = disksize
}
...
func (c *ComputerBuilder) Build() Computer {
    comp := c.computer
    c.Reset()
    return comp
}
```

Listing 4.11 Concrete Builder Implementation of the ComputerBuilder (Go)

The concrete Builder can be used either directly or via a corresponding ComputerDirector, which contains special methods for creating the configuration of the objects in a simplified manner. In the following example, a standard office PC is configured entirely in the director. Figure 4.8 shows the process in a sequence diagram. The implementation is shown in Listing 4.12.

```go
package model

type ComputerDirector struct {
}
```

```go
func NewComputerDirector() ComputerDirector {
    return ComputerDirector{}
}

func (d ComputerDirector) BuildStandardOfficePC(builder Builder) {
    builder.Reset()
    builder.SetRamInGB(128)
    builder.SetHardDisksizeInGB(1000)
    builder.SetCPU("latest greatest")
}
```

Listing 4.12 Implementation of the Director: ComputerDirector (Go)

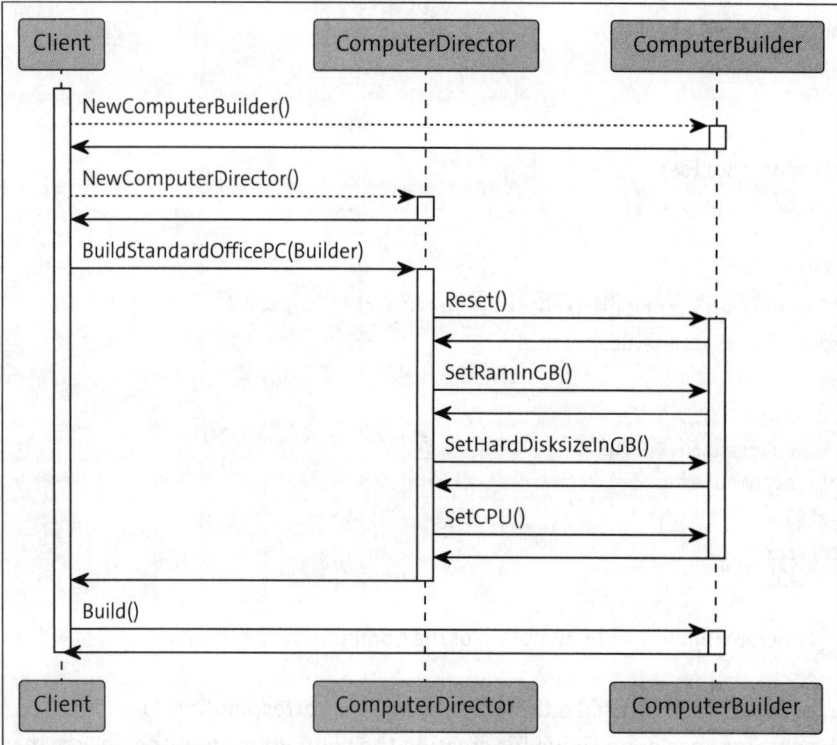

Figure 4.8 Builder Pattern for Computer: Sequence Diagram

Inside the client, the `ComputerBuilder` and `ComputerDirector` classes can then be instantiated and used, as shown in Listing 4.13.

```
builder := model.NewComputerBuilder()

director := model.NewComputerDirector()
```

```
director.BuildStandardOfficePC(builder)

computer := builder.Build()
```

Listing 4.13 Calling the Director and the Builder: The Client (Go)

> **Fluent API**
>
> Many builder implementations are based on an application programming interface (API) called a *fluent API*, in which method calls can be chained together. Methods that do not require a return value return their own called object instance in a fluent API.
>
> Without a fluent API, the object instance or reference must be specified anew for each call, as in the following Go example:
>
> ```
> ship := model.Ship{}
> ship.SetName("Unsinkable")
> ship.SetPosition(...)
> ```
>
> However, if the setter methods return their own object instance, as shown in the following implementation, the calls can be chained:
>
> ```
> func (s *Ship) SetName(name string) *Ship {
>     s.name = name
>     return s
> }
> ```
>
> The call then changes accordingly and becomes a more compact, chained fluent API expression:
>
> ```
> ship.SetName("Unsinkable").SetPosition(...)
> ```

4.2.4 When To Use the Pattern

Some examples of when to use the pattern include the following:

- **For creating complex objects**

  The *builder pattern* enables you to keep complex creation logic separate from the objects to be created. With many optional parameters that can be transferred when the objects are created, the approach allows these properties to be set gradually and only when necessary. This strategy results in a clear, lean, and simple interface for the construction of objects.

  Using a director implementation, the interface for API clients can also be simplified through predefined configurations.

- **For creating objects with different representations**

  You can always use this pattern if different representations of an object to be created must be generated and the same steps must always performed in the creation process.

4.2.5 Consequence

Some consequences of using these patterns are as follows:

▶ **Better control and readable code**
Objects can be created step by step and with optional parameters without needing a separate constructor for each permutation (called *telescope constructors*) and without needing a separate subclass in each case.

By extracting the logic, a clearer, more transparent, and more flexible interface is in place for creating complex objects.

▶ **Interchangeability of representation**
The abstract definition of the `Builder` interface, in which the individual steps of the creation process are defined, allows multiple specific implementations to be realized. The internal structure of the objects remains hidden as an internal detail of the respective builder.

The creation logic of the client or of the director can be used for several representations or builder implementations.

▶ **Separates complex creation logic from the object**
By separating the logic that creates the objects from the business logic of these objects, dependencies between the code parts can be minimized. This separation promotes maintainability and at the same time implements the *single responsibility principle*. Each component has a clear responsibility: either the creation of objects or their use in business operations.

▶ **The complexity of the creation increases**
Using the *builder pattern* requires the development of additional classes and methods to replace the constructor. This additional code can provide advantages in complex systems by simplifying the object construction and making it easier to read. For simpler use cases with few attributes and less complexity, however, the overhead caused by the pattern can be perceived as unnecessary.

▶ **Possible inconsistencies during the creation**
The ability to set configuration options only optionally can lead to different—possibly inconsistent—internal states if not all the necessary configuration steps have been carried out. Different developers can potentially use the interface differently and thus unintentionally cause errors.

4.2.6 Real-World Example in Open-Source Software

With the *Apache HttpComponents*, the *Apache Software Foundation* provides libraries for simplified, low level communication using HTTP and associated protocols. The libraries can be used on both the client and server side.

The components are divided as follows:

- *HttpCore* for the low level communication on the transport layer
- The *HttpClient* library based on it, which simplifies HTTP/1.1-based communication with a clear and easy-to-use interface

A `HttpClient` instance is used to set up HTTP communication with the help of the library. This can be created via `HttpClientBuilder`, either directly or indirectly via the `HttpClients` class, which in this case acts as a *director* and in turn contains methods for creating preconfigured clients, as shown in Figure 4.9.

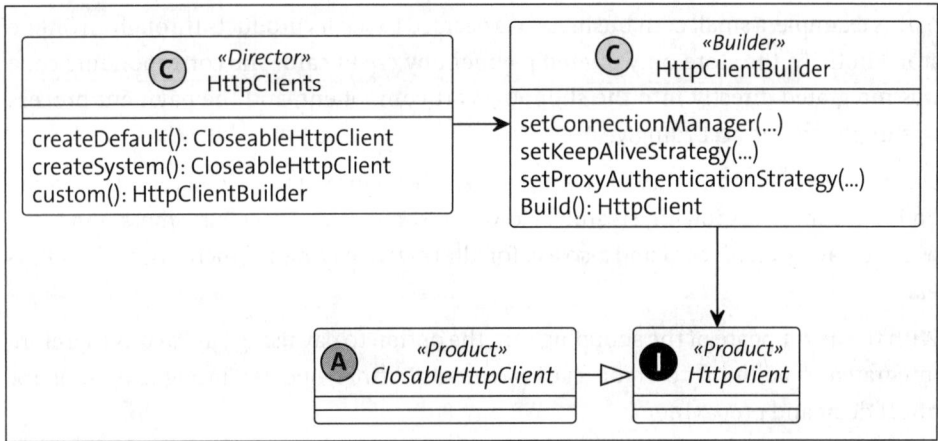

Figure 4.9 Structure of the HttpClient Library

The example in Listing 4.14 shows the creation of an `HttpClient` with the `HttpClients` *director* and the subsequent use of the new instance. Notice how the required request objects are also created using separate builders.

```
try (CloseableHttpClient httpclient = HttpClients.createDefault()) {
    ClassicHttpRequest httpGet =
            ClassicRequestBuilder.get("http://httpbin.org/get").build();

    httpclient.execute(httpGet, response -> {
        final HttpEntity entity1 = response.getEntity();
        //... do something
        EntityUtils.consume(entity1);
        return null;
    });

}
```

Listing 4.14 Sample Use of the HttpClient Library (Java)

4.3 Strategy

Origin: Gang of Four—Behavioral Pattern

The *strategy pattern* belongs to the category of behavioral patterns. This pattern allows the implementation of a family of algorithms, each of which is encapsulated in a separate class. This encapsulation enables these algorithms to be changed independently, making them interchangeable.

4.3.1 Problem and Motivation

In this example, a small craft business has decided to sell its products through an online store. Initially, the store only offered payment by credit card. The corresponding code was integrated directly into the shopping cart component, and the payment process was implemented accordingly.

Due to the excellent quality of the products, sales figures are now increasing, and the online business is growing steadily. However, some of the new customers don't just want to pay by credit card and also ask for alternative payment functions, such as PayPal.

With the next update of the shopping cart, the option to pay using PayPal was therefore integrated in addition to credit card payment. Further updates brought options for direct debit and prepayment.

From management's point of view, online sales are a great success: More and more products have been sold, and the customer base has become increasingly larger.

In the source code of the shopping cart, however, the logic has become increasingly complex and therefore more error prone. Many `if` conditions check which payment variant is currently involved and how it should be handled.

Each additional payment function leads to code changes in the shopping cart and potential problems with the existing functions. Changes to an algorithm of an individual payment method might jeopardize the accuracy or functionality of all the processes in the shopping cart.

In addition, this bundling of different payment algorithms in one central source code location makes it more difficult for multiple developers to collaborate since changes are made in exactly one place and may have to be laboriously brought together via an error-prone *merge process*.

4.3.2 Solution

With the *strategy pattern*, you can use different algorithms or procedures without having to differentiate between them within a common class by means of condition checks.

The specific algorithms or procedures are extracted into separate classes (the *concrete strategies*) and called by the using class (the *context*) via an abstract algorithm interface (the *strategy*). Figure 4.10 shows the structure of the *strategy pattern*.

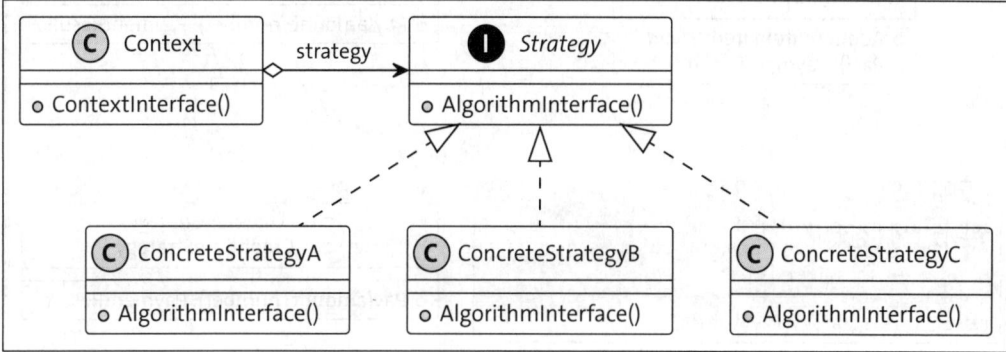

Figure 4.10 Structure of the Strategy Pattern

A reference to a specific strategy is saved within the context object and used accordingly in the logic. The strategy actually used is not selected within the context object, but via its caller.

The abstract definition of the algorithm in the Strategy interface means that the specific implementations are interchangeable and can be managed independently of each other. Later inclusion of new algorithms is simpler and more error tolerant because no central point requires adaptation.

The structure and participants in this pattern are as follows:

▶ Strategy: Defines an abstract interface for the algorithm that can be used within the context object.
▶ Concrete strategy: Implementation of a concrete algorithm.
▶ Context: Implementation with reference to a strategy that is used in the logic. The strategy is selected by the user of the context object.

4.3.3 Sample Solution

Let's now use the *strategy pattern* to solve the problem with the complex logic in the shopping cart described earlier. The example whose class diagram is shown in Figure 4.11 is implemented in *TypeScript*.

The logic for the various payment options previously contained in the shopping cart gets extracted into separate classes. The PaymentStrategy interface is introduced as a common interface which contains the Pay method as an abstract description for the payment process. The newly introduced PaymentResult object provides additional information on the outcome of the actual payment process.

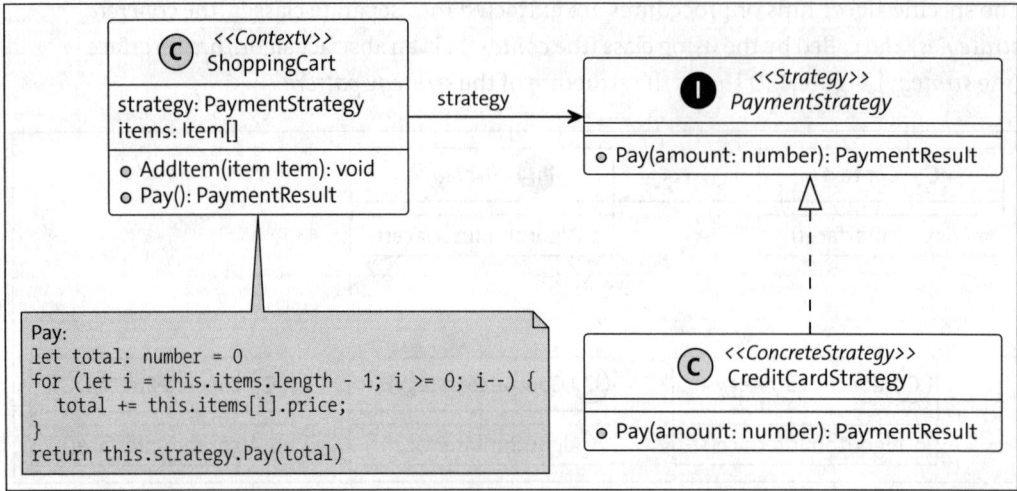

Figure 4.11 Different Payment Options with TypeScript

Only one specific strategy, the CreditCardStrategy, is shown in the class diagram from Figure 4.11 and in the *TypeScript* source code in Listing 4.15. Other implementations for payment processes can then be added using corresponding classes.

```
interface PaymentStrategy {
    Pay(amount: number): PaymentResult
}

class PaymentResult {
    success: boolean
    //...more attributes for more details
    constructor(success: boolean) {
        this.success = success;
    }
}

class CreditCardStrategy implements PaymentStrategy{
    Pay(amount: number): PaymentResult {
        console.log("pay with CreaitCard")
        //... Logic for payment
        return new PaymentResult(true);
    }

}
```

Listing 4.15 A Strategy with an Implementation in TypeScript

The `ShoppingCart` class contains the logic for managing the individual products to be purchased (`Item`) and a reference to a `PaymentStrategy` with which the payment process is to be carried out.

The user of the shopping cart (`ShoppingCart`) decides which strategy is used via the API. The specific strategy can either be passed as a parameter via the constructor or through a setter method.

Within the `Pay` method of the shopping cart, only the total amount to be paid for the products is calculated, and the payment process is started, as shown in Listing 4.16. The specific details of the payment process undertaken are then left to the individual strategy.

```typescript
class Item {
    price: number;

    constructor(price: number) {
        this.price = price;
    }
}

class ShoppingCart {

    private items: Item[];
    public strategy: PaymentStrategy;

    constructor(strategy: PaymentStrategy) {
        this.strategy = strategy;
        this.items = [];
    }

    AddItem(item: Item) {
        this.items.concat(item);
    }

    Pay() : PaymentResult {
        let total: number = 0
        for (let i = this.items.length - 1; i >= 0; i--) {
            total += this.items[i].price;
        }
        return this.strategy.Pay(total)
    }
}
```

Listing 4.16 Implementing a Shopping Cart: Context (TypeScript)

The shopping cart is easy to use. After instantiating the strategy and setting the reference in a new shopping cart, products can be added to it and paid for, as shown in Listing 4.17.

```
let strategy: PaymentStrategy = new CreditCardStrategy();
let shoppingCart: ShoppingCart = new ShoppingCart(strategy)

let item1: Item = new Item(20)

shoppingCart.AddItem(item1)
let result:PaymentResult = shoppingCart.Pay()
console.log(result)
```

Listing 4.17 Using the TypeScript PaymentStrategy

4.3.4 When To Use the Pattern

Some examples of when to use the pattern include the following:

- **There are multiple variants of an algorithm available**
  You can always use the *strategy pattern* if you have multiple versions of a partial behavior or algorithm that must be developed or maintained independently of each other.
- **There are classes that differ only in parts of the implementation**
  The *strategy pattern* can be used to break down complex inheritance hierarchies that only serve to separate different sub-behaviors. In our example, as shown in Figure 4.12, the different behavioral variants could also be mapped in an inheritance hierarchy. An abstract ShoppingCart class contains the higher-level logic, and the respective derivatives contain the additional code for the corresponding payment process.

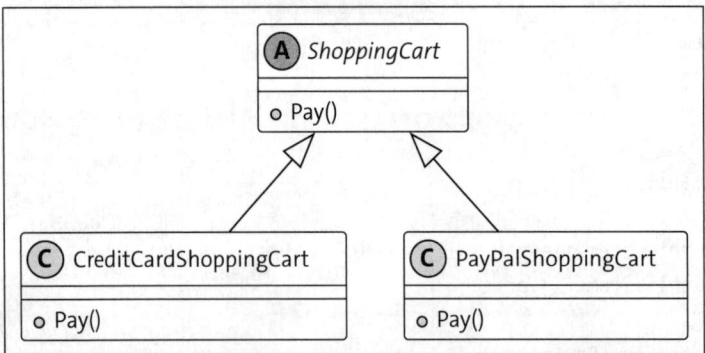

Figure 4.12 When Should You Use the Strategy Pattern? Definitely Here!

Such an implementation variant leads to complex class hierarchies however and also lacks a clear separation of responsibilities. The `ShoppingCart` class contains code to control the general process as well as details on specific behavioral variants.

- **You want to hide implementation details of the algorithms**
  By using the pattern, an interface can be created via the context class in which no internal details or dependencies of the individual algorithms are known. In other words, the internal data structures of the algorithms, for example, do not need to be passed on to callers and then published.

- **Many code branches for behavioral differences**
  If all behavioral differences are implemented in just one class, as in our online store example, conditions and branches must be built into the code so that the right behavior is selected and executed. If the number of behavioral variants increases, the complexity of the class increases at the same time.

  Using the *strategy pattern*, different variants of a behavior can be extracted into separate classes, which thus reduces the differentiation logic. The extracted behavioral differences are addressed via an abstract interface.

- **The logic of behavioral differences is distributed**
  The pattern can be used if the logic for various individual algorithms is not implemented centrally, but instead at different code locations, thus limiting maintainability.

4.3.5 Consequences

Some consequences of using these patterns are as follows:

- **Selecting and changing an algorithm**
  By implementing the *strategy pattern*, different versions of an algorithm can be used and, under certain circumstances, even swapped at runtime.

- **Alternative to inheritance hierarchies**
  Complex inheritance hierarchies with unclear separation of responsibilities for process control and algorithm implementation are prevented. No abstract base classes are created that are extended by behavior-dependent classes. The context contains the process control, and the implementation of the individual algorithms takes place in the specific strategies. Responsibilities are therefore clearly defined.

- **Strategies minimize condition checks**
  By extracting the individual strategies into separate classes and introducing a corresponding abstract interface for their execution, the conditional check for strategy selection is omitted in the calling class. The sequence control becomes clearer and can therefore be tested more easily.

- **Inheritance and polymorphism**
  The *strategy pattern* is a classic application of object-oriented polymorphism but can also be implemented without inheritance, as described in the next information box.
- **Strategies must be known to the client**
  The responsibility for selecting the right strategy lies with the caller of the context class and must therefore also be known. The client must know the relevant strategies and also be able to make sensible selections.
- **Increasing complexity**
  Implementing the pattern may result in a more complex application structure and is less useful if you only have a few variants in the algorithm.

Alternative Implementation through Functional Types

As an alternative to an inheritance hierarchy, you can implement strategies in some programming languages by using *functional types* where a type is defined to which a function can be assigned. The following example shows this approach using Go:

```go
//Type definition - Functional type
type PaymentStrategy = func(amount int) error
```

```go
//Implementation of the type
func CreditCardPayment(amount int) error {
    fmt.Println("pay with CreditCard")
    return nil
}
```

The use for the pattern is then analogous:

```go
ShoppingCart := model.ShoppingCart{}
shoppingCart.SetStrategy(model.CreditCardPayment)
shoppingCart.AddItem(item)
err = shoppingCart.Pay()
```

The effort required to use the pattern is thus reduced, and responsibilities are still separated.

4.3.6 Real-World Example

In Java, probably one of the best-known use cases for the *strategy pattern* can be found in the *Java Collections Framework*.

All instances of the `java.util.List` interface can be sorted, either via the `java.util.Collections` class or using the `sort(...)` method directly on the instance itself. However, the sorting process is determined by a `Comparator`, which defines the rules for comparing elements within the list and thus the order of elements after sorting, as shown in Figure 4.13.

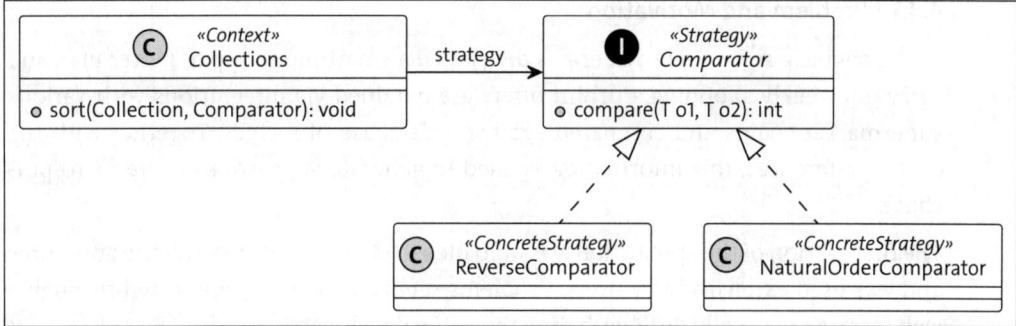

Figure 4.13 Sample Strategy Pattern of the Java Collections

The standard *Java Development Kit (JDK)* library contains multiple implementations. You can also add your own implementations by implementing the interface. The example shown in Listing 4.18 uses the `Comparator.reverseOrder()` and the `Comparator.naturalOrder()` to sort a list of strings; the output is shown in Listing 4.19.

```
List<String> values = new ArrayList<>();
values.add("A");
values.add("C");
values.add("B");

Collections.sort(values, Comparator.reverseOrder());
values.sort(Comparator.reverseOrder());
System.out.println(values);
values.sort(Comparator.naturalOrder());
System.out.println(values);
```

Listing 4.18 Strategy Implementation in the Java Collections Framework

```
[C, B, A]
[A, B, C]
```

Listing 4.19 Output of the Application

4.4 Chain of Responsibility

Origin: Gang of Four—Behavioral Pattern

The *chain of responsibility pattern* is a behavioral pattern originally from the Gang of Four.

You can use this pattern to separate the creator of a request from the actual executor of the request. The executing objects can be linked to form a chain, and the request is passed along this chain until one object processes the request.

4.4.1 Problem and Motivation

Let's imagine a software provider has developed a smartphone app to better plan and carry out weekly shopping. Current offers are obtained via integrations with various supermarket chains and compared against a database of recipes. Together with the user's preferences, this information is used to generate suggestions of items to purchase.

The first version of the application was uploaded to the relevant smartphone app stores and was used extensively by users. Problems reported by users, submitted through a web form, were usually dealt with directly by the developers.

Over time, the number of downloads increased, and the app became popular and successful. New functions were added so that the application became increasingly comprehensive.

The increasing popularity and greater complexity led to an increasing number of reported problems, which were forwarded directly to the developers. In many cases, however, the problems reported were not bugs but user issues. As a result, the developers were so busy providing support for the application that they were unable to implement any new functions.

An AI-based solution was introduced to process customer questions and relieve the burden on developers to answer incoming questions. Inquiries that could not be answered here were forwarded to the developers. The corresponding logic was built directly into the implementation of the web form for receiving support requests.

This significantly reduced the number of support requests to the developers, allowing them to implement additional functions. However, the problem reappeared: The new functions attracted new users and inquiries increased again. The AI-based solution worked well, but the support requests received by the developers increased again. Therefore, the company decided to create a separate support group that would take on the requests that couldn't be processed by the AI and only pass them on to developers if absolutely necessary.

All logic for controlling the process and the logic for forwarding the requests was incorporated directly into the web form again, which resulted in complex, error prone, and poorly maintainable code.

When the time to switch from email-based communication to instant messaging-based transmission arrived, this changeover led to problems, failures, and high costs.

4.4.2 Solution

Using the *chain of responsibility* pattern, more complex processing can be divided into multiple individual steps, and these steps are each converted into independent objects. The behavior of the steps is defined via an abstract interface and a `Handler` interface and then implemented by the individual steps, as shown in Figure 4.14.

4.4 Chain of Responsibility

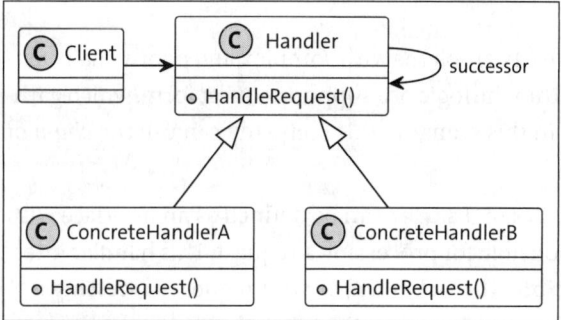

Figure 4.14 Structure of the Chain of Responsibility Pattern

A reference to the next concrete handler, which each handler implementation contains, is used to set up a processing chain that takes over the processing of the request.

A client only knows the first link in the chain. If a request is not or cannot be processed within a handler, the corresponding handler redirects the request to its `successor`, which then also attempts to process this request. As soon as the request is processed, processing stops, and the response is returned to the caller.

Not all handler implementations in the chain are necessarily called, as shown in Figure 4.15. In this example, the `ConcreteHandlerB` can process the request and return a result. Each handler therefore has the option of canceling further processing.

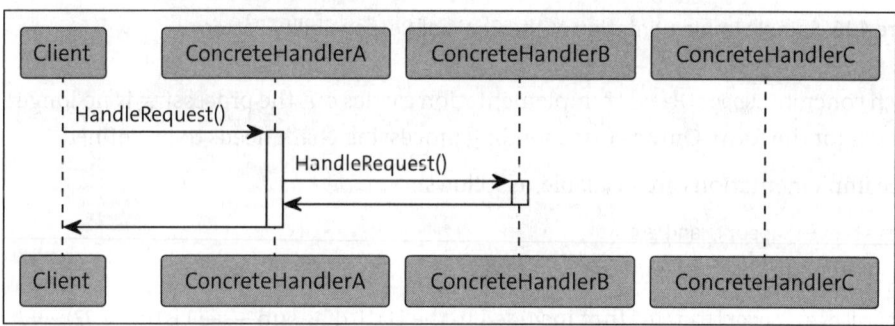

Figure 4.15 Sequence Diagram of the Chain of Responsibility

The structure and participants in this pattern are as follows:

- Processor/handler: Defines an abstract interface for the individual steps that will be linked together.
- Concrete handler: Implementation of the abstract interface in a single concrete step. Each specific processor contains a reference to its successor, which it calls up if it does not want to or cannot process the request itself.
- Caller/client: This calling code starts the first concrete handler in the processing chain and expects a result in return. The client only knows the abstract interface and is decoupled from the concrete implementations.

4.4.3 Sample Solution

Let's say our software company solves its problems with complex and poorly maintainable code for support processing, within the logic of a support request form, which gathers from the problem descriptions. In this scenario, this company can use the *chain of responsibility pattern*.

As shown in Figure 4.16, an abstract SupportHandler can be defined as an interface with a handleRequest method that is responsible for processing a request. This handler interface is then used within the SupportForm (the web form) to start request processing.

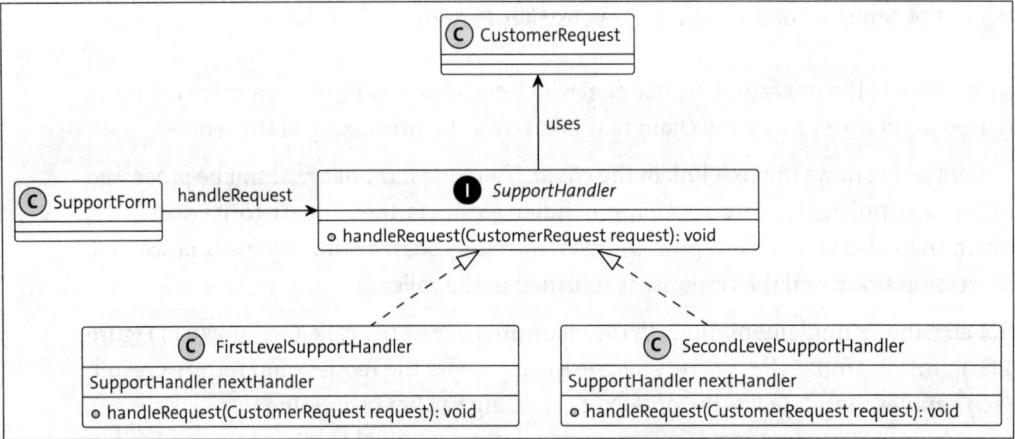

Figure 4.16 Sample Implementation of the Chain of Responsibility Pattern

Which concrete SupportHandler implementation carries out the processing is no longer relevant for the form. Only a corresponding processing chain needs to be defined.

Three implementations are available, as follows:

- FirstLevelSupportHandler
- SecondLevelSupportHandler
- ThirdLevelSupportHandler (not included in the UML diagram—see Listing 4.22)

An implementation of the SupportHandler interface in Java is shown in Listing 4.20.

```
package com.sourcefellows.patterns.cor;

public interface SupportHandler {
    void handleRequest(CustomerRequest request);
}
```

Listing 4.20 Sample Implementation of the Handler in Java

In this example, the CustomerRequest class represents a request to be answered. However, the interface of the SupportHandler interface does not define a response type for the `handleRequest` method, as the response of the corresponding handler should be sent to a reply address contained in the `CustomerRequest` object, for example, to the customer's email address.

Each specific handler implementation must have a reference to its successor in the chain. In our example, this reference is set for `FirstLevelSupportHandler` via the constructor. The `handleRequest` method determines whether the request should be processed or whether it should be passed on to the next handler in the chain. Listing 4.21 shows the implementation.

```
package com.sourcefellows.patterns.cor;

public class FirstLevelSupportHandler implements SupportHandler {

    private SupportHandler nextHandler;

    public FirstLevelSupportHandler(SupportHandler nextHandler) {
        this.nextHandler = nextHandler;
    }

    @Override
    public void handleRequest(CustomerRequest request) {
        if (...) { // Handler does not want to or cannot take over
            this.nextHandler.handleRequest(request);
        }
        ...
    }
}
```

Listing 4.21 Implementation of a Concrete Handler in Java

The implementation of the `SecondLevelSupportHandler` is quite similar and also contains a reference to its successor.

The processing chain can be created and used within the form. Three handler implementations are shown in Listing 4.22. The third implementation, `ThirdLevelSupportHandler`, no longer has a successor and must therefore take care of all requests that have not yet been answered.

```
SupportHandler third = new ThirdLevelSupportHandler();
SupportHandler second = new SecondLevelSupportHandler(third);
SupportHandler handler = new FirstLevelSupportHandler(second);

handler.handleRequest(new CustomerRequest());
```

Listing 4.22 Call Within the Web Form (Java)

As an alternative to this implementation using an interface and then direct concrete implementations, you can also use an abstract class to manage a subsequent handler, as shown in Figure 4.17.

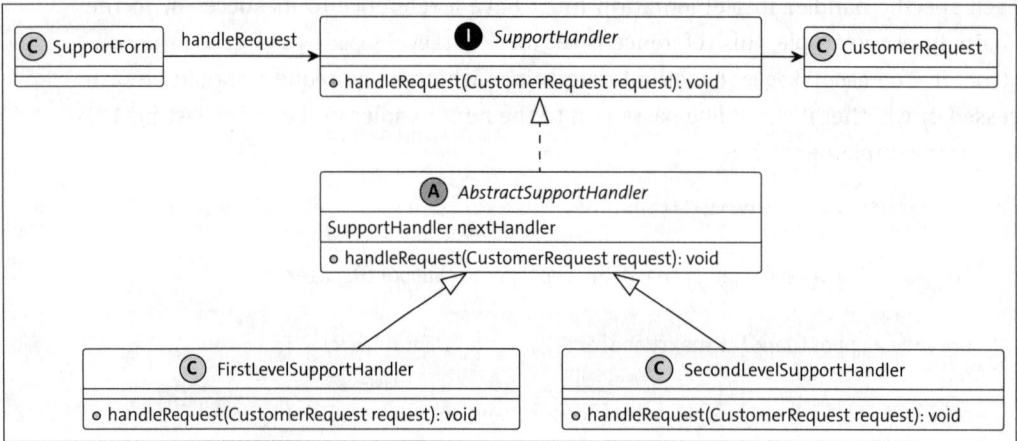

Figure 4.17 Implementation Using an Abstract Base Class

4.4.4 When To Use the Pattern

Some examples of when to use the pattern include the following:

- **Multiple handler implementations with unknown sequence**
  You can always use the *chain of responsibility pattern* if you have multiple different implementations for processing a request. Another scenario is when these implementations must be added later, but their execution sequences have not yet been determined at the start.

- **Decoupling between caller and performer desired**
  By using the pattern, the calling code can be decoupled from the processing code that is actually called. The caller does not know the specific implementation used to process the request.

- **Enabling a dynamic sequence**
  The structure of the processing chain is flexible and can also take place at runtime under certain circumstances. With an appropriate configuration, handlers can be adapted or exchanged dynamically.

4.4.5 Consequences

Some consequences of using these patterns are as follows:

- **Flexible order and distribution of responsibilities**
  Separating the processing of requests into their individual steps also leads to a distribution of responsibilities for the overall process. Each individual step takes on a

corresponding share of the responsibility and, with the flexible configuration, enables a fine-grained control of the specific process.

- **Decoupling of the calling and executing code**
  The pattern decouples the *calling code* from the *called code*: The client code works with the abstract interface and does not know about the concrete implementations. Conversely, the specific handler implementation does not know the client either. Both components work independently of each other, and none of the specific implementations need to know the structure of the processing chain or of another implementation.

  As a result, using this pattern can reduce dependencies within the implementations and increase testability.

- **No execution guaranteed**
  There are no guarantees for the receipt or processing of a request. An error in a handler implementation can cause the entire processing chain to become faulty, and requests can potentially not be processed. If, for example, the chain is configured incorrectly, the last handler in the chain may no longer be able to pass on the request, and the request might be discarded.

4.4.6 Real-World Example

Event bubbling is a mechanism for event propagation within a *Document Object Model (DOM)*, as used in XML and HTML. Each object in a document is represented as a node within a hierarchical tree structure and can receive *events*.

If new events occur, these events are first sent to the innermost element and then successively to the surrounding parent objects. This process continues until the outermost structure or the root node of the tree structure has been reached.

In the example shown in Figure 4.18, an event that occurs on element 3 is first forwarded to element 2 and then to element 1. All elements have the ability to react to the event.

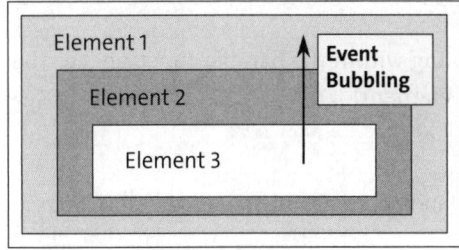

Figure 4.18 Event Bubbling in a DOM

In web browsers, for example, a user's actions, such as clicking a button or typing in an input field, trigger events, which is the only way to handle interaction in a browser.

You can register *event handlers* for each element of a document, which enables you to respond to specific results. JavaScript provides the `addEventListener` method for this purpose, which is available in every implementation of the `EventTarget` interface, as shown in Listing 4.23.

```
addEventListener(type, listener, options)
```

Listing 4.23 Method for Registering an EventListener (JavaScript)

A function, called a *listener* or a *handler*, can be passed as a parameter of the method with a `handleEvent` function. These functions in turn receive the event that has occurred as a parameter, which they can then evaluate. Listing 4.24 shows a corresponding function signature.

```
function handleEvent(event) {
...
}
```

Listing 4.24 Listener or Handler Implementation (JavaScript)

> **Event Bubbling: What's Missing from the Chain of Responsibility Pattern?**
>
> Strictly speaking, in JavaScript, event bubbling does not correspond completely to the *chain of responsibility pattern*. The handler implementations lack reference to the next handler in the chain. However, you can cancel further processing by calling a method.
>
> If the HTML page itself were built using JavaScript, the processing chain could also be adapted dynamically at runtime. However, these details are beyond the scope of this example.
>
> The *Gang of Four (GoF)* book, in which the pattern was first described, also uses a user interface as an example. Help texts for elements are displayed in an application. Each element has the option of displaying a text or forwarding the processing to its parent element in order to use the higher-level help text.

As shown in Listing 4.25, you can cancel bubbling within the handler function (i.e., the automatic forwarding of events to the parent element).

```
event.stopPropagation();
```

Listing 4.25 Canceling Event Bubbling Within a Handler (JavaScript)

If the code fragments shown so far are brought together, a handler implementation can be created for the following HTML page and registered as shown in Listing 4.26.

```
<!DOCTYPE html>
<html lang="en">
<head>...</head>
```

```
<body id="body">
    <h1>Chain of Responsibility</h1>
    <p id="paragraph">
        <img id="logo" src="...">
    </p>
    <script src="index.js"></script>
</body>
</html>
```

Listing 4.26 Sample HTML Document with Elements with ID Attributes

---

ID Attributes?

For the sake of clarity, our HTML page uses ID attributes for its elements, even if it's not necessary in this case.

In this example, the JavaScript code shown in Listing 4.27 is integrated into the HTML page shown earlier via an external file. The corresponding handlers can be registered and used in this file in order to access the HTML structure defined in the HTML file.

```
// Reference elements in document
const img = document.getElementById("logo");
const paragraph = document.getElementById("paragraph");
const body = document.getElementById("body");
// Handler implementations
function imgHandler(event) {
    console.log("Image Handling:", event);
}
function paragraphHandler(event) {
    console.log("Paragraph Handling:", event);
    event.stopPropagation();
}
function bodyHandler(event) {
    console.log("Body Handling:", event);
}
// Handler registrations for click events
img.addEventListener("click", imgHandler);
paragraph.addEventListener("click", paragraphHandler);
body.addEventListener("click", bodyHandler);
```

Listing 4.27 Implementation of Event Listeners (JavaScript)

If the HTML page is loaded in the browser and a user clicks the displayed image, first the `imgHandler` function is called, and then the `paragraphHandler` function.

By calling `stopPropagation` within the `paragraphHandler` function, the bubbling of events is interrupted, and processing ends.

4.5 Command

Origin: Gang of Four—Behavioral Pattern

The *command pattern* encapsulates a command that can be parameterized by the client in what are called command objects. These commands can be used as parameters for methods, for example, to register their execution for a later time or to enable them via a queue. Commands can also be implemented in such a way that their execution can be undone later.

4.5.1 Problem and Motivation

In 2014, a large online department store introduced electronic buttons, known as *dash buttons*, which could customers click to purchase predefined products directly from its online store. Each button was configured for a specific product. When the button was pressed, an order was triggered, and the goods were delivered to the customer. Communication with the internet worked by integrating the button into the customer's Wi-Fi network.

Since the internet department store service has been discontinued, goods can no longer be ordered via the buttons, but the Wi-Fi signals can still be used to trigger other actions. When the button is pressed, signals are generated that can be assigned to the respective dash button via a unique switch ID.

A savvy developer then began implementing software that could be used to perform various actions in his home. As the functionality of the switches is similar, he summarized the similarities in an abstract basic implementation (`Button` class) and created a separate derivation of the abstract class for each switch, as shown in Figure 4.19.

This resulted in multiple implementations: one for switching the living room light (`LivingRoomLightSwitch` class), another for starting the dishwasher (`StartDishwasherButton` class), and one for muting the doorbell (`MuteDoorbellButton` class).

Each of those concrete implementation uses the shared functionality of the `Button` base class for the receipt of messages, and contains details of the triggering action or switch (such as the switch ID).

With this approach, the inclusion of additional switches in the environment leads to an increasing number of subclasses, which doesn't need to be a problem. However, as soon as changes to the base class must be made, you run the risk that the functionality in the specific implementations will become faulty, and problems will arise.

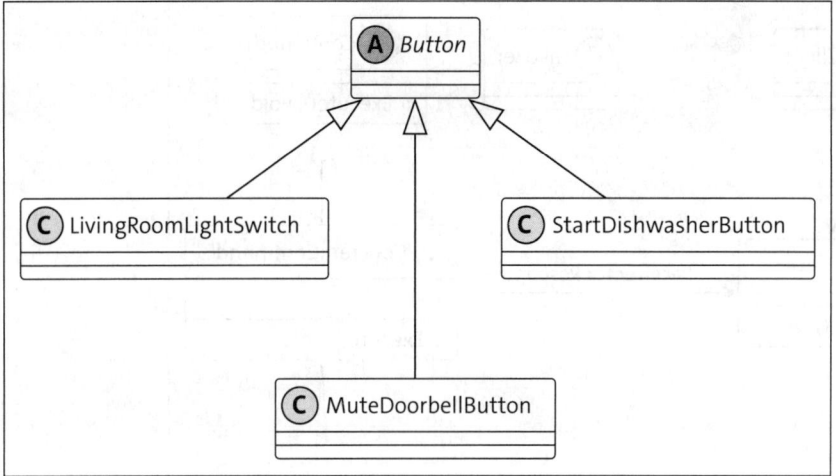

Figure 4.19 Implementing the Commands as a Class Hierarchy

Each implementation also mixes the logic for communication with the switch with the actual business logic, that is, the execution of the action. These implementations lack a clear separation of responsibilities.

After the service was shut down, cheap buttons kept popping up for sale on various online exchanges, and the developer came up with the idea of acquiring and using more buttons.

However, these new buttons didn't take on any new functions, only trigger existing actions. From then on, not only should it be possible to switch off the doorbell in the hallway, but also from the bedroom.

As the specific implementations have code for exactly one switch (namely via the switch ID), you cannot continue using the action logic for a second switch without problems. Another subclass would solve the problem but would result in redundant and confusing code.

4.5.2 Solution

The *command pattern* decouples the action to be executed from the calling logic by implementing the individual actions or commands as implementations of an abstract interface description, the Command interface.

The concrete implementation of a command only contains the logic of an action. Changes or method calls that are necessary during a call are carried out at a Receiver. In the structure of the *command pattern*, shown in Figure 4.20, the ConcreteCommand executes the Action action of the Receiver. The execution is initiated via a caller, the Invoker.

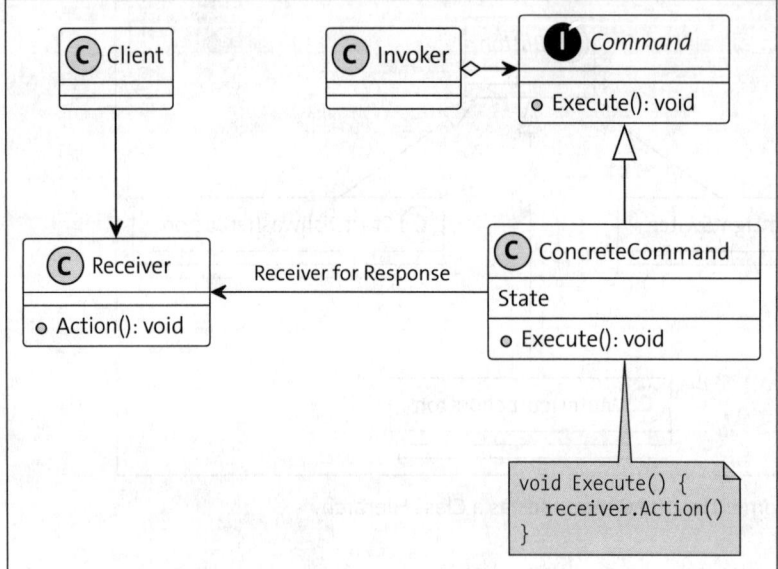

Figure 4.20 Structure of the Command Pattern

The entire sequence, shown in Figure 4.21, follows these steps:

- A Client creates an instance of a concrete command (ConcreteCommand).
- The Client then passes the command to the Invoker, by calling a method (in the example, the DoSth method) with the command as parameter.
- The Invoker starts the specific command by calling the Execute method of the concrete command.
- The command executes the action logic (Action) by using methods of the Receiver.

The structure and participants in this pattern are as follows:

- Command: Defines an abstract interface for the operations or actions to be executed. In most cases, the method has no parameters.
- Concrete command: Implements the abstract command interface and defines the connection between the action and the receiver. The command executes the corresponding methods on the receiver object to perform the action.
- Client: Creates a specific command and defines the corresponding recipient on which the action is to be executed.
- Caller/invoker: Uses a command to execute an action.
- Receiver: The object on which the actual action is to be carried out. Each class can act as a receiver.

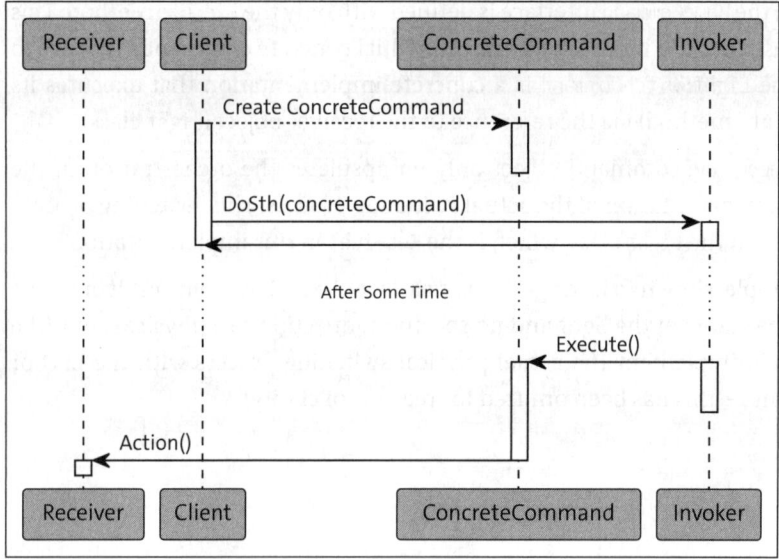

Figure 4.21 Sequence Diagram of the Command Pattern

4.5.3 Sample Solution

Using our example of the *command pattern* for connecting the Wi-Fi switches for home automation separates the action logic from the execution logic for the switch. The developer has the option of integrating additional switches into the system and continuing to use the execution logic or creating more complex scenarios by linking several commands. For example, the dishwasher could be started, the living room light dimmed, and the doorbell switched off at the touch of a single button.

In concrete terms, an implementation for controlling a light switch is shown in Figure 4.22.

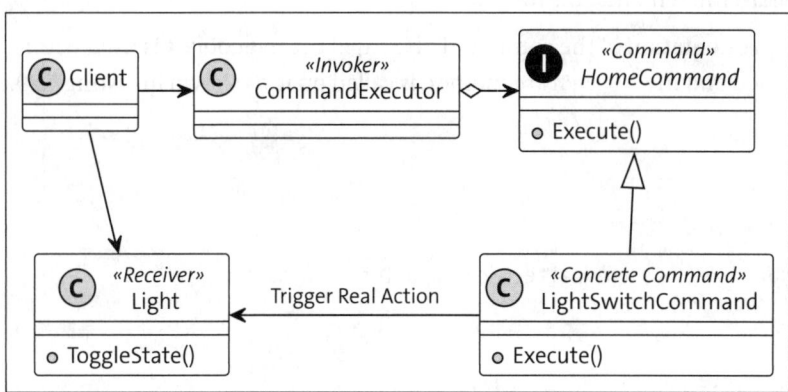

Figure 4.22 Solving the Problem Using the Command Pattern

In this example, the HomeCommand interface is defined with only the Execute method. This abstract type describes the interface for the individual concrete commands that are to be executed. The LightSwitchCommand is a concrete implementation that executes its action (ToggleState method) via the reference to the receiver object (Light class).

The LightSwitchCommand command object only encapsulates the orchestration of the action, not the implementation of the actual switching process. This switching process is performed within the Light class, which is the Receiver in this implementation.

A Go-based example, shown in Listing 4.28, is clearly laid out: The Light receiver object only contains the status of the light and no specific connection to a physical light (the Light class should implement the actual physical switching process with the LED or incandescent lamp—this has been omitted for reasons of clarity).

```go
package model

type Light struct {
    onOffState bool
}
//Constructor/factory method
func NewLight(onOffState bool) *Light {
    return &Light{onOffState: onOffState}
}
// ToggleState switches the state of the light.
func (l *Light) ToggleState() {
    l.onOffState = !l.onOffState
}
```

Listing 4.28 The Receiver Object (Go)

As described earlier, the HomeCommand interface defines the Execute method, and the LightSwitchCommand implements this interface.

In the actual implementation of the command, the Light receiver object is used within the Execute method and the ToggleState method is called on it, as shown in Listing 4.29.

```go
package internal

import (
    "fmt"
    "golang.source-fellows.com/pattern/command/internal/model"
)

type HomeCommand interface {
    Execute()
}
```

```go
type LightSwitchCommand struct {
    light *model.Light
}

func NewLightSwitchCommand(light *model.Light) *LightSwitchCommand {
    return &LightSwitchCommand{light: light}
}

func (l LightSwitchCommand) Execute() {
    l.light.ToggleState()
}
```

Listing 4.29 Example of the Command Pattern (Go)

This example code, `LightSwitchCommand`, only contains action steps that are called on a receiver object. The receiver object or receiver is specified as a parameter of the `NewLightSwitchCommand` function when the command object is created.

The actual execution of the commands takes place according to the pattern via what's called the *invoker*. In the example shown in Listing 4.30, this invoker was implemented with the name `CommandExecutor` and contains a list of all commands to be executed. Within the `ExecuteAllCommands` method, the individual concrete commands are called in a loop, and the list is reinitialized after their execution. New commands can be added.

```go
package internal

import "fmt"

type CommandExecutor struct {
    commands []HomeCommand
}

func (i *CommandExecutor) AddCommand(command HomeCommand) {
    i.commands = append(i.commands, command)
}

func (i *CommandExecutor) ExecuteAllCommands() {
    for _, command := range i.commands {
        command.Execute()
    }
    i.commands = []HomeCommand{}
}
```

Listing 4.30 Sample Invoker of the Command Pattern (Go)

In a client, the configuration and the call are shown in Listing 4.31.

```
//Create the invoker/executor
invoker := &internal.CommandExecutor{}

//Create the receiver/receiver object
livingRoomLight := model.NewLight(false)

//Create the command with the reference to the receiver
cmd1 := internal.NewLightSwitchCommand(livingRoomLight)

//Add the command to the list
invoker.AddCommand(cmd1)
//Execute the command list
invoker.ExecuteAllCommands()
```

Listing 4.31 Sample Client of a Command Pattern Implementation (Go)

Additional commands can be implemented in the same way, and each command can use separate receiver objects.

For our home automation example, the developer can generate a new command with every switch interaction and pass it to the CommandExecutor. Depending on how the command passed is configured, it can, for example, pass the light to be switched as a parameter. The corresponding implementations are decoupled from each other and can be used flexibly.

4.5.4 When To Use the Pattern

Some examples of when to use the pattern include the following:

▶ **Parameterizable objects for execution**
The *command pattern* allows you to express the callback functions, which you may be familiar with from procedural programming, in an object-oriented syntax. Method calls are converted into independent objects that can, for example, be passed on as method parameters or saved for later execution. Parameterization of the objects enables flexible use of the commands.

The pattern is therefore suitable for cases in which tasks or commands need be configured and executed later.

▶ **Different times for definition, application, and execution**
The Invoker is used to separate the times at which a command is created and executed. The Invoker receives the commands and decides when the actual execution starts. This pattern can be used whenever such a time division is required. Accordingly, the pattern is often used for asynchronous processing of orders, which in this case are transferred as a command to a queue. In addition to local execution, scenarios are conceivable in which a command is transmitted via a remote interface and executed in a remote instance.

▶ **Undoing actions**
Even if no action could be undone in our example, the pattern is suitable for this purpose. Commands can provide a compensation method or Unexecute method for such a case, and the corresponding action can be reversed if necessary. If the command has not yet been executed, it can be removed from the queue. Commands that have already been executed can be stored in a separate history object with a corresponding list.

▶ **Logging of actions**
If a log of the individual commands executed is required, you can, for example, create an audit log within the Invoker instance with a manageable amount of effort.

4.5.5 Consequences

Some consequences of using these patterns are as follows:

▶ **Decoupling between execution and definition**
This pattern defines two objects: an object that controls the execution and an object on which the action is executed. This approach decouples the tasks of the two objects according to the *single responsibility principle*.

▶ **Commands are first-class objects**
Method calls become what are called *first-class objects* through encapsulation in independent objects. They can be treated like any other object and passed on as parameters, for example.

▶ **Combination of commands possible**
Individual commands can be combined to perform more complex tasks. In word processing software, this sequence is known as a macro command, where smaller independent commands are combined to form a larger, more complex sequence.

▶ **New commands can be extended**
According to the *open-closed principle*, new commands can be added without having to change the existing code.

4.5.6 Real-World Example

After significant work by Doug Lea, the `java.util.concurrent` package was included in the *Java Development Kit (JDK)* version 1.5. The package was initially maintained as an external library and provides helper classes that facilitate concurrent programming in Java.

In simple terms, tasks can be executed with one of the main components, the *executors*. On the one hand, interfaces for execution and, on the other hand, suitable concrete implementations are provided.

The tasks to be executed are passed to an Executor and must implement the java.lang.Runnable interface. Multiple implementations are already available in this case as well. Figure 4.23 shows the basic structure of the Executor component.

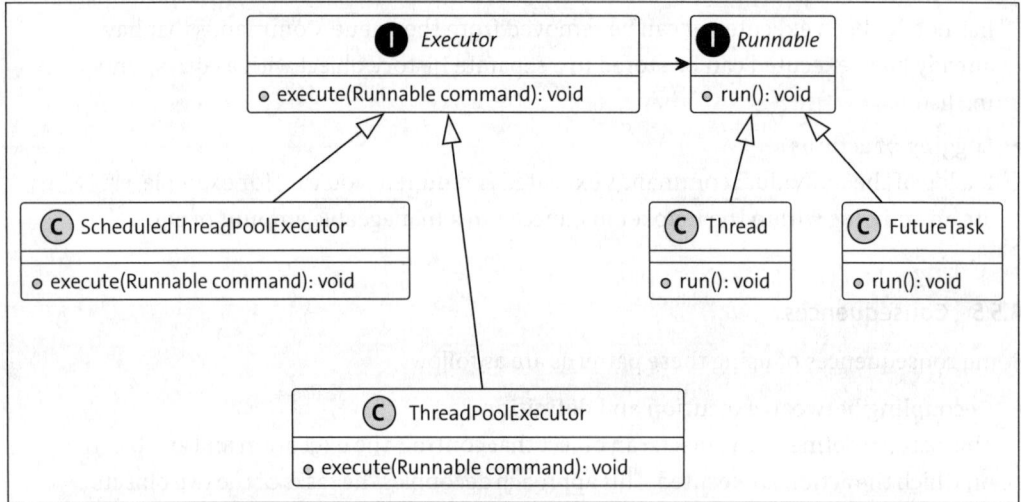

Figure 4.23 Main Executor Component of the Concurrent Package

The concurrent package was further expanded in the subsequent Java versions, and additional subclasses were added for the execution of tasks. In the following example, however, we'll work with the basics, which is why I will not introduce the newer classes any further.

The class diagram shows the basic structure of a *command pattern*. The Executor corresponds to the Invoker, and the Runnable interface corresponds to the command.

In a concrete example, the WebCrawler class is used to retrieve and analyze web pages. The results of the analysis are stored in a database and can be called there later. Using the *command pattern* enables better control the volume of data retrieved and distributes it throughout the day.

Figure 4.24 shows the class diagram for our example. The CrawlCommand class implements the Runnable interface and can therefore be transferred to a ThreadPoolExecutor.

The implementation is kept simple, as shown in Listing 4.32.

```
package com.sourcefellows.patterns.command;

public class CrawlCommand implements Runnable{

    private String url;
    private final WebCrawler webCrawler;
```

```
public CrawlCommand(String siteToBeCrwawled, WebCrawler collector) {
    this.url = siteToBeCrwawled;
    this.webCrawler = collector;
}

@Override
public void run() {
    webCrawler.crawl(this.url);
}
}
```

Listing 4.32 Command Object for Retrieving Internet Pages (Java)

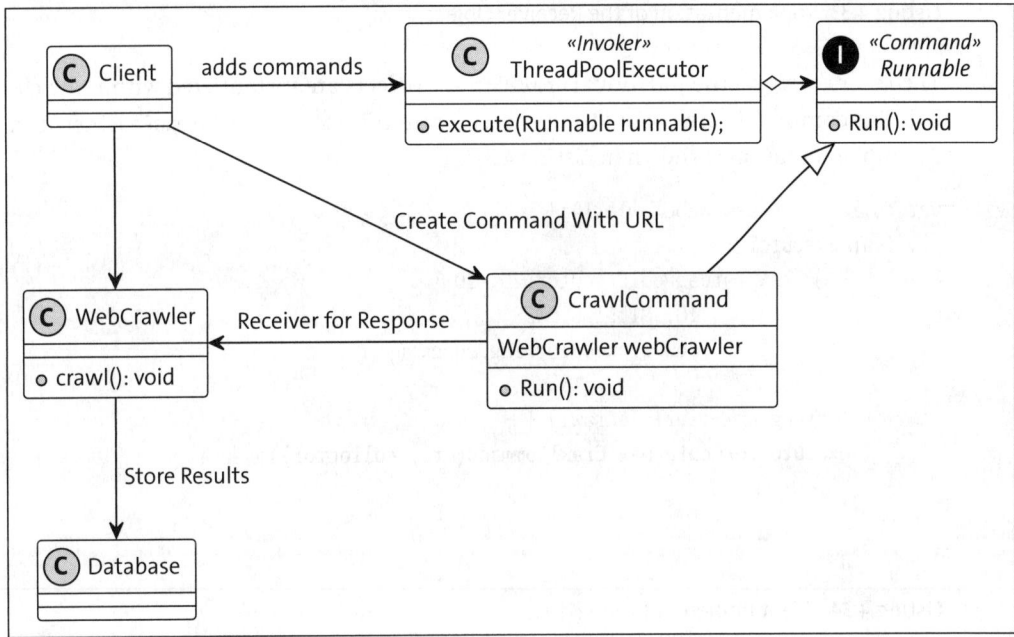

Figure 4.24 Example Command Pattern with the Java Concurrent Package

The task of the `CrawlCommand` class is to call the `crawl` method with the corresponding URL on the `WebCrawler` receiver object. Receiver and URL to be retrieved are transferred as parameters of the constructor when the object is created.

The `WebCrawler` implementation receives the URL, retrieves it, and saves the result of the evaluation, as shown in Listing 4.33.

```java
package com.sourcefellows.patterns.command;

public class WebCrawler {

    public WebCrawler() {}

    public void crawl(String url) {
        //Crawl URL
        //Save result in DB
    }

}
```
Listing 4.33 Implementation of the Receiver Object (Java)

The `WebCrawler` and the individual `CrawlCommand` objects are instantiated within the client. The commands are passed to the `executor`, in this example a `ThreadPoolExecutor`. An implementation is shown in Listing 4.34.

```java
var collector = new WebCrawler();
try (var executor =
            new ThreadPoolExecutor(20, 20,
                    10, MILLISECONDS,
                    new LinkedBlockingDeque<>())) {

    for (String url : urlsToCrawl) {
        executor.execute(new CrawlCommand(url, collector));
    }

}
```
Listing 4.34 Client Implementation (Java)

> **Lambda Expressions or Method Reference**
>
> In some programming languages, the *command pattern* can also be implemented using *lambda expressions* or alternatively with the transfer of method or function references.
>
> Our Java-based example can, for example, be implemented with a lambda expression without implementing a specific command object. The setting of the command within the loop would be shortened accordingly:
>
> `executor.execute(() -> collector.crawl(url));`
>
> With more extensive logic within the command class, lambda expressions can become more confusing. Alternatively, you could also use references to methods or functions.

4.6 Observer

Origin: Gang of Four—Behavioral Pattern

Using the *observer pattern*, you can establish a connection from one object to multiple dependent objects so that these objects are automatically informed whenever the state of the central object changes and can react accordingly.

4.6.1 Problem and Motivation

An online news portal specializes in publishing important information on commodities trading. Customers all over the world receive the latest reports and detailed dossiers on a wide range of topics at all times. In addition to a traditional website, the portal also provides the content as a PDF or in EPUB format so that customers can also consume the information on devices, which aren't always connected to the internet.

After logging into the portal, customers can select the appropriate data format with a click, whereupon the latest version of the documents is generated for the customer and provided for download.

Over time, as the number of subscribers increases, so do the costs of the cloud environment in which the online news portal is operated, and solutions are sought.

After initial investigations, clearly, generating the documents places a heavy demand on resources and therefore accounts for a large proportion of the costs incurred in the cloud: On one hand, the latest reports and messages must be loaded from a database, and on the other hand, the creation of the documents requires a lot of computing time. Both aspects can be regarded as cost drivers in a cloud environment.

The developers then set up a cache for documents in which they store pregenerated documents that they deliver directly to customers. Not only does doing so mean that customers no longer wait for the documents to be generated, but also that the documents don't need to be generated anew for each customer, which saves time and computing resources and therefore money.

Over time, however, customer complaints increase as they realize that the different formats contain different information and that they may be provided with outdated news, which had not been the case so far.

The reason is the implementation of the document update mechanism: The developers have implemented different mechanisms depending on the data format. The HTML page is still created directly from the database and therefore contains the latest data. PDFs and the EPUB format, on the other hand, are generated via processing jobs that are executed at regular intervals. Unfortunately, the generation times are different and documents with different content are created.

If no new articles are published, job processing and document generation continue. Even if nothing has changed, new documents are created and published.

4.6.2 Solution

The *observer pattern* can distribute state changes that occur on an object, called the *subject*, to registered *observers* or *subscribers*. These observers can react to changes and adjust their own states accordingly.

Objects that want to receive change information for a specific object must implement or extend the abstract interface description, Observer. This interface defines a method that is called whenever there are changes to the monitored object. In the class diagram in Figure 4.25, this is the Update method of the abstract Observer class or the concrete implementation; ConcreteObserver.

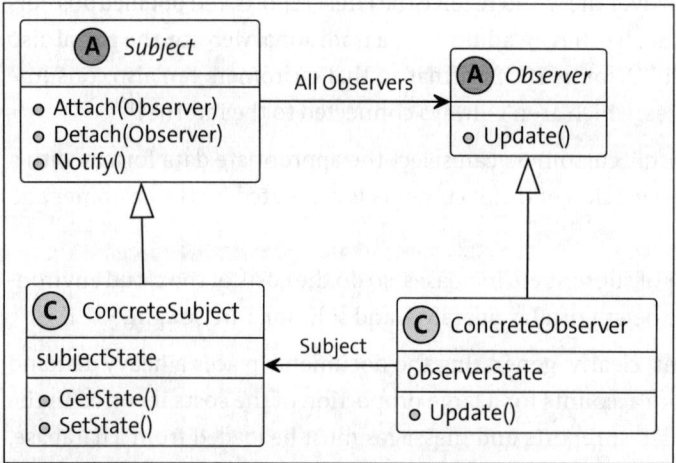

Figure 4.25 Structure of the Observer Pattern

To ensure that the Observer implementations are informed about status changes to the monitored object (Subject), these subscribers are registered or deregistered in the subject via method calls. In the class diagram, the Attach and Detach methods each expect an observer as a parameter.

The ConcreteSubject class holds the actual status to be monitored and informs the registered and saved observers when status changes occur. This step is performed by calling the Notify method of the abstract Subject class. This method informs all registered Observer instances of a status change. For this task, the method iterates over the saved list of observers and calls the Update method for each observer.

If the updated data is not provided as a parameter of the Update method, the individual observers can retrieve this data via a status request, which you can see in the sequence diagram shown in Figure 4.26, and adjust their internal status accordingly if necessary.

The structure and participants in this pattern are as follows:

▶ Subject: Describes an abstract interface where the parties interested in status changes can be registered. A list of all registered interested parties, the *observers*, is stored internally and they are informed of any changes. An interface is provided for notifying interested parties. The subject is often referred to as the *publisher* and is implemented as a separate class and not as a base class.

- Concrete subject: Stores and manages the relevant status to be observed. All interested parties and observers will be informed of any changes to this status. If no data is contained as a parameter in the notification method, the specific subject has a method for querying the corresponding status.
- Observer: Defines an interface for notification of status changes.
- Concrete observer: Concrete implementation of the notification interface. A reference to the monitored object is kept internally so that the current status can be called there in the event of status changes. Concrete observers are logged on to the subject and receive status changes accordingly.

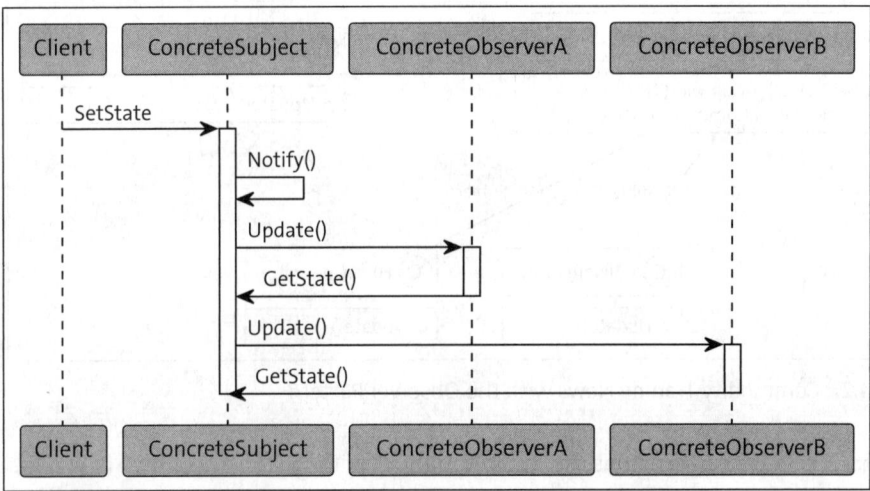

Figure 4.26 Sequence Diagram of the Observer Pattern

Observer and Messaging

The observer pattern can also be used in a messaging environment. The *enterprise integration patterns* contain, for example, the *publish-subscribe channel pattern*, which transfers the *observer pattern* to a messaging-based environment by using an *event channel* for communication. However, the basic idea remains the same.

4.6.3 Sample Solution

The online news portal from the problem description can solve the non-uniform generation of the individual data formats and the non-optimal use of resources by using the *observer pattern*.

The individual document generators are implemented as observers that are called when new messages are created. For the observers, this message is a signal that new output documents must be created. Since all document generators are always called in a loop, the newly received information is immediately available in all document formats, and the regular processing jobs are also eliminated. Figure 4.27 shows a UML diagram of the application.

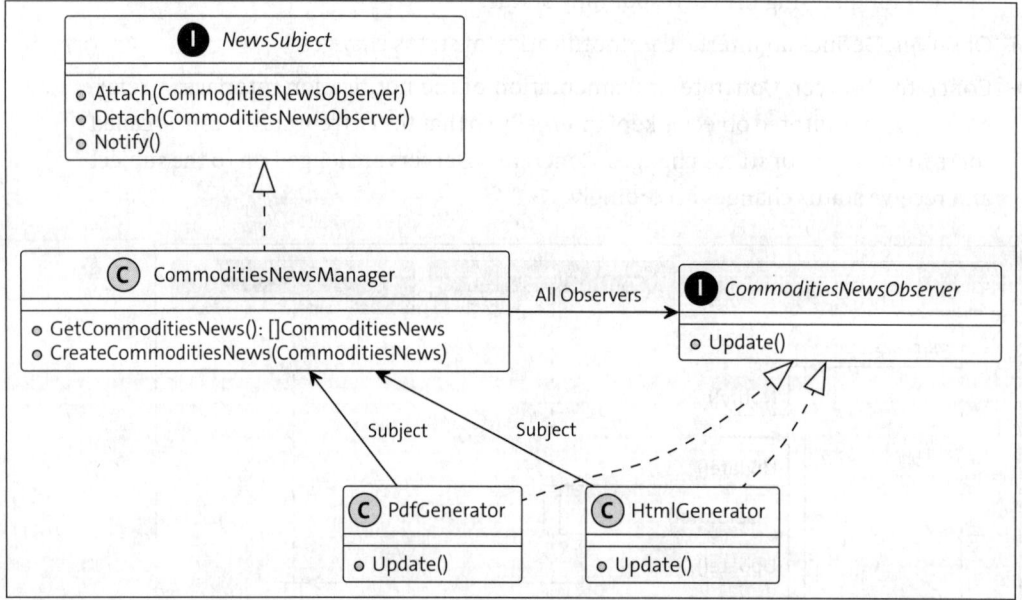

Figure 4.27 Commodity Trading News with the Observer Pattern

A CommoditiesNewsManager acts as the concrete subject of the pattern, in which new news objects (the CommoditiesNews) are added via the CreateCommoditiesNews method and stored internally. Once the messages have been stored, the Notify method is called to inform all registered CommoditiesNewsObservers about news.

In the Go-based implementation shown in Listing 4.35, the Subject interface NewsSubject is defined, which is implemented by CommoditiesNewsManager as a concrete subject.

```
type NewsSubject interface {
    Attach(observer CommoditiesNewsObserver)
    Detach(observer CommoditiesNewsObserver)
    Notify()
}

type CommoditiesNewsManager struct {
    observers          []CommoditiesNewsObserver
    allCommoditiesNews []*model.CommoditiesNews
}
```

```go
func (m *CommoditiesNewsManager) Attach(
    observer CommoditiesNewsObserver) {
    m.observers = append(m.observers, observer)
}

func (m *CommoditiesNewsManager) Detach(observer CommoditiesNewsObserver) {
    ...
}

func (m *CommoditiesNewsManager) Notify() {
    for _, concreteObserver := range m.observers {
        concreteObserver.Update()
    }
}

func (m *CommoditiesNewsManager) CreateCommoditiesNews(commoditiesNews
*model.CommoditiesNews) {
    m.allCommoditiesNews = append(m.allCommoditiesNews, commoditiesNews)
    m.Notify()
}

func (m *CommoditiesNewsManager) GetCommoditiesNews() []*model.CommoditiesNews
{
    return m.allCommoditiesNews
}
```

Listing 4.35 Sample Implementation for a Concrete Subject (Go)

To use the implementation, the concrete subject (CommoditiesNewsManager) must first be created. The individual observers, the PdfGenerator and the HtmlGenerator, can then be attached to the specific subject using the Attach method, as shown in Listing 4.36.

```go
newTraining := model.NewCommoditiesNews("GO")
manager := &internal.CommoditiesNewsManager{}

pdfGenerator := pdf.NewGenerator(manager)
htmlGenerator := html.NewGenerator(manager)

manager.Attach(pdfGenerator)
manager.Attach(htmlGenerator)

manager.CreateCommoditiesNews(newTraining)
```

Listing 4.36 Using the Observer Pattern (Go)

Both generators and observers each receive a reference to the specific subject so that they can query the current status from the subject, as shown in Listing 4.37. Then, to implement the `CommoditiesNewsObserver` interface, you'll need to implement the `Update` method.

```go
type Generator struct {
    manager *internal.CommoditiesNewsManager
}

func NewGenerator(manager *internal.CommoditiesNewsManager) *Generator {
    return &Generator{manager: manager}
}
func (c *Generator) Update() {
    commoditiesNews := c.manager.GetCommoditiesNews()
    for _, news := range commoditiesNews {
        //...
    }
}
```

Listing 4.37 Implementation of a Concrete Observer with Reference to the Subject (Go)

4.6.4 When To Use the Pattern

Some examples of when to use the pattern include the following:

- **Distributing state changes of an object**
  You can always use the *observer pattern* when state changes on an object must be passed on to several dependent objects or when these dependent objects must be informed of the change.

- **State changes must be monitored**
  Often, monitoring state changes to objects in applications is necessary so that other components can update themselves accordingly.

  The "monitoring direction" can be reversed using the *observer pattern*. In other words, interested parties no longer need to actively request information (*polling*) but instead can register with the object to be monitored so that they are notified whenever a state changes (*push*).

- **The recipient of state changes is unknown at the start**
  At the beginning of the development of an application, you may not know which objects or components are interested in state changes of specific objects. The pattern provides flexibility in this case and allows the list of recipients to be adapted dynamically.

4.6.5 Consequences

Some consequences of using these patterns are as follows:

- **Decoupling between sender and receiver**
  By introducing an abstract interface for notifying interested parties or the observers, the subject no longer has a direct dependency on the concrete implementations. The coupling between subject and observer is abstract and minimal.

- **The list of recipients is variable**
  The list of observers is variable and does not necessarily have to be determined in advance. Observers can be registered and deregistered. Dynamic adjustment of the list is possible in the application. The recipient or observer list can be adjusted without changing the subject, corresponding to the *open-closed principle*.

- **No sequence when calling observers**
  The observers can be logged in and out of the subject. However, the pattern does not define the order in which the registered observers are called.

- **Unwanted cascades of state updates**
  The subject is decoupled from the observers and does not know the concrete implementations. Changes will be passed on to all registered observers. If observers in turn act as subjects (i.e., pass on their state changes themselves), unwanted cascades of state changes can occur, which may generate inconsistent states or, in the worst case, lead to endless loops during updates.

- **The lapsed listener problem**
  Observers must be deregistered when they are no longer required. As the subject continues to hold a reference to the observer object, it continues to be informed. In programming languages with garbage collection, these objects cannot be removed from memory. This problem, called the *lapsed listener problem*, is one of the most common sources of *memory leaks*.

4.6.6 Real-World Example

The *Spring Framework* is widely used in the Java environment. What's called an *Inversion of Control (IoC)* container is an integral part of the framework and is responsible for managing components (the *beans*). Beans can be registered either in the code via the `ApplicationContext` interface, via corresponding annotations, or via configuration files.

State changes of the `ApplicationContext`, such as starting or stopping the container, are published and can be processed by registered `EventListener` implementations.

In the following example, an `ApplicationListener` implementation was created that will be informed whenever changes occur to the `ApplicationContext`. Figure 4.28 shows the structure of the `ApplicationListener` of the Spring Framework in a simplified class diagram. For the sake of clarity, derived classes and components that handle the distribution of events are not shown in this illustration.

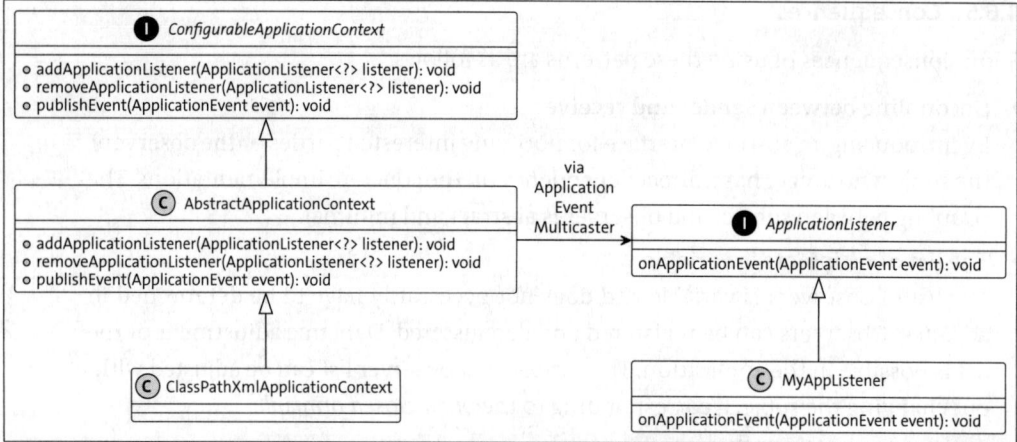

Figure 4.28 The Observer Pattern in the Spring Framework

In the Spring Framework, the `ApplicationListener` interface corresponds to the observer of the *observer pattern*, and the `ConfigurableApplicationContext` represents the subject. The concrete implementations of the subject correspond to the `AbstractApplicationContext` or the `ClassPathXmlApplicationContext`.

Listing 4.38 shows the implementation of the specific `MyAppListener` observer.

```
import ...context.ApplicationEvent;
import ...context.ApplicationListener;

public class MyAppListener implements ApplicationListener {
    @Override
    public void onApplicationEvent(ApplicationEvent event) {
        System.out.println("Event " + event);
    }
}
```

Listing 4.38 Implementation of the ApplicationListener Interface (Java)

The application shown in Listing 4.39 instantiates an `ApplicationContext`, registers a specific observer or listener, and then closes the context again.

```
// The long package names have been shortened.
import ...context.ConfigurableApplicationContext;
import ...context.support.ClassPathXmlApplicationContext;

public class Main {
    public static void main(String[] args) {
        ConfigurableApplicationContext context =
            new ClassPathXmlApplicationContext("services.xml");
```

```
        var myAppListener = new MyAppListener();
        context.addApplicationListener(myAppListener);

        context.start();
        context.stop();

        context.close();
        context.removeApplicationListener(myAppListener);
    }
}
```

Listing 4.39 ApplicationContext and ApplicationListener in the Spring Framework (Java)

A shortened console output (when running the example) is shown in Listing 4.40.

```
Event org.springframework.context.event.ContextStartedEvent[source=org.spring-
framework.context.support.ClassPathXmlApplicationContext@5ce65a89,
    started …]
Event org.springframework.context.event.ContextStoppedEvent[source=org.spring-
framework.context.support.ClassPathXmlApplicationContext@5ce65a89,
    started …]
Event org.springframework.context.event.ContextClosedEvent[source=org.spring-
framework.context.support.ClassPathXmlApplicationContext@5ce65a89,
    started …]
```

Listing 4.40 Output of the Sample Application

4.7 Singleton

Origin: Gang of Four—Creational Pattern

The *singleton pattern* is a creational pattern that originally comes the Gang of Four. This pattern is intended to ensure that only one instance of a type exists and also defines a public access point for that object.

4.7.1 Problem and Motivation

In this example, a manufacturer of high-performance machine tools wants to enable its customers to use the machines it produces with as few faults as possible. During production operation, however, the mechanical load on all components can repeatedly cause problems that result in failures.

Some of these problems become apparent days or hours before the machine actually breaks down. If certain key figures deviate from normal operations, the manufacturer wants to use a predictive maintenance approach to inform the customer at an early stage and be able to react together with the customer.

A newly installed *Internet of Things (IoT)* device is designed to inform the manufacturer and the customer of a potential fault via a cellphone message.

The software department creates an HTTP client for this purpose, which handles communication on the end device with the backend system developed by the company. Internally, the client uses the built-in cellular modem. Once the HTTP client was available, various software components began to use this HTTP client to send data to the server at regular intervals.

After the initial tests, the team realized that communication with the backend system was error prone. In many cases, communication fails because data from different components is obviously mixed up.

When the source code was checked, as shown in Figure 4.29, it was found that each component uses its own instance of the HTTP client and that concurrent communication via the mobile modem is prone to errors if used by multiple components at the same time. The currently implemented HTTP client was not designed for this situation. Data arrived incomplete or incorrect on the server side. In addition, each instance of the HTTP client reauthenticated itself when it was called, which led to an increased data volume and therefore to higher transmission costs.

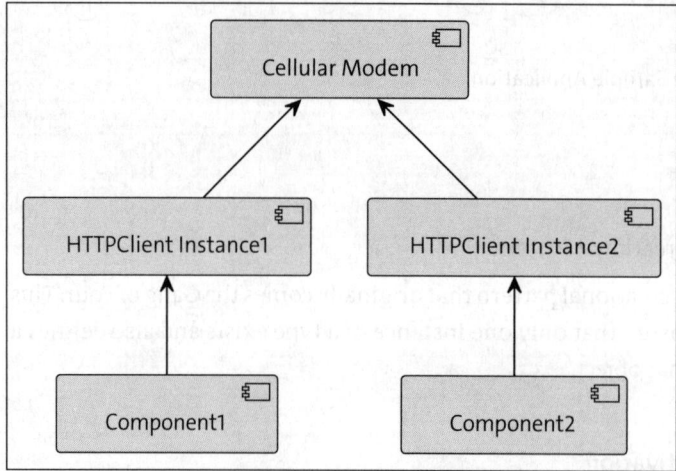

Figure 4.29 Problem Without the Singleton Pattern

4.7.2 Solution

The *singleton pattern* ensures that only one instance of a type or class is created and that only one access interface can be used for this object. The pattern enforces the *single responsibility principle*.

When using the pattern, the instance is not created via a publicly accessible constructor of the respective type, but via a separate static method since constructors must always return a new instance of the corresponding type according to the object-oriented definition.

For the implementation of the pattern, the constructor of the type to be created is no longer made publicly available so that clients cannot use it to create new instances. In many programming languages, the effect is that instances can no longer be created using the new keyword.

A new static method takes over the task of the constructor for creating the object; in our example shown in Figure 4.30, this method is the getInstance method.

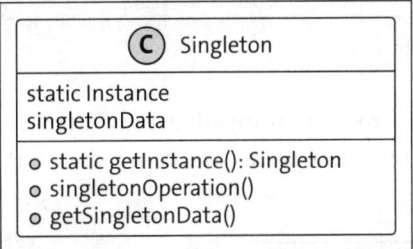

Figure 4.30 Structure of a Singleton

After the instance has been created via the private constructor within the static method, the instance is saved in a static attribute and returned on subsequent calls. The access interface is defined directly via the returned object.

The structure and participants in this pattern are as follows:

▶ Singleton: Defines the singleton with a corresponding interface. Depending on the programming language, the type also defines the method for instance creation.

4.7.3 Sample Solution

The problem with a cellular modem or with the HTTP clients from the problem description can be avoided by using the *singleton pattern*. The HTTP client is generated in the user components using a static method, and access to the modem is synchronized within the HTTP client so that communication with the backend system is not disrupted.

Figure 4.31 shows the structure of such an application. The HTTClient class has the static getInstance method, which is used to create a new instance of the class. Communication with the cellular modem is encapsulated via the sendData method and should also be synchronized.

4 Software Patterns

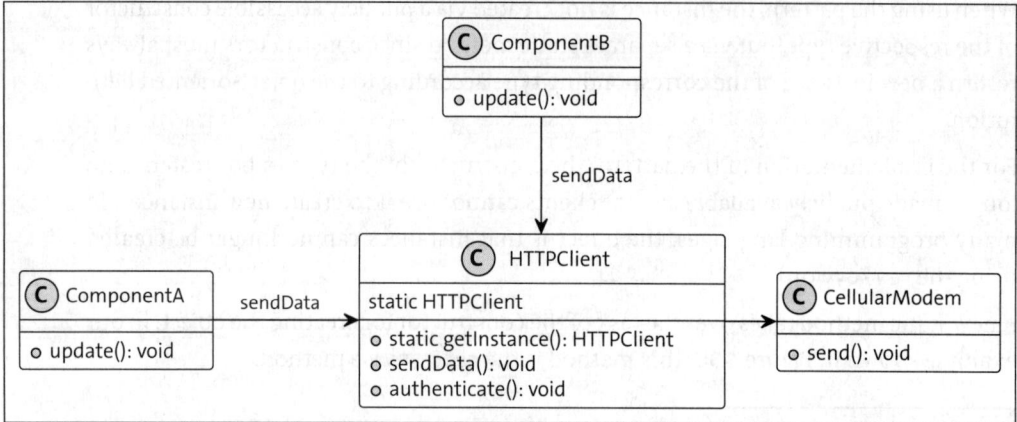

Figure 4.31 Sample Structure with the Singleton Pattern

Within the getInstance method, the authenticate method is called directly after the instance is created so that the client is only authenticated once, as shown in Listing 4.41. Figure 4.32 shows the process in a sequence diagram.

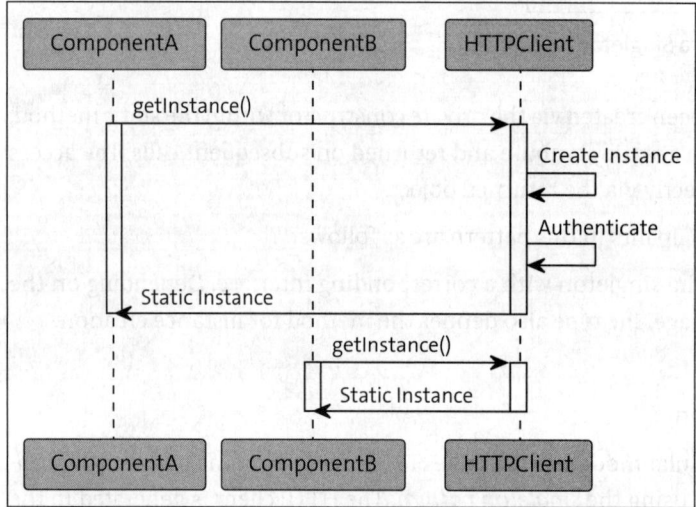

Figure 4.32 Sequence of Our Example Singleton Pattern

```
public class HTTPClient {

    // Static instance
    static HTTPClient instance;

    // Private constructor
    private HTTPClient() {
    }
```

```java
    public static HTTPClient getInstance() {
        // see the box later in this section for mulithreading
        if (instance == null) {
            // Instance generation
            instance = new HTTPClient();
            // direct execution
            instance.authenticate();
        }
        return instance;
    }

    public void authenticate() {
        //...
    }
    public void sendData(Object data) {
        //...
    }
}
```

Listing 4.41 HTTPClient Implementation as a Singleton (Java)

The individual components can use the client and send data, as shown in Listing 4.42, within their update methods.

```java
public class ComponentA {
    public void updateData() {
        HTTPClient.getInstance().sendData(...);
    }
}
```

Listing 4.42 Using the HTTPClient (Java)

4.7.4 When To Use the Pattern

Some examples of when to use the pattern include the following:

- **Only one instance of an object**
  The pattern can always be used if only one instance of a type should be used. This instance is usually an object that encapsulates access to resources (e.g., databases or files).

- **Restricting the creation of objects**
  Due to the *singleton pattern*, the creation of instances is encapsulated in a method. The implementation can therefore determine the number of objects created, making creating multiple objects easy. A pool implementation could be used for this purpose, for example.

▸ **Derived singleton classes can be created consistently**
Classes that are derived from the singleton class can also be returned by the static method, and the corresponding clients don't need to be changed. Extensions to the singleton implementation can be introduced without having to adapt the clients themselves.

4.7.5 Consequences

Some consequences of using these patterns are as follows:

▸ **Restriction to one instance**
By using the pattern, the creation of instances of a type and access to them are controlled. You can ensure that only one instance of a type is created and used.

▸ **Delayed creation of objects (lazy init)**
Since object creation takes place within a special method, instantiation can take place as late as possible.

▸ **Multithreaded accesses must be taken into account**
One of the major problems when using the *singleton pattern* is concurrent access and ensuring that only one instance is created. In our example, in an environment with several concurrently executed threads, we cannot be sure that only one instance is created.

If two clients perform the zero check "simultaneously," as shown in Listing 4.43, both clients will create their own HTTP client instances.

```
if (instance == null) {
   // Instance generation
   instance = new HTTPClient();
}
```

Listing 4.43 Problematic Null Check for Instance Creation (Java)

This block must either be synchronized somehow, or the delayed generation must be dispensed with.

Instantiation directly in the class is possible with one variant, shown in Listing 4.44.

```
public class HTTPClient {

   // Static instance
   static HTTPClient instance = new HTTPClient();
   ...
```

Listing 4.44 Creation Without Lazy Init (Java)

4.7 Singleton

Multithreaded Access: Always an Issue

Which programming language is used doesn't matter: When working in a concurrent environment, you must always ensure that only one instance of an object can be created. This rule is usually not trivial.

In Java, for example, you can use the `synchronized` keyword, but doing so can lead to rather extensive locks in the memory when using larger code blocks and therefore contributes to slower execution. One possible use could look as follows:

```java
public static Singleton getInstance(String value) {
    Singleton result = instance;
    if (result != null) {
        return result;
    }
    synchronized(Singleton.class) {
        if (instance == null) {
            instance = new Singleton(value);
        }
        return instance;
    }
}
```

Listing 4.45 Synchronized Creation of a Singleton Object (Java)

In Go, the use of the `sync.once` function is often recommended because doing so ensures that a code block is only executed once:

```go
var instance *Singleton
var once sync.Once
func GetInstance() *Singleton {
    once.Do(func() {
        instance = &Singleton{}
    })
    return instance
}
```

Listing 4.46 Singleton Implementation in Go

- **Difficult tests**

    In many test frameworks, dependencies are replaced by generated mock objects to achieve better control of the behavior. The use of *dependency injection* is also widespread to deal with these dependencies during testing.

    In our examples, the components have a direct dependency on the `HTTPClient` and generate a corresponding instance themselves. During testing, replacing the concrete instance with a mock implementation becomes difficult.

4.7.6 Real-World Example

In the `java.lang` package of the *Java software development kit (SDK)*, you'll find a very prominent example of the implementation of the *singleton pattern*: the `java.lang.Runtime` class. This class has methods that you can use to communicate with the *Java Virtual Machine (Java VM)* or to execute commands directly on the target operating system.

The `Runtime` implementation is saved as a static attribute within the class itself and instantiated directly when the class is created. With the corresponding code shown in Listing 4.47, creation is not delayed.

```java
public class Runtime {
    private static final Runtime currentRuntime = new Runtime();
    public static Runtime getRuntime() {
        return currentRuntime;
    }
    private Runtime() {}
    ...
}
```

Listing 4.47 Runtime Implementation from the Java API

You can use the `Runtime` implementation in your own code by using the `getRuntime` method and then working with the instance. Listing 4.48 shows an example.

```java
Runtime r = Runtime.getRuntime();
r.addShutdownHook(new Thread(() -> {
    System.out.println("Bye bye");
}));

// Call garbage collector
r.gc();

// Call command
String[] cmd = new String[]{"sh", "-c", "echo Hello"};
String[] params = new String[]{};
Process exec = r.exec(cmd, params);
BufferedReader br = exec.inputReader();
System.out.println(br.readLine());

// End process
r.exit(0);
```

Listing 4.48 Example with Java Runtime

> **Singletons in Dependency Injection Frameworks**
>
> *Contexts and dependency injection* frameworks (such as the Spring Framework for Java) usually manage registered services as *singleton beans* by default. As a result, only one instance of the *bean* exists in the entire Spring container, and this instance is shared by all dependencies.
>
> In the world of CDI, an object defined by the application and managed by the framework is referred to as a *bean*.
>
> The following code snippet shows the implementation of a Spring `Service` class:
>
> ```
> @Service // Singleton by default
> public class MyService {
>     public String greet(String name) {
>         return "Hello, " + name + "!";
>     }
> }
> ```

4.8 Adapter/Wrapper

Origin: Gang of Four—Structural Pattern

The *adapter pattern*, also referred to as the *wrapper pattern*, is a structural patterns used to convert or translate the interface of an object so that clients expecting a different interface can use this object without any adjustments. This pattern is intended to compensate for incompatibilities between the interfaces.

4.8.1 Problem and Motivation

Let's say an insurance company specializing in life insurance offers its customers a list of all the contracts they have concluded and corresponding interaction options via its own internet portal.

The relevant contract data is stored in a database and then retrieved and displayed directly by the portal software. The developers have designed a corresponding interface for accessing the database and used it in a list component, as shown in Figure 4.33.

For various internal reasons, the company creates a new, independent division to sell property insurance in addition to selling life insurance. This data is managed in a completely independent system and cannot yet be displayed on the internet portal.

With the next update of the internet portal, customers should see their property insurance policies in addition to their life insurance policies and will also have suitable interaction options.

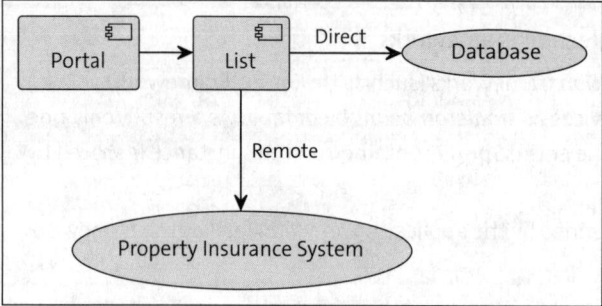

Figure 4.33 Aggregation of Data Via Two Interfaces

Effort must be kept to a minimum, and the property insurance system must be connected via an existing HTTP-based remote interface.

4.8.2 Solution

In this case, we can use the *adapter pattern* to make a non-compatible interface usable within the client code. A special object, known as an *adapter*, translates the method or interface calls that the client makes to the interface it knows into the interface that is incompatible with the client.

All calls from the client to the target object (the *adaptee*), are packaged by the adapter, and the necessary complexity is encapsulated in the adapter. Ideally, the called *adaptee* is not aware of this encapsulation or the call by the adapter.

A complete protocol conversion or parameter type conversion can take place within an adapter. If, for example, parameters are transferred as string values but are required as integer values in the target interface, it is the task of the adapter to perform such a conversion.

The compatible interface used by the client is defined for the implementation of the pattern in the form of an abstract interface description and implemented by the adapter. The following illustrations show the Target interface.

The *adapter pattern* can be implemented in two different variants:

- As a *class adapter*, as shown in Figure 4.34
- As an *object adapter*, as shown in Figure 4.35

When implemented as a class adapter, the Adapter implements the compatible Target interface and extends the object to be addressed (i.e., the Adaptee). Within the adapter, the functionality of the extended object addressed can be accessed via the subclass implementation.

In the second variant, the adapter implements the Target interface and has a reference to the object to be called, the Adaptee. Both adapter implementations can be used by the client without compatibility problems.

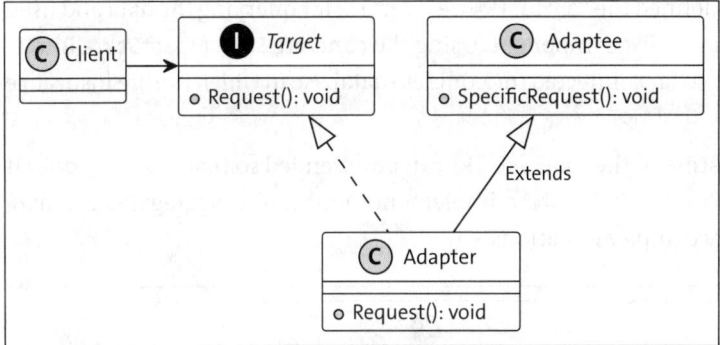

Figure 4.34 The Adapter Pattern Via a Class Adapter

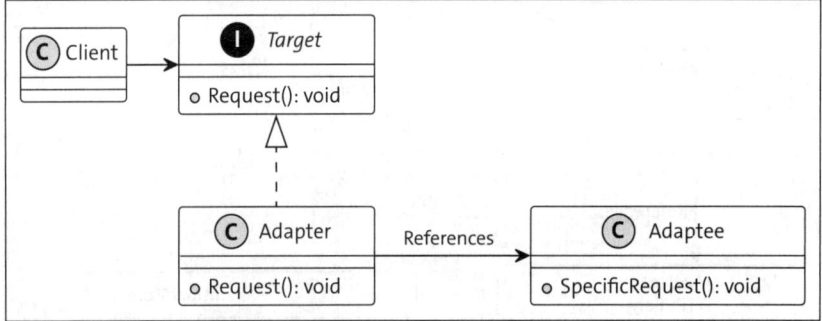

Figure 4.35 The Adapter Pattern Via an Object Adapter

The structure and participants in this pattern are as follows:

- Target: The `Target` interface describes the protocol via which the client communicates. This interface must be implemented by all corresponding implementations, regardless of whether they are adapters or not.
- Client: The client uses the `Target` interface and has no knowledge of the specific implementations used. Client and adapter are decoupled by the `Target` interface.
- Adapter: The adapter acts as an intermediary that provides the client with an interface that is completely independent of the actual implementation of the target object (the adaptee). Internally, it translates the client calls into the specific interface of the adaptee.
- Adaptee: The object to be called with an interface that is incompatible with the client. In some pattern descriptions, the adaptee is also referred to as a *service*. The adaptee is called by the adapter.

4.8.3 Sample Solution

The challenge described earlier for the insurance company in operating its portal can be solved by using the *adapter pattern*.

The developers have defined the ContractStore interface for querying the data and used it within the ContractList list component. Using the concrete SQLContractStore implementation, they have so far connected the concrete database in which the life insurance contracts are stored.

With a minimal adjustment, the ContractList can be extended so that, not only does it obtain its data from one ContractStore implementation; it also aggregates the data from multiple interface implementations.

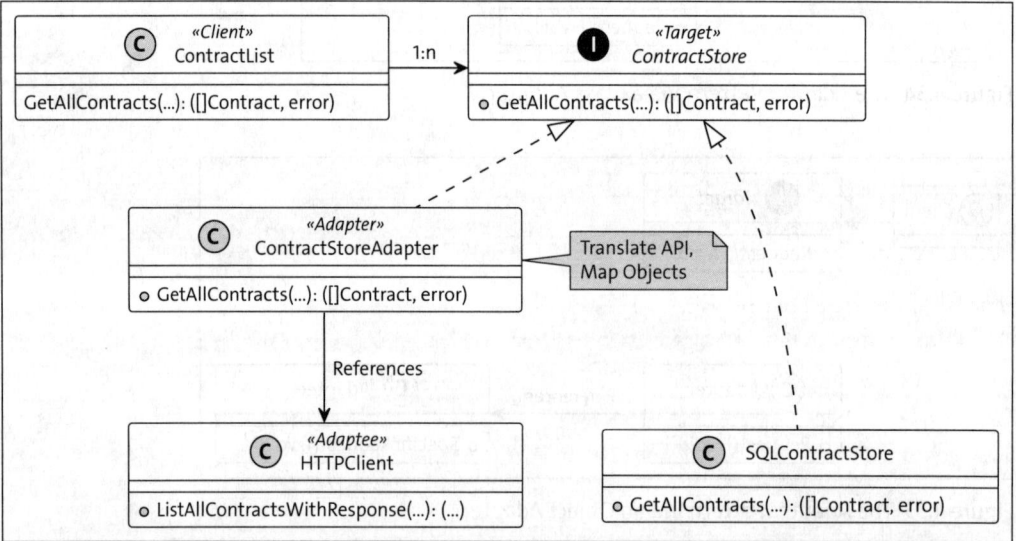

Figure 4.36 Structure of an Example for the Adapter Pattern

The HTTP-based interface for property insurance is connected using an HTTPClient implementation, which is generated from a corresponding interface description in the form of an openapi.yaml. This is provided by the team responsible for the property insurance system.

As shown in Figure 4.36, the ContractStore interface used within the ContractList differs from the interface that was generated for the HTTPClient: The ContractStore interface has the GetAllContracts method, and the HTTPClient provides a list of all contracts via the ListAllContractsWithResponse method.

A translation of the interface is necessary. This task is performed by the ContractStoreAdapter. It converts the parameters and the return values into the required format. The code example shown in Listing 4.49 shows the adapter implementation for our example in Go.

```go
// ContractStore corresponds to the Target interface.
type ContractStore interface {
    GetAllContracts(ctx context.Context,
        customer model.Customer) ([]model.Contract, error)
}
```

Listing 4.49 Target Interface for the Sample Implementation (Go)

The associated implementation of the `ContractStore` interface is carried out with the `ContractStoreAdapter`. The generated `HTTPClient` is called within the `GetAllContracts` method, and the data is transferred to the required data model.

I have deliberately omitted any detailed error handling in this example, shown in Listing 4.50.

```go
type ContractStoreAdapter struct{}

func (c ContractStoreAdapter) GetAllContracts(
    ctx context.Context,
    customer model.Customer) ([]model.Contract, error) {

    client, err := NewClientWithResponses("..")
    if err != nil {
        return nil, err
    }
    // Query data via HTTPClient
    response, err :=
        client.ListAllContractsWithResponse(ctx, customer.CustomerNo)
    if err != nil {
        return nil, err
    }
    // Check for errors
    if response.JSON200 == nil {
        return nil, errors.New("something wrong")
    }
    // Convert result data into the correct format
    var returnValue []model.Contract
    for _, contract := range *response.JSON200 {
        returnValue = append(returnValue,
            model.Contract{Name: contract.Name})
    }

    return returnValue, nil
}
```

Listing 4.50 The ContractStoreAdapter Implementation (Go)

> **Generating an OpenAPI Client in Go**
>
> In this example, I used the *oapi-codegen* library to generate the `HTTPClient` in Go. This library helps to generate suitable clients from *OpenAPI* documents and to deploy them in a time-saving manner.
>
> Both the generator and the library can be found at *https://github.com/deepmap/oapi-codegen*.

Within the `ContractList` component, an iteration over several implementations of the `ContractStore` interface takes place, and the data is aggregated for a complete list, as shown in Listing 4.51, which can then be displayed in the portal.

```go
type ContractList struct {
    stores []ContractStore
}

func (l ContractList) GetAllContracts(
    ctx context.Context,
    customer model.Customer) ([]model.Contract, error) {

    var allContracts []model.Contract
    for _, store := range l.stores {
        contracts, err := store.GetAllContracts(ctx, customer)
        if err != nil {
            return nil, err
        }
        allContracts = append(allContracts, contracts...)
    }
    return allContracts, nil
}
```

Listing 4.51 Implementation of the Client Code in the ContractList (Go)

4.8.4 When To Use the Pattern

Some examples of when to use the pattern include the following:

- **The interface of an existing class is incompatible**
  You can use this pattern in cases where an existing class should be used in an existing object, but its interface is not compatible with the calling code. The adapter adapts the interface.

- **Reusable class with unknown dependencies**
  In some cases, a reusable class is created whose dependencies are not fully known at the beginning and are likely to have incompatible interfaces.

4.8.5 Consequences

Some consequences of using these patterns are as follows:

- **Transformation separated from the logic**
  The introduction of an adapter separates the calling business logic from the need to transform the interface.

- **Extension with new adapters**
  The abstraction of the target interface makes it possible to replace or extend the application with alternative adapter implementations. The calling code does not need to be changed for this.

- **Increasing complexity**
  The introduction of abstraction and the splitting of the code can result in a higher complexity of the application.

4.8.6 Real-World Example

For a long time, the *Java Development Kit (JDK)* contained classes for reading and writing XML documents. From JDK version 9, these classes for *Java XML Binding (JAXB)* were successively deleted and finally removed from the standard scope with JDK version 11. The reason was the changed requirements and the modularization of the entire JDK.

XML documents can still be read or written in Java. However, since version 9 at the latest, external libraries for the APIs and a specific implementation must be integrated.

The serialization or deserialization of the XML documents and the transfer to object structures is performed by the JAXB framework whose behavior can be adapted using annotations. You can also create your own `XmlAdapter` implementations for complex structural adaptations.

If, for example, a `java.util.HashMap` is serialized via JAXB, it is generated as XML, as shown in Listing 4.52.

```
<hashmap>
    <entry>
        <key>1</key>
        <value>Test 1</value>
    </entry>
    <entry>
        <key>2</key>
        <value>Test 2</value>
    </entry>
</hashmap>
```

Listing 4.52 Standard Serialization of java.util.HashMap

A more compact version is shown in Listing 4.53.

```xml
<hashmap>
    <entry key="1">Test 1</entry>
    <entry key="2">Test 2</entry>
</hashmap>
```

Listing 4.53 Alternative XML Structure

The implementation of a separate XmlAdapter means you can adapt the structure of the data according to the second XML example. Only two methods need to be implemented in the XMLAdapter adapter class for this purpose:

- marshal is called by the framework when serializing the data.
- unmarshal is used by the framework during deserialization.

The XmlAdapter class is a generic abstract class that defines two generic types: the BoundType, which represents the type used in the application object model, and the ValueType, which is used for serialization/deserialization in the framework. Both methods receive or return the corresponding data types, as shown in Listing 4.54.

```java
public abstract class XmlAdapter<ValueType, BoundType> {
    protected XmlAdapter() {
    }

    public abstract BoundType unmarshal(ValueType var) throws Exception;
    ...
}
```

Listing 4.54 Abstract Definition of the XmlAdapter in Java

In our example of HashMap serialization, a HashMap can be used in the application object model (i.e., in your own Java classes), and an alternative data type is used within the adapter to be used in the framework. For the compact serialization described, as shown in Figure 4.37, these types are the MyHashMapType and MyHashMapEntryType types, which have a customized object structure and suitable XML annotations.

The MyHashMapType does not save the data in a map but instead in a list of MyHashMapEntryType objects, as shown in Listing 4.55.

```java
@XmlAccessorType(XmlAccessType.FIELD)
public class MyHashMapType {
    List<MyHashMapEntryType> entry;
}
```

Listing 4.55 MyHashMapType Implementation (Java)

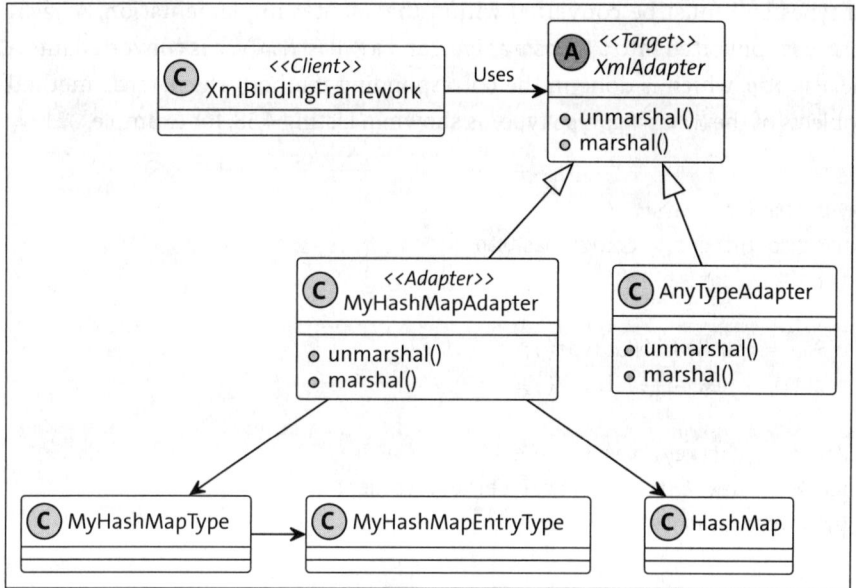

Figure 4.37 Java JAXB Adapter for Data Conversion

MyHashMapEntryType is configured to the desired XML structure via XML annotations. Listing 4.56 shows the implementation.

```java
public class MyHashMapEntryType {
    @XmlAttribute
    public Integer key;

    @XmlValue
    public String value;

    public MyHashMapEntryType(Integer key, String value) {
        this.key = key;
        this.value = value;
    }
}
```

Listing 4.56 MyHashMapEntryType Implementation (Java)

The serialization of MyHashMapType then leads analogously to the XML structure, shown in Listing 4.57.

```
<entry key="1">Test 1</entry>
```

Listing 4.57 MyHashMapEntryType Serialized as XML

The data types still must be converted within the adapter implementation. A `java.util.HashMap` is converted into a `MyHashMapType`, and a `MyHashMapType` is converted into a `java.util.HashMap`, which is done in the corresponding method. The `marshal` method returns objects of the `MyHashMapType` type, as shown in Listing 4.58, for example.

```java
@Override
public MyHashMapType marshal(
        HashMap<Integer, String> hashMap)
        throws Exception {

    var myMap = new MyHashMapType();
    myMap.entry = new ArrayList<>();

    hashMap.forEach((key, value) -> {
        var et = new MyHashMapEntryType(key, value);
        myMap.entry.add(et);
    });

    return myMap;
}
```

Listing 4.58 Marshal Method of a Custom Adapter Implementation (Java)

The adapter can be registered directly on the attribute of a class, as shown in Listing 4.59.

```java
@XmlRootElement
@XmlAccessorType(XmlAccessType.FIELD)
public class Foo {
    @XmlJavaTypeAdapter(MyHashMapAdapter.class)
    HashMap<Integer, String> hashmap;
}
```

Listing 4.59 Registering the Adapter for the Serialization/Deserialization of Foo Objects (Java)

The XML format of the generated documents changes accordingly as a result of the adjustments described.

4.9 Iterator

Origin: Gang of Four—Behavioral Pattern

The *iterator pattern* is a behavioral pattern that enables sequential access to object structures without knowing the underlying structure. Lists or tree structures, for example, can be traversed using a standardized interface.

4.9.1 Problem and Motivation

One department of a large consulting company wants to inform its customers about news and trends by sending emails to achieve better customer loyalty and thus provide reasons for interaction. However, the addresses of the relevant customers are not maintained centrally in a company system but in a database managed by the department itself.

For this purpose, the department created a small piece of software that iterates over a list of addresses from the database and sends the corresponding messages using an email component. The messages are each based on a text template created by the department. Figure 4.38 shows the simple structure of the application.

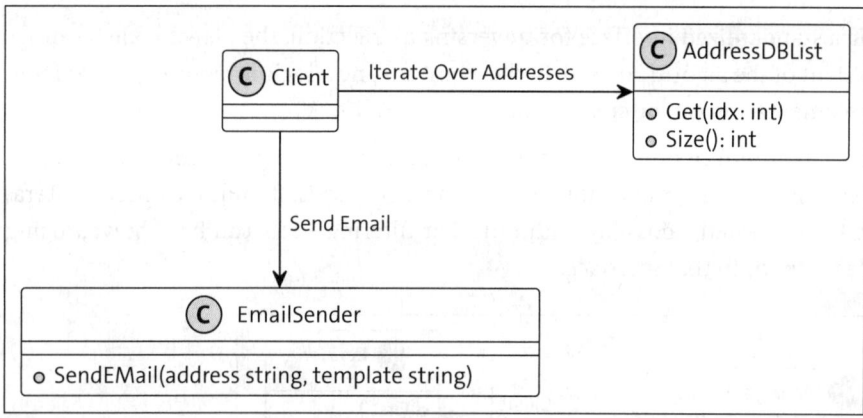

Figure 4.38 Initial Solution for Sending Emails

The first mailing campaigns are proving successful, and customers and the department are satisfied with the emails and the responses to them. However, as not all information is relevant for every customer, the software is supposed to be expanded so that different mailings can be created and sent for different interest groups. The selection of members of the interest groups is to be realized via a filter for the data.

Another department has also shown interest in the software. However, their customer data is stored in a text file and not in a database.

Both departments agree to expand the software. Despite the different requirements for the filtering logic and different list implementations, the logic for creating and sending emails should be retained.

4.9.2 Solution

With the *iterator pattern*, the logic for iterating or traversing collections of objects or collections is extracted into a separate implementation. The internal structure of the collection is irrelevant; whether it is a list or a tree structure doesn't matter.

An abstract interface, the Iterator interface, creates a standardized access option for traversing the objects in the collection (called the Aggregate). Concrete implementations of the Iterator interface are instantiated and implement the algorithm required for traversing. For example, those concrete implementations hold the current position in the object list or manage filter criteria.

Figure 4.39 shows the structure of the *iterator pattern* as a class diagram. Note how the Iterator interface, in most implementations, has at least one method to go to the next element in the collection. In our example, this method is the Next method. If the method no longer returns any values, the end of the collection has been reached.

Iterators can also provide methods for positioning a pointer which points to the current object in the list, such as a First or Last method.

By using a standardized interface for traversing a collection, the Client code becomes independent of the algorithm used. In the event that new logic is required, a new Iterator implementation can be created and used.

The Aggregate interface provides a simple interface for creating new concrete Iterator instances. This interface can contain one or more methods. If multiple different iterators are to be created—possibly with different filtering—this can be achieved using multiple methods in that interface.

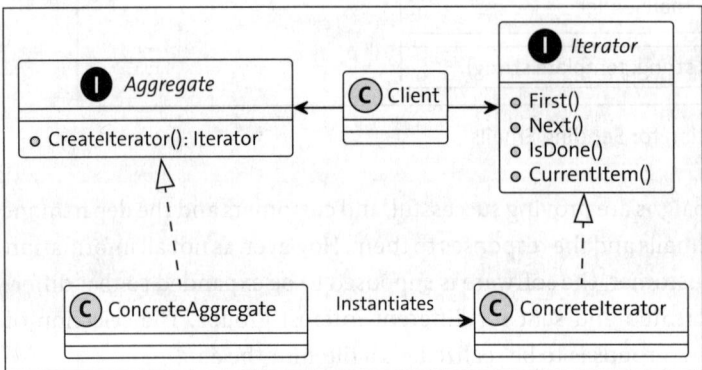

Figure 4.39 Structure of the Iterator Pattern

The structure and participants in this pattern are as follows:

- Iterator: Defines the abstract interface for accessing the aggregate and traversing its elements.
- Concrete iterator: Implements the interface defined by the iterator. The implementation saves the status required for traversing.
- Aggregate: Defines the abstract interface for creating one or more different iterator objects.
- Concrete aggregate: Implements the interface defined by the aggregate and returns a corresponding concrete iterator implementation.

4.9.3 Sample Solution

The extension of the consultancy company's Go-based software can be implemented, as shown in Listing 4.60 and Figure 4.40.

A `CustomerIterator` interface is introduced for the client, which is used to iterate over the list of customers. Different implementations of the interface can provide different filter options. In our example, an `InterestFilteringIterator` is created whose filter criteria only returns customers with certain interests.

```go
type CustomerIterator interface {
    First() *Customer
    Next() *Customer
    IsDone() bool
    CurrentItem() *Customer
}
```

Listing 4.60 Iterator Interface for Customers (Go)

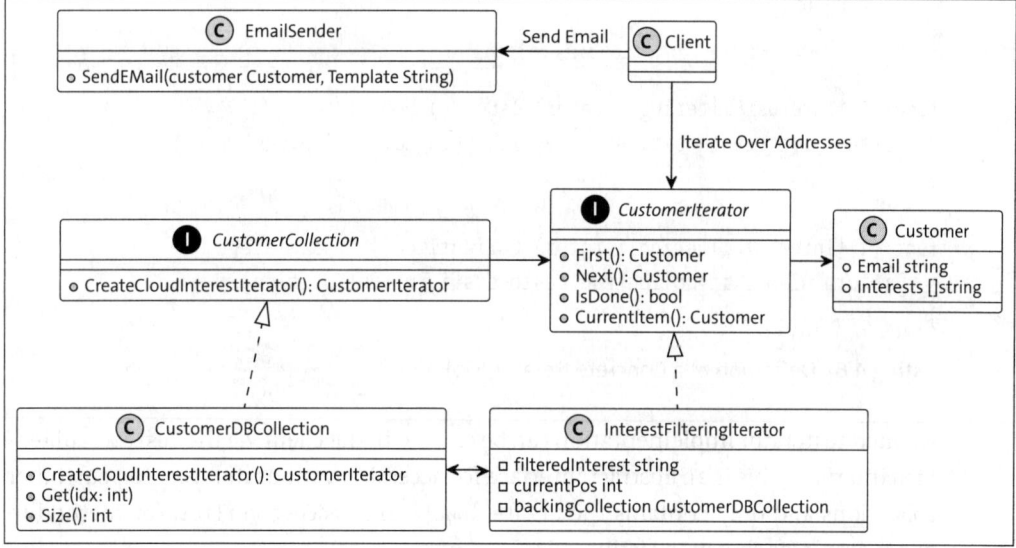

Figure 4.40 Sample Solution Using the Iterator Pattern

The `InterestFilteringIterator` type which implements the iterator interface saves a reference to the specific collection in the `backingCollection` field as well as the current position of the iterator in the collection (`currentPosition`). The details are shown in Listing 4.61.

```go
type InterestFilteringIterator struct {
    filteredInterest  string
    currentPosition   int
```

```go
    backingCollection *CustomerDBCollection
}
func (i *InterestFilteringIterator) First() *Customer {
    i.currentPosition = -1
    return i.Next()
}

func (i *InterestFilteringIterator) Next() *Customer {
    i.currentPosition = i.currentPosition + 1
    for newIdx, customer :=
       range i.backingCollection.customers[i.currentPosition:] {

        if customer.HasInterest(i.filteredInterest) {
            i.currentPosition = newIdx + i.currentPosition
            return &customer
        }
    }
    return nil
}

func (i *InterestFilteringIterator) IsDone() bool {
    return i.currentPosition == len(i.backingCollection.customers)
}

func (i *InterestFilteringIterator) CurrentItem() *Customer {
    return &i.backingCollection.customers[i.currentPosition]
}
```

Listing 4.61 Definition of a Concrete Iterator (Go)

A concrete iterator implementation can be created in the client via the `CustomerCollection` interface. This is an abstract interface for accessing various iterators for customer collections and only contains the `CreateCloudInterestedCustomerIterator` method in our example, as shown in Listing 4.62.

This method is used to create a preconfigured iterator with filter functionality. Concrete collection implementations must implement this interface accordingly.

```go
type CustomerCollection interface {
    CreateCloudInterestedCustomerIterator() CustomerIterator
}
```

Listing 4.62 Interface Definition (Go)

For the first department, which stores its data in a database, the `CustomerDBCollection` type is implemented. This type implements the `CustomerCollection` interface and returns a preconfigured iterator with a filter for customers who are interested in cloud topics.

For the example shown in Listing 4.63, I have omitted an actual database connection for the sake of clarity. The data is saved locally in a slice.

```go
type CustomerDBCollection struct {
    customers []Customer
}

func (cc *CustomerDBCollection) CreateCloudInterestedCustomerIterator() 
    CustomerIterator {

    return &InterestFilteringIterator{
        currentPosition:  -1,
        filteredInterest: "Cloud",
        backingCollection: cc}
}
```

Listing 4.63 CustomerCollection Implementation for a Department (Go)

The corresponding interfaces and implementation can be used within the client implementation, as shown in Listing 4.64.

```go
//use Collection as CustomerCollection
var col internal.CustomerCollection = ...
sender := internal.EMailSender{}
iter := col.CreateCloudInterestedCustomerIterator()

for iter.Next() != nil {
    sender.SendEMail(iter.CurrentItem(), "cloud-template")
}

fmt.Println(iter.IsDone())
```

Listing 4.64 Implementation of a Client for the Iterator (Go)

The second department can then create its own `CustomerCollection` and `CustomerIterator` implementation and use the client code unchanged.

4.9.4 When To Use the Pattern

Some examples of when to use the pattern include the following:

▸ **Easy access to underlying structure required**
You can use the *iterator pattern* to create a standardized and simpler interface for traversing object structures. The iterator interface contains only a few clear methods in

order to keep the internal structure and complexity away from the calling code. The client doesn't need to deal with the possibly extensive API of the collection.

- **Central logic for traversing**
  For some collections, the logic for traversing is complex and time consuming to implement. The *iterator pattern* encapsulates the logic in a central location so that it does not need to be implemented in several places.

- **Adding additional collection implementation later**
  Often, not all possible characteristics of a collection or of a traversing logic are known from the start of the implementation. The *iterator pattern* can be used to add new iterators or aggregate implementations to the application at any time.

4.9.5 Consequences

Some consequences of using these patterns are as follows:

- **Central logic for traversing**
  As I have already described, the code for traversing is bundled in one place and does not need to be implemented multiple times by the client.

- **Extensions possible**
  The implementation of the pattern makes it possible to work according to the *open-closed principle*. New implementations can be added at any time without having to change the existing client code.

- **Multiple iterators possible for one collection**
  As the status of an iterator, such as the position in the collection, is stored in a separate instance, you can use multiple iterators simultaneously for a collection.

- **Effort and speed**
  The use of the pattern leads to more complex code and may prevent the use of special access methods optimized for the collection due to the uniform interface.

4.9.6 Real-World Example

In JDK 1.2, the `java.util.Iterator` interface and some corresponding implementations were introduced for the *Java Collections Framework*. Since then, the Java Collections Framework has been successively expanded, making it increasingly powerful and comprehensive.

As shown in Listing 4.65, the `Iterator` interface consists of only four methods, of which `hasNext()` and `next()` are probably the most important. The structure within the JDK is shown in Figure 4.41.

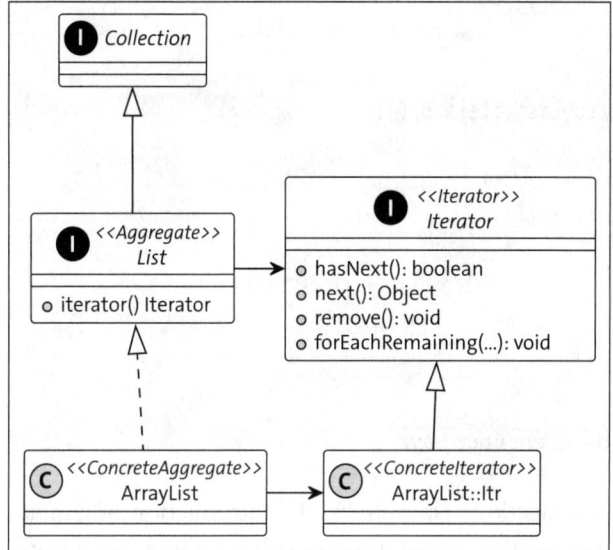

Figure 4.41 Iterator in the Java Software Development Kit (SDK)

To provide a better overview, I have omitted further list and iterator implementations.

```
package java.util;

import java.util.function.Consumer;

public interface Iterator<E> {
    boolean hasNext();

    E next();

    default void remove() {
        throw new UnsupportedOperationException("remove");
    }

    default void forEachRemaining(Consumer<? super E> action) {
        Objects.requireNonNull(action);
        while (hasNext())
            action.accept(next());
    }
}
```

Listing 4.65 The Java Iterator Interface

In the client code, the Iterator methods for traversing the objects can be used for every List implementation—even independently of the internal structure.

```
List<String> arrayList = new ArrayList<>();
arrayList.add("Hello World");

Iterator<String> iterator = arrayList.iterator();

while (iterator.hasNext()) {
    System.out.println(iterator.next());
}

for (String s : arrayList) {
    System.out.println(s);
}
```

Listing 4.66 Use of an Iterator in the Client Code (Java)

With the extensions to the Java Collections Framework, the introduction of lambda expressions, and so on, shorter expressions are now also possible, such as the for loop shown in Listing 4.66. The Iterator interface is still used internally.

> **Iterators in Other Programming Languages**
> The iterator pattern is a widespread design pattern used in many programming languages. This pattern is used in C++, for example, in the form of iterators, which serve as an abstraction for pointers. Other modern programming languages (including Java, Python, and JavaScript) use it for collections and loops.

4.10 Composite

Origin: Gang of Four—Structural Pattern

The *composite pattern* is a structural patterns for organizing objects in a tree structure and for mapping *part-whole hierarchies*. Every object in a tree structure has a standardized interface and can be used in the same way.

4.10.1 Problem and Motivation

In this example, a vehicle manufacturer wants to determine the weight of its new electric car early on in the development phase. However, vehicle development is at such an early stage that no vehicle exists to put on a scale.

The software department then develops an application with which a list of components can be recorded and the total weight of the vehicle calculated.

In the first step, the software works well, as the development department only records larger assemblies with corresponding weights as components and uses them to determine an approximate vehicle weight. However, additional steps will be taken to refine the assemblies and break them down into their individual components. A steering wheel, for example, consists of an airbag, turn signals, and a horn.

Since the software can currently only manage a list of components, the assemblies must be replaced by their individual parts, and the important information about which assembly a component belongs to is lost. This delinking detracts from the clarity of the overall list and makes replacing components with alternative parts difficult because you never know whether the part belongs with a certain assembly. Assemblies can also consist of other assemblies.

By extending the software, mapping a hierarchy of the components is possible so that their relationships to each other or to the assemblies are retained. As mentioned earlier, each module can consist of individual components or assemblies.

For the weight calculation, whether the object is an assembly or a component shouldn't matter. In both cases, the weight results from the total of the individual parts or from the weight of the component itself.

4.10.2 Solution

You can use the *composite pattern* to implement hierarchical object structures and map part-whole hierarchies. Each node and each leaf of the tree has a uniform interface and can therefore be treated uniformly.

Figure 4.42 shows the structure of the *composite pattern* in a class diagram. The abstract Component class defines the uniform interface and functionality of all elements within the tree.

Trees have leaves, and in our example we have the Leaf class, which contain no further elements, and CompoundComponent objects, which in turn can contain other Component objects as containers.

Clients can use a uniform interface (the abstract Component class)—regardless of the specific characteristics of the objects contained in the tree—and use each object in the same way. Whether the object is a leaf or a node doesn't matter.

If a method is called on them, composite objects—the composites—can call this method on the objects they contain, which is transparent for the caller.

The design pattern can be implemented in different ways. In the class diagram shown, the Component class is implemented as an abstract class with the AddChild and RemoveChild methods for managing the child objects it contains.

As the AddChild method is only required in CompoundComponents, the corresponding abstract implementation always returns an error, and only the concrete implementation within the CompoundComponent implements the logic.

4 Software Patterns

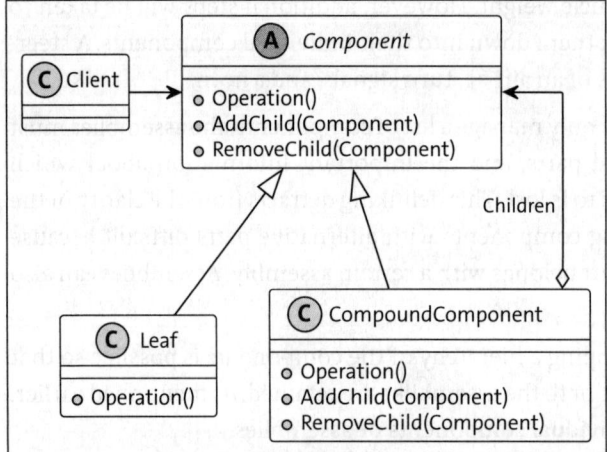

Figure 4.42 Structure of the Composite Pattern

As an alternative, often only an interface for accessing the individual components is created, and the tree structure is managed in parallel. In our example, the Component interface would only contain the Operation method.

The structure and participants in this pattern are as follows:

- Component: Defines the interface for standardized access to all objects in the tree and may implement shared functionality.
- Leaf: A single tree element that contains no other elements and implements the interface of the component. The logic is usually implemented in these objects because it can no longer be delegated.
- Composite: This element, often referred to as a *container*, manages other tree elements as its children. In addition to the component interface, a composite provides operations for the management of its children. The composite usually delegates the logic that goes beyond managing the children to the objects it contains.
- Client: A client uses the elements within the tree via the standardized interface of the component.

4.10.3 Sample Solution

Our car manufacturer can address its challenges by using the *composite pattern*. The implementation allows the hierarchy of components and assemblies to be managed and a uniform weight calculation to be carried out.

Individual components are represented by the SimpleComponent class, while the CompoundComponent class, as a composite of the pattern, combines these components into assemblies. It contains a list of all the components that make up the assembly. These components can be individual parts or entire assemblies.

Both classes implement the Component interface with the getWeight method for weight calculation. Figure 4.43 shows the structure as a class diagram.

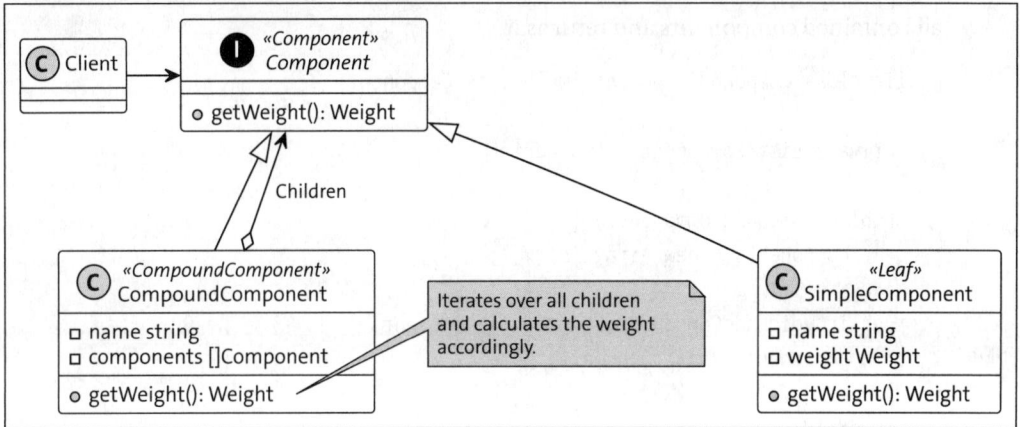

Figure 4.43 Example of the Composite Pattern

In Java, you can implement this example and the corresponding interfaces and classes using the code shown in Listing 4.67.

```java
public interface Component {
    Weight getWeight();
}
```

Listing 4.67 The Component Interface for Weight Calculation (Java)

The component implementation contains a weight and a name. The getWeight method returns the weight of the component, as shown in Listing 4.68.

```java
public class SimpleComponent implements Component {
    private final Weight weight;
    ...
    public SimpleComponent(String name, Weight weight) {
        this.weight = weight;
        ...
    }
    @Override
    public Weight getWeight() {
        return weight;
    }
    ...
}
```

Listing 4.68 Implementation of a Concrete Component Within the Tree Structure (Java)

The compound or assembly is implemented by the CompoundComponent class. This class manages a list of all contained Component objects internally, as shown in Listing 4.69. When the getWeight method is called, the implementation calculates the total weight of all contained components and returns it.

```java
public class CompoundComponent implements Component{

    private List<Component> components;

    public CompoundComponent() {
        components = new ArrayList<>();
    }
    public void addComponent(Component component) {
        components.add(component);
    }
    @Override
    public Weight getWeight() {
        var weight = Weight.createWeight(0, Weight.Unit.MILLIGRAMS);
        for (var component : components) {
            weight = Weight.sum(weight, component.getWeight());
        }
        return weight;
    }
}
```

Listing 4.69 Assembly Implementation (Java)

For the client, the use and weight calculation is simple. The assemblies of the vehicle can be put together and the total weight of the "Car" assembly can be determined at the end, as you can see in Listing 4.70.

```java
var car = new CompoundComponent();
var engine =
        new SimpleComponent("Engine", Weight.createWeight(200,
        Weight.Unit.KILOGRAM));

var seats = new CompoundComponent();
seats.addComponent(new SimpleComponent("Left front seat",
        Weight.createWeight(28, Weight.Unit.KILOGRAM)));
seats.addComponent(new SimpleComponent("Right front seat",
        Weight.createWeight(28, Weight.Unit.KILOGRAM)));
seats.addComponent(new SimpleComponent("Backseat",
        Weight.createWeight(62, Weight.Unit.KILOGRAM)));
```

```
car.addComponent(engine);
car.addComponent(seats);

System.out.println(car.getWeight());
```

Listing 4.70 Example of Using the Tree Structure for Weight Calculation (Java)

4.10.4 When To Use the Pattern

Some examples of when to use the pattern include the following:

- **A tree structure of objects must be implemented**
  You can use the *composite pattern* to implement tree structures with a part-whole hierarchy. The individual elements in the tree can be leaves or nodes at any point.

- **Handling objects uniformly in a tree structure**
  The *composite pattern* helps you use different elements uniformly in a tree structure, regardless of whether the contained object is a complex, composite, or simple object.

4.10.5 Consequences

Some consequences of using these patterns are as follows:

- **Using tree structures becomes simpler and more standardized**
  Access to tree structures is made easier for the client since all objects contained in the tree can be accessed uniformly via the common interface. Recursions are executed by the composite objects themselves, and whether they are simple or complex objects in the tree is irrelevant to the client.

- **New components can be added**
  The structure of the pattern enables the expansion and inclusion of new components and thus corresponds to the *open-closed principle*.

- **Generalized design as a result**
  You can use the pattern to standardize and generalize access to the objects contained in the tree. The result might be, however, rather abstract interfaces that are difficult for a client to work with and objects that must be converted or cast into a corresponding type before they can be used.

4.10.6 Real-World Example

The `http` package from the standard library is used in Go for HTTP-based communication. It provides implementations for both the client and the server side.

If a separate server with HTTP communication is to be created, it always consists of a `Server` instance and a `Handler` which is responsible for processing the requests. The `Handler` interface of the same name exclusively defines the `ServerHTTP` method with a

ResponseWriter and a pointer to Request as parameters. Individual use cases for communication are usually implemented within these handler implementations.

When a Server instance is created, a handler can be specified directly to process the HTTP requests:

```
server := &Server{Addr: addr, Handler: handler}
```

If a single Handler implementation is registered directly, the server uses only this implementation to respond to each request and can only ever take care of the one use case implemented in this handler. This may make sense in some cases, but usually several functions should be provided under different URIs for HTTP servers.

One component of the standard library is the ServerMux type which handles *multiplexing* (i.e., the assignment and redirecting of incoming HTTP requests to a corresponding handler). ServerMux itself implements the Handler interface and can be registered to the Server instance accordingly. Internally, an assignment of URI specifications to Handler implementations is managed. These handlers can also be ServerMux instances. Figure 4.44 shows the structure of the *composite pattern*.

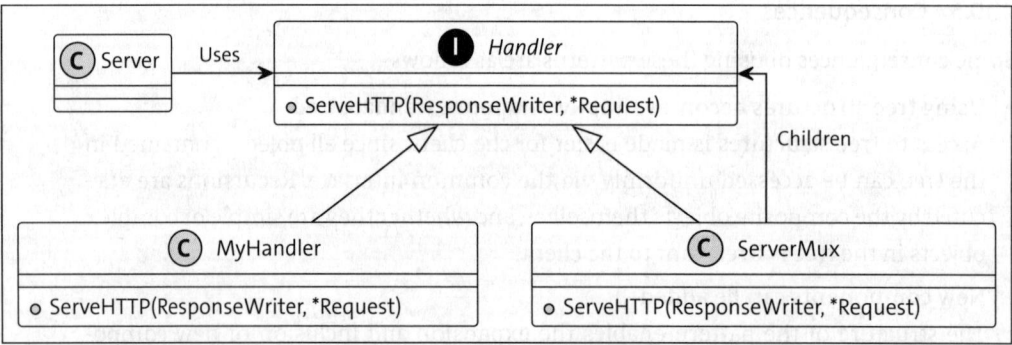

Figure 4.44 An HTTP Server in Go Implemented with the Composite Pattern

The Server and Handler implementations can be used via *convenience methods*. With their help, more complex functionality is made available via a simple interface.

For example, the ListenAndServe function of the http package takes a binding specification and a Handler implementation as parameters and internally creates a new Server instance with a corresponding network binding configuration.

If no Handler parameter is given to the ListenAndServe function, the internal default multiplexer is automatically used, and additional handlers are automatically registered with the multiplexer, for example, via the HandleFunc function with a specification of URI and implementation.

The example shown in Listing 4.71 is a complete HTTP server implementation in Go, in which the internal ServerMux implementation is used. The Hello World text is returned under the URL *http://localhost:8081/nix*.

```
package main

import "net/http"

func main() {
    http.HandleFunc("/nix", handle)
    http.ListenAndServe(":8081", nil)
}
func handle(writer http.ResponseWriter, _ *http.Request) {
    writer.Write([]byte("Hello World"))
}
```

Listing 4.71 Complete HTTP Server Example (Go)

4.11 The Concept of Anti-Patterns

In the previous sections, I presented some of the most common design patterns.

In addition to the design patterns, however, their opposites, called *anti-patterns*, also emerged. This term describes procedures and solutions that are frequently used but have proven to be inefficient, problematic, or counterproductive. Anti-patterns are like errors in the construction plan of a house, which often arise due to insufficient knowledge, experience, or time pressure.

The issues caused by anti-patterns usually manifest themselves in the form of increased maintenance effort, poor performance, or complex code.

Looking at anti-patterns can thus provide valuable insights into which approaches and practices you should avoid to prevent common pitfalls and difficulties in software development.

For this reason, I present five of the "most popular" anti-patterns in the following sections.

4.11.1 Big Ball of Mud

The *big ball of mud* anti-pattern describes an unstructured code base with no recognizable structure or clear separation of responsibilities. This anti-pattern usually results from disorganized development in which no precise definition of an architecture or no design principles have been adhered to.

Many reasons can cause the formation of a *big ball of mud*. Either a system has already been started without clear communication or without the existence of a software design, or an existing system has been extended in an ill-considered manner (i.e., without taking the architecture into account). A lack of communication or a lack of clear design guidelines are often the decisive factors.

The negative effects of this anti-pattern include the following:

- Insufficient or poor readability and comprehensibility of the source code
- Low developer productivity, as efficient work is made more difficult
- Higher susceptibility to errors due to the unnecessarily complex and opaque structure of the software
- Difficult to maintain and the team may be afraid of change, as changes to one part of the software may affect other parts

Some countermeasures against this anti-pattern include the following:

- Defining and documenting the architecture
- Applying design principles
- Performing code reviews and refactoring
- Communication and information exchange within the team

4.11.2 God Object

A *God object* is a class or component that combines a large number of responsibilities and is therefore virtually "omnipotent" but also confusing and difficult to maintain. *God objects* usually play a central role in a system and have many dependencies on other components, which makes changes more difficult.

Listing 4.72 shows an example of a *God object* with many mixed responsibilities.

```java
public class ApplicationManager {
    public void loginUser(String user, String password) {
        ...
    }
    public void parseXML(String document) {
        ...
    }
    public String renderHTML() {
        ...
        return "<html/>";
    }
    public void logError(String message) {
        ...
    }
    public void executeSQLQuery(String query) {
        ...
    }
}
```

Listing 4.72 Example of a God Object in the ApplicationManager Class (Java)

The negative effects of this anti-pattern include the following:

- **Strong coupling**
  The many dependencies on and to the God object lead to a strong coupling of actually independent program parts. Changes potentially affect large parts of the application.

- **Difficult testability**
  Components that use the God object also require a corresponding reference during execution within a unit test. Testing often requires extensive configurations or complex mocking procedures.

- **Difficult to maintain**
  Bug fixes and changes are more complex and risky due to the many component dependencies. Each change can impact many other parts of the application.

- **Not reusable**
  Due to the many different tasks and dependencies of the God object, it is difficult to reuse it in other contexts. The referencing components cannot be reused without the God object either.

Some countermeasures against this anti-pattern include the following:

- **Decomposition or modularization**
  The God object should be broken down into multiple specialized classes with clearly defined responsibilities.

- **Compliance with the single responsibility principle**
  Compliance with this principle is required when splitting into multiple classes.

- **Abstraction**
  The use of interfaces and abstract classes can further reduce the dependencies between the components.

- **Regular refactorings**
  In the source code, recognizing the emergence of a central omnipotent object at an early stage is not easy. Regular refactorings can help you discover God objects and divide them into smaller units.

4.11.3 Spaghetti Code

The term *spaghetti code* describes incomprehensible source code in which the control flow is difficult to understand and follow. The name is reminiscent of the noodle, as its functional blocks are chaotically intertwined and have no recognizable structure.

The signs of spaghetti code are long methods with a high nesting depth and many dependencies. If there are also nested loops and jump statements such as GOTO statements, the program flow becomes almost impossible to follow.

The negative effects of this anti-pattern include the following:

- **Complexity and confusion**
  The code becomes more complex and confusing with every nesting depth and every jump statement, until it is no longer comprehensible.
- **High susceptibility to errors**
  An untraceable control flow increases the likelihood of bugs and also makes it difficult to fix them.
- **Maintenance effort**
  Adaptations or extensions to the code are time-consuming and laborious due to the complexity. Any change can have unwanted side effects.

Some countermeasures against this anti-pattern include the following:

- **Regular refactorings and code reviews**
  By reviewing code (for example, via *Extract Method* refactoring), code blocks are created that are smaller and therefore easier to maintain and more manageable. Regular code reviews can help to identify areas where spaghetti code threatens to emerge at an early stage.
- **Use of design patterns**
  Using design patterns, recurring problems can be solved in a standardized way.
- **Modularization**
  Smaller, logically related classes or components are easier to maintain and clearer.
- **Compliance with the single responsibility principle**
  As with the other anti-patterns, make sure that a code block focuses on a single task and that there is only one reason for change.

4.11.4 Reinventing the Wheel

The *reinventing the wheel* anti-pattern describes the practice of someone developing a solution to a problem for which there are already proven solutions. Instead of using existing libraries, for example, this approach involves developing something from scratch with unnecessary effort.

Developers are usually technicians and love to master technical challenges. For this reason, they often prefer to implement technical and specialist topics in projects with a high level of detail. However, as most detailed technical solutions take a long time to actually work, they cannot usually be created to their full extent in practice, resulting in errors and inadequacies in the unfinished solution.

The negative effects of this anti-pattern include the following:

- **Waste of resources**
  Developing your own solution requires time and effort that could be used for other tasks. The time required is therefore not available for implementing other business logic or new features.

- **Cost factor**
  The development costs of software can rise quickly, especially if external specialist knowledge or technical expertise is required.
- **Lower quality**
  Existing solutions are usually tried and tested. As a rule, in-house implementations must first reach a certain level of maturity.
- **Higher maintenance effort**
  If you use an existing open-source library, for example, a lot of the support effort is taken over by the library's community. You have to maintain your own implementations completely yourself.

Some countermeasures against this anti-pattern include the following:

- **Research!**
  Before any new implementation, you should search for existing alternatives and evaluate whether you can use them in your project.
- **Use open-source projects**
  High-quality, open-source solutions have now been developed for most requirements, which are available free of charge and which you can adapt to your own needs. However, licenses and usage restrictions might exist that you should be aware of. You should also think about actively participating in the community.
- **Consider commercial services**
  For certain tasks, powerful commercial services are available that can save you a lot of time.

4.11.5 Cargo Cult Programming

Cargo cult programming is an anti-pattern where programmers blindly adopt or apply code without fully understanding its purpose or functionality.

The *cargo cult* concept goes back to a religious and political movement in Melanesia, a Pacific island group northeast of Australia. Locals lived in expectation that imitating certain actions they had previously observed in Europeans and Americans during World War II would convince them to bring Western goods (see *https://en.wikipedia.org/wiki/Cargo_cult*).

For example, airplanes and airports were imitated on the islands in the hope that *cargo* would be brought to them again, as it had been during the war. The inhabitants had no technical knowledge of airplanes or logistics.

In software development, a cargo cult often starts with Stack Overflow or ChatGPT when solutions are copied without reflection or when new, popular, and "hip" technologies are used in projects even though they are not suitable for the actual problem.

The negative effects of this anti-pattern include the following:

- **Missing problem solution**
  The adopted code may not be able to solve the current problem, or not effectively, as it was intended for use in a different context. It may also be used or configured incorrectly.

- **Technical debt**
  Copying code without reflection can lead to the accumulation of technical debt and make software confusing and difficult to maintain. The copied code must also comply with the architectural decisions made.

- **Limited learning curve**
  If code is only copied superficially, no underlying concepts or principles are understood, and the solution of future problems is made more difficult.

Some countermeasures against this anti-pattern include the following:

- **Learning the basics**
  Books, training courses, conferences, and workshops help developers understand the concepts and principles behind the technologies and patterns used.

- **Questioning topics and solutions**
  Code reviews and pair programming can help to better understand the concepts, solutions, and implementations used. Question solutions in order to better understand them.

- **Debugging and familiarization**
  Using tools such as debuggers to analyze code step by step and understand its behavior will help you to mentally penetrate the code and understand how it works in detail.

Chapter 5
Software Architecture, Styles, and Patterns

Software architecture forms the basic framework for the structure of a software solution. The architecture defines an application's overall structure and the relationships between its various components, while the software design specifies certain details and implementations. An architecture lays the foundation for the efficient development, maintenance, and scaling of applications. In this chapter, you'll learn about different architectural styles and architectural patterns as well as patterns that help with implementation.

Creating high-quality software is a demanding task that becomes increasingly difficult as system complexity increases. Even if all best practices and design patterns are implemented at the lowest possible level in your source code and if your source code is readable, you're still not guaranteed a good final product.

For high-quality software, a clear definition of its structure through hierarchical decomposition into components at the highest level is essential. Software lacking clear structure usually requires strong coupling among the individual components. Such an interwoven structure makes an application's maintenance and later development more difficult because changes in one area can have unforeseen effects on other areas of the system.

Software architectures define the fundamental framework and basic structure of an application and break down the application's complexity into manageable parts. They lay the foundation for the efficient development, maintenance, and scaling of applications.

Martin C. Fowler, a leading expert and author on software architecture, names two aspects that he believes characterize an architecture:

- Architectures break down a system into components at the highest level.
- Architectures are used to make decisions that are difficult to change.

Patterns can also play a decisive role in the development of software architectures. As with software design at the source code level, they make a significant contribution to defining central properties and the basic behavior of applications, albeit at a higher level of abstraction.

Some of these *software architecture patterns* are suitable for highly scalable applications, for example, while others are more suitable for agile applications.

To choose a suitable architecture for your own application and its specific technical requirements, an important task to know the characteristics, strengths, and weaknesses of various architectural approaches.

Architectural patterns are classified using *architectural styles* based on the structure or organization of the code used and the interactions between the various software components.

In most projects, usually one person plays the role of the *software architect*. This person makes the decisions on software architectures and ensures that the software architecture is implemented.

Different architectures can coexist within an application. Their use is not static, and it should be possible to adapt them to changing requirements. Even though Fowler emphasizes that architectural changes can be difficult, architectures must be flexible enough to adapt to new challenges or changing requirements.

In this chapter, you'll learn about the various tasks of the software architect as well as different architectural styles, architectural patterns, and patterns to support the implementation of these architectures.

5.1 The Role of the Software Architect

Software architects are responsible for the definition and successful implementation of a *software architecture*. They ensure that the architecture meets the requirements of the project and remains successful in the long term.

Simon Brown, author of *Software Architecture for Developers*, defines six areas of responsibility of the software architect role in a practical and pragmatic way:

- Understanding the professional and technical context
- Defining a viable software architecture
- Dealing with technical challenges and risks
- Further developing the architecture
- Actively participating in development
- Conducting quality assurance tasks

Brown emphasizes that not only must software architects make decisions, they must also (co-)implement them. In his view, nobody needs an *ivory tower architect* who won't collaborate with the team on solutions and on the overall architecture.

Let's now look at each of these points in detail.

- **Understanding the professional and technical context**
  In a project, one task of a software architect is to know, at least in an overview, of the functional and technical requirements of the software. These requirements form the basis for all subsequent decisions in the architectural process.

  Although some technical framework conditions or non-specialist requirements can have a major influence on a software architecture, these concepts are often only formulated vaguely or not in specific terms. A software architect must therefore identify and record these restrictions, evaluate them in relation to the software architecture, and communicate their effects.

- **Defining a viable software architecture**
  A software architect designs the structure of an application by using their contextual knowledge of the application to effectively solve underlying business challenges. In this task, they must always consider technical limitations and the potential for further development of the architecture.

  In a corporate environment, technical restrictions often include specifications for the programming language used, for the frameworks to be used, or for the execution environment.

  Some server-based software systems run in on-premise data centers, while others are operated by one of the large hyperscalers. Many different aspects must be considered and incorporated into your decision in favor of any particular architecture.

  Every decision about the architecture (and the architecture itself) must be documented in an adequate format so that decisions can be discussed and communicated at any time. I presented the relevant topics in Chapter 3.

- **Dealing with technical challenges and risks**
  Every decision in a software architecture can potentially turn out to be unfavorable or wrong and lead to problems in retrospect. These decisions might relate to the chosen technological basis or the chosen software architecture. Not every new technology, approach, or product purchased fulfills all the promises advertised.

  For this reason, the selection of a technology or architectural approach requires careful consideration and should take many different influencing factors into account. Regardless of the decision-making process, some residual risk always exists that must be dealt with effectively.

  The architect's job is to make these decisions, question them, validate them, and adapt them if necessary. Many issues won't arise until the respective technology or architectural approach is employed, even though they sound promising in theory.

  Decisions about software architectures are not static, and software architects must proactively identify potential issues and mitigate them as early as possible in an application's life cycle.

▶ **Further development of the architecture**
As I have already described, an architecture should not be perceived as static and inflexible. Requirements may change during the course of a project, or earlier assumptions may prove to be impractical or impossible to implement.

In some cases, adjusting decisions that have already been made makes sense. Therefore, not only should architects play an initial role in projects and specify architectures; they should also be continuously involved in development and, if necessary, drive and support further development of the architecture.

Sometimes new, creative solution concepts are sought because using the same solution template for every application is often not effective.

▶ **Active participation in development**
Software architects should be actively involved in software development to understand exactly how the chosen architecture is perceived and implemented in real life by developers. Architects don't need to be fully integrated into the day-to-day work of developers, but they should be involved in the development process and be encouraged to actively participate.

▶ **Quality assurance**
No technology or architectural approach can guarantee error-free software. Quality is an explicit goal that must be constantly monitored and improved. The responsibility for the quality of software does not lie with a single person or role but instead is a joint task that must be performed by multiple parties. The tasks of a software architect include, among other things:

- Documentation and communication of the software architecture and the decisions on which it is based
- Regular code reviews and technical support for the development team
- Specifying technical standards and guidelines and checking that these are adhered to
- Continuous review and validation of the selected software architecture
- Making design decisions that ensure, for example, the performance, security, and scalability of the application

5.2 Software Architecture Styles

An *architectural style* defines a basic structure for organizing software components and capturing their relationships with each other at a higher level of abstraction. Each style is based on specific principles and procedures designed to fulfill certain requirements, such as scalability, modularity, or maintainability.

Architectural styles can be grouped in different ways. The following software architectures represent different styles that focus on the basic structure and interaction between the individual components.

5.2.1 Client-Server Architecture

One of the fundamental concepts in distributed applications and services is the *client-server model*. In this distribution model, applications or services are provided by a *server* component for a *client* component. Communication among the components takes place via a network. Communication is based on a request-response mechanism in which a client sends a request to the server, and the server then answers.

A distinction can be made between different types of clients, the most important of which include the following:

- **Thin clients**
  A thin client is a client that consumes only minimal resources and mainly uses the resources of the server. Common thin clients include web browsers like Mozilla Firefox or Google Chrome, which display classic server-side web apps without much JavaScript.

- **Thick/fat clients**
  A thick client (or fat client) is a powerful client that requires fewer resources from the server, as processing and storage tasks are performed locally on the terminal device. Examples of fat clients include Microsoft Word running locally or video games that require a lot of local computing power, but still communicate with a server.

> **Single Page Applications: Thin Clients or Thick Clients?**
>
> A *single page application (SPA)* is a web application in which all content and functions are displayed within a single HTML page. Instead of repeatedly loading the entire page, the required data and interface elements are dynamically reloaded and displayed via JavaScript.
>
> Since a large part of the application logic is usually swapped out from the server to the client, SPAs are often regarded as thick clients. However, making a clear distinction between thick and thin clients can be difficult and depends heavily on which component does the majority of the processing.

In the client-server model, a server actively waits for client requests and responds to each request with its specific *services*. A service-specific protocol is used for communication between the partners.

Most modern software architectures are based on the client-server model, and components can often act as both server and client. As shown in Figure 5.1, `Service A` is called by the user as a *service*. However, `Service A` in turn calls `Service B` and acts as a *client* in this communication.

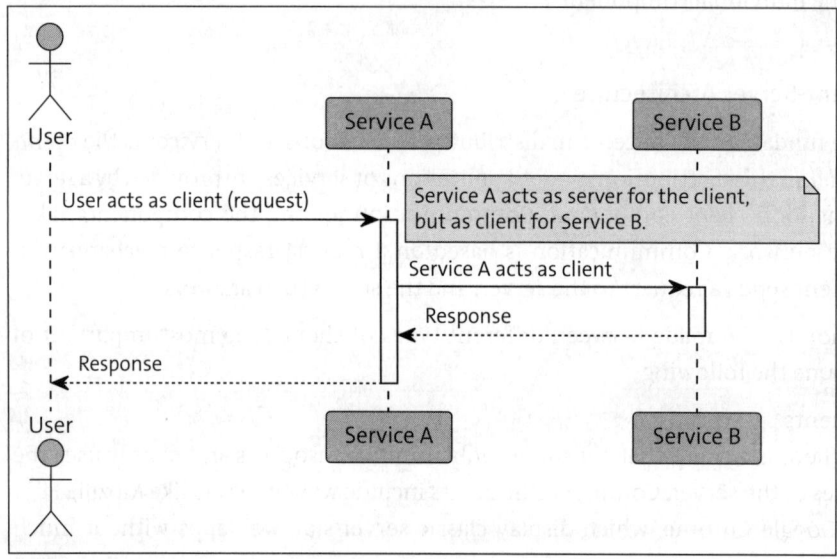

Figure 5.1 Client-Server Model with Multiple Clients

5.2.2 Layered Architecture and Service Layers

The *layered architecture*, also known as *n-tier architecture*, is one of the most common architectural patterns in information technology, especially in enterprise applications.

Due to its proximity to traditional corporate communication and organizational structures, this approach is a familiar standard for most developers and architects. In larger companies, you'll often find database teams, business logic or service teams, and front-end teams, for example, which each take care of the needs of a specific area or layer.

Dividing software into individual layers that correspond to these structures is therefore a simple technique for subdividing a complex system into smaller, more manageable units. As shown in Figure 5.2, each layer is based on an underlying layer and uses the services or interfaces provided. The call is transparent for the called layer. However, the called layer does not know the layers that build upon it. The dependency is only ever defined in one direction: from a higher to a lower layer.

Each layer provides its own separate interface and encapsulates the layers below it. In this example, layer 2 uses the services of layer 3, which in turn accesses the database in layer 4. For layer 2, access to layer 4 is transparent, and the layer is not aware of this.

The task of isolating and separating individual layers results in *layers of isolation* and promotes the maintainability and expandability of your software. Changes in one layer do not directly affect other layers due to the strict separation and clear interfaces.

5.2 Software Architecture Styles

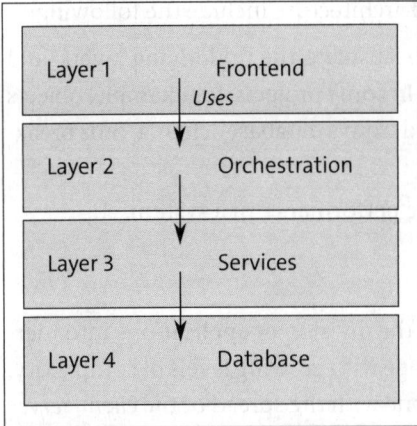

Figure 5.2 Layered Architecture

Some further advantages of adopting this concept also include the following:

- The clear definition of responsibilities for each layer and thus the implementation of the *separation of concerns* principle. The individual layers can be better tested and managed separately.
- Individual layers can be understood separately without knowing the "big picture." When saving a file via an operating system, which is also divided into different layers, you don't need to deal with the details of the file system or the management of data on an SSD hard disk, for example. The application programming interface (API) storage can be understood individually and used without detailed knowledge of the rest of the storage.
- Layers can be exchanged using adequate implementations with the same interfaces. An application that saves data via the operating system API, for example, can use SSD hard disks or mechanical HDD hard disks without any adjustments as long as the interface remains the same.
- Dependencies between the layers are minimized. With the abstraction and encapsulation between the layers, the individual components become more independent, and complex dependencies across multiple layers are prevented.
- The definition of layers with explicit interfaces provides opportunities for standardization. For example, the *Open Systems Interconnection Model (OSI layer model)* defines a layered architecture for network protocols that can be useful for understanding some standards like TCP or IP.
- Multiple layers can be built on a single layer. In the example shown in Figure 5.2, exactly one additional layer was built on top of a layer. In a layered architecture, however, multiple layers can also be based on a single layer. If, for example, an application requires multiple user interfaces, multiple parallel frontend layers can be created, for example, one for web browser-based interfaces and one for mobile apps.

However, the disadvantages of creating a layered architecture include the following:

- Often, individual layers do not sufficiently encapsulate the underlying layers, and changes must be made across multiple layers. In some projects, for example, objects that are used in the user interface are generated from a database schema, thus breaking up the layer model.
- Many layers can have a negative impact on the performance of a system.

Three-Layer Architecture

Even in the early days of software development, the division of applications into hierarchically arranged modular layers was used to cope with growing levels of complexity. In the 1990s, this development reached a milestone with the spread of the *client-server architecture*, which separated tasks between clients and servers and often used a database as the central data source, on the server side.

Client applications were mostly desktop applications designed as *rich clients* that processed data locally on a client device and/or cached large amounts of data.

Many manufacturers offered component libraries or even complete development environments for a wide variety of operating systems for the implementation of these architectures. With these resources, you could create an application via drag-and-drop graphical tools. The component libraries included user interface components that enabled direct SQL database connections to be configured and applications to be created quickly. Examples of these development environments include *Borland Delphi*, *Microsoft Visual Basic*, and *PowerBuilder* from Sybase.

However, this approach resulted in some problems with the reuse of business logic code. Often, this logic was often implemented directly—and therefore isolated from other logic—in the individual frontend pages or forms of an application. Thus, reusing existing components is difficult, and the effort required for maintenance work and extensions increases because changes must be made in multiple places. The more complex an application becomes, the more obvious this problem.

In some cases, logic blocks were then implemented directly in the database as *stored procedures*, while other blocks were extracted to special components or libraries and shared. Stored procedures are always database dependent, and by using them, you lose the option of simply porting to an alternative persistence technology. Managing the central libraries and supporting multiple versions simultaneously also became a challenge.

As a result, *three-tier architectures* became increasingly popular, facilitated above all by the spread of object-oriented programming languages and technologies like *Java* and *Java, Enterprise Edition (Java EE)*. In addition, with the rise of the internet, applications suddenly required web interfaces as well as desktop applications. Logic that was only implemented within a desktop application was a hindrance because this logic could not be reused in another client.

Classically, three-layer architectures, whose individual layers are sometimes also physically separated, can be described as having the following three layers:

- Presentation layer: Services or information are made available to the user via this layer. This layer can consist of rich clients, HTML-based interfaces, or programming interfaces.
- Business logic layer or domain layer: The core of the application, containing the actual logic as well as all processes and rules essential for business operations. This layer is where data from the database is processed, calculations are carried out, checks are made for correctness or validity, and decisions are made on the basis of implemented rules.
- Data storage layer or data source: This layer handles communications with external systems that perform various tasks for the application or that store data. Databases are often used as external systems.

Considering the dependencies between the layers is just as important as defining their individual responsibilities. In a layered architecture, dependencies should only ever run in one direction, never bidirectionally. In the three-tier architecture this rule means, for example, that the presentation layer should access the business logic layer, but not vice versa. With this rule, a presentation layer or a data storage layer can potentially be easily replaced by an alternative implementation without having to reimplement business logic again.

> **N-Layer Architectures**
>
> Layered architectures don't need to consist of exactly three layers. In his book *Software Architecture Patterns*, Mark Richards always adds a fourth layer to the classic three-layer architecture:
>
> - Presentation
> - Business logic
> - Storage/persistence
> - Database
>
> The number of layers is not relevant as long as the layers are separated and its hierarchical structure is adhered to. These architectures are often simply called *n-layer architectures*.

The Principle of Open and Closed Layers

According to Mark Richards in *Software Architecture Patterns*, a key principle of a layered architecture is the division of the individual layers into open and closed units. This use of "open" and "closed" does not refer to the mandatory *open-closed principle* in which a layer or its interface should be open to extensions but closed to changes.

Richards is describing which layers are involved in processing a request and which layer can be skipped, for example.

As shown in Figure 5.3, we have layered architecture consisting of four *closed layers*. Thus, a request received by layer 1 is processed or passed through all the layers below it until it reaches layer 4 (the database layer). Each layer is involved in the processing of the request and ensures the separation of the individual layers.

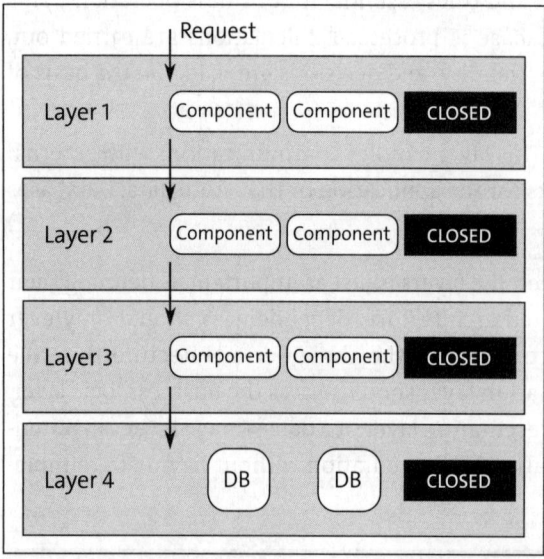

Figure 5.3 Closed Layers

Strict separation and isolation of the layers is essential for good maintainability of the application and ensures that changes to one layer do not spread throughout the entire system and that all layers have to be updated.

In some situations, however, defining *open layers* where requests may not be routed through them, but past them, might be a good idea. As shown in Figure 5.4, for example, layer 2 is marked as open, and a request from layer 1 could in this case be forwarded directly to layer 3 without being processed in layer 2.

If, for example, shared business components must be integrated into the architecture that are to be used by both the existing business logic layer and the existing presentation logic, then an additional *service layer* can be defined. From the point of view of the architecture, this layer is located below the business logic layer and is therefore not directly accessible for the presentation logic. By defining the new layer, an access restriction for the new layer is already achieved via the layer architecture.

However, with the introduction of the new layer, which is located below the business logic layer, every call would then have to be processed by this layer, even if no further processing of the request would be carried out in the new layer. In this case, it makes sense to mark the layer as open and pass calls directly to the next layer down.

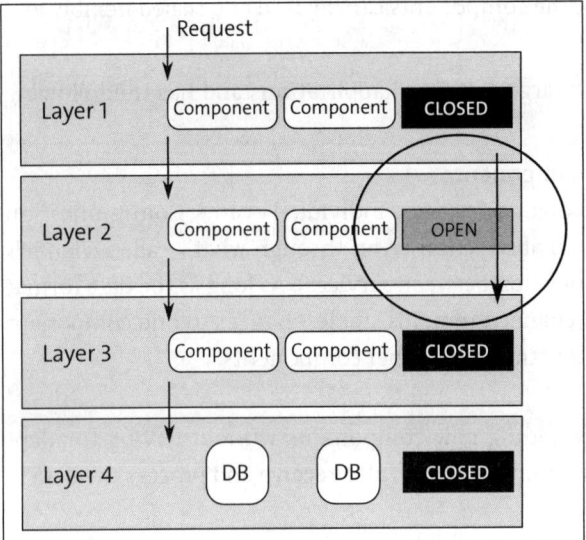

Figure 5.4 A Combination of Closed and Open Layers

The use of open layers must be carefully considered. Too many open layers and direct access to other layers can make a system and its components confusing and highly coupled. As a result, the layered architecture loses its advantages and systems become more susceptible to rampant changes and therefore more difficult to test and maintain.

> **Open-Closed Layers and Patterns**
>
> Open and closed layers are sometimes defined within the framework of patterns. For Java EE, for example, *Design Patterns for Optimizing the Performance of J2EE Applications* was published in 2001. This list of patterns for optimizing Java EE applications contains, for example, the *fast lane reader pattern*, which describes a kind of fast lane to bypass several layers for faster database access. In this case, all layers are considered open for read access.
>
> However, this pattern is now considered more of an anti-pattern: Completely bypassing all layers is not recommended because it undermines the entire layer architecture.

5.2.3 Event-Driven Architecture

An *event-driven architecture (EDA)* is an architectural style based on the processing of *events*. "Processing" in this context means the way in which we react to certain events. Each event represents a state change within a system, such as a change to a customer address.

This kind of architecture consists of independent components that communicate with each other via events, with each component receiving and processing data. Because communication is asynchronous, the components can be used and scaled flexibly and independently of each other.

This approach is suitable for both large and small applications and has the following advantages:

- **Decoupling of the individual components**
  The individual components receive and process individual events. Communication between them takes place via an abstraction layer, through what's called *channels* (see also Chapter 6 on communication between services). As long as the data format of the events that are being exchanged remains stable, changes to one component won't require any additional adjustments to other components.

- **Functionality enhancements**
  New features can be added by adding new components without having to adapt existing components. The new components can also receive and process the data.

- **Scalability**
  In an EDA, events are processed asynchronously, which means they can be processed in parallel or with a time delay during peak loads.

- **Flexibility**
  Components can be developed, expanded, and combined independently of each other, which allows functionality to be flexibly adapted, added, or removed.

- **Real-time processing**
  Events can be processed directly, thus enabling the creation of what's called *reactive applications*.

Communications between components can be implemented in an event-driven architecture with two topologies or patterns:

- *Message processor*, or *mediator topology*
- *Message broker*, or *broker topology*

In the *mediator topology*, a central *mediator* or *message processor* controls the coordination and processing of events and thus orchestrates the message processing of the components. This approach is particularly suitable for more complex workflows within an application.

In the *broker topology*, on the other hand, events are distributed via a central *event broker* that is exclusively responsible for forwarding and distributing events.

An event-driven architecture is often implemented with messaging technologies, which we'll look at in more detail in Chapter 6.

Message Processor

The orchestration of a process by a central mediator or *process manager* always makes sense if an event must be processed by multiple components in a more complex workflow.

For example, let's say a customer triggers an order in an online store system, the central process manager can orchestrate the next steps, such as the following:

- Validating the order by checking the availability of the item and reserving it for the customer
- Communicating with the payment service provider for payment processing
- Preparing the shipment
- Invoicing and dispatching to customers
- Mailing customers with invitations to a feedback form

In our case, the sequence of individual steps and the dependencies between them can be decisive; for instance, shipping should only take place after successful payment. The process manager ensures that the individual steps are coordinated and executed correctly before the event is forwarded to other components.

The use of a central process manager or mediator is called the *hub-and-spoke* message flow pattern, which is shown in Figure 5.5.

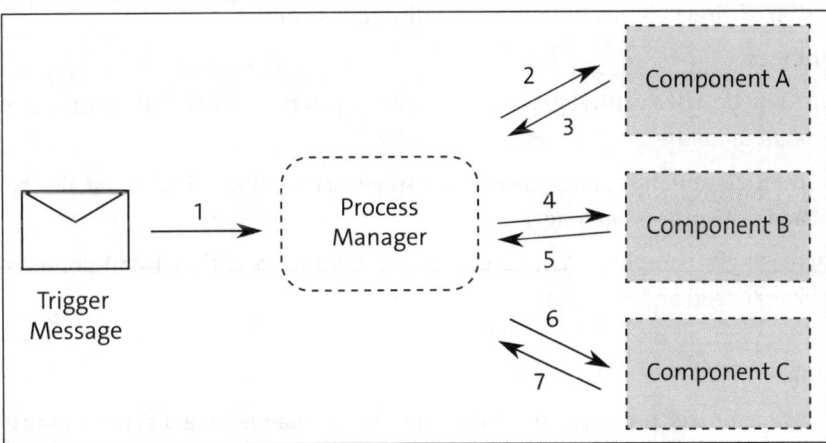

Figure 5.5 Event-Driven Architecture with Process Manager

In this pattern, a *trigger message* starts a new process within the process manager. The process manager then analyzes the message and uses the logic and rules it contains to determine which component should take over the processing. The process manager then redirects the message to the corresponding component. As soon as the component has completed its task, the component sends its response back to the process manager. Then, the process manager uses the message to decide which component should perform the next processing step and forwards the message accordingly.

The process manager does not execute its own business logic on the messages. Its sole task is to correctly forward messages to the loosely coupled, independent components that fulfill quite specific tasks.

A central process manager or mediator can either be implemented in-house, or you can use existing software products and libraries. Various commercial and open-source products are available for this purpose, such as the following:

- *Spring Integration*
- *Apache Camel*
- *IBM App Connect Enterprise*
- Middleware like *Redux Thunk*

The pros and cons of adopting a message processor include the following:

- **Benefits**
  - Independent, loosely coupled components that can be integrated
  - Flexible and diverse application options for the process manager, such as the execution of parallelized processes, which means that most interaction sequences can be implemented
  - Centralized management of individual business processes and better monitoring options
  - Reduction of direct connections between different systems
- **Disadvantages**
  - The introduction of a central process manager, which runs the risk of becoming a *single point of failure*
  - Extensive logic and numerous rules for message distribution, which must also be managed in the process manager.
  - An increasingly complex architecture in use because additional components must be managed and monitored.

Message Broker

With the *message broker* topology, messages are also exchanged via a central point, called the *message broker*. However, the message broker does not contain any logic or rules for mapping a process; it is only responsible for message transfers between individual components. The corresponding logic for the message flow is distributed across the components involved.

In the message broker topology, the individual components process events and in turn generate events that reflect the result of the executed action, as shown in Figure 5.6.

For example, when a customer triggers an order in an online store system, the shopping cart component generates the "New order" event and transmits it to the message broker. All components that are interested in this event receive the event and can process it. As a result, the validation component returns the "Order successfully validated" event and thus enables the next interested component to continue working. This process is repeated continuously until the order process is completed or no more components are interested.

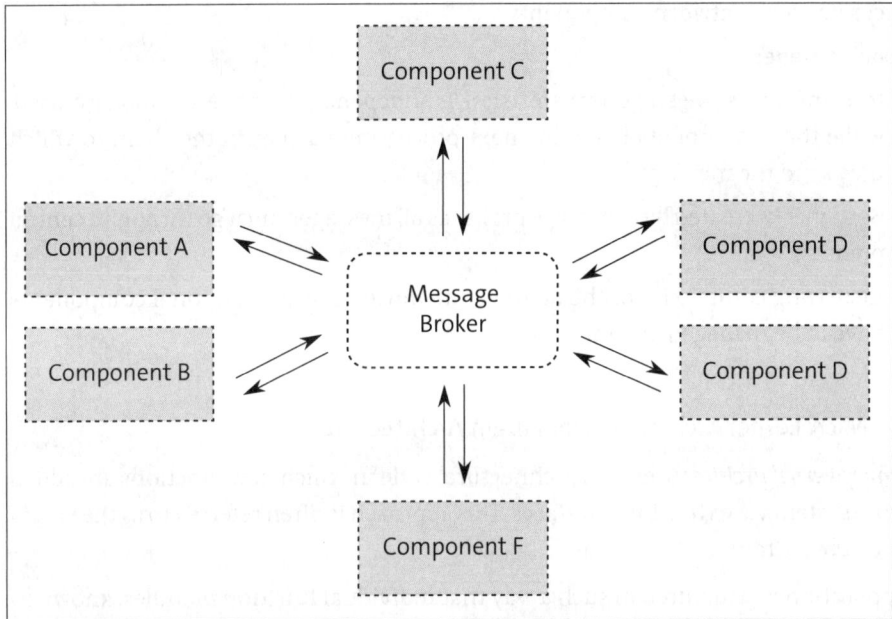

Figure 5.6 Setting Up an EDA with a Message Broker

The way this topology works can be compared to a relay race: Each component passes the event to the next component in the chain after it has processed it.

This topology is suitable for use cases in which only limited logic is required for message distribution and the central orchestration of the processes is not desired.

Some proven software products and libraries available for implementing a message broker topology include:

- Apache Kafka
- RabbitMQ
- Google's Cloud Pub/Sub
- Amazon Simple Notification Service (SNS) and Amazon Simple Queue Service (SQS)

The pros and cons of adopting a message broker include the following:

- **Benefits**
  - Loose coupling of individual components since the message broker handles the message transmission

- Ability to change and add new components, without making adjustments to existing components
- Asynchronous and flexible processing of messages by the components
- Good options for scaling the components, which can also run in parallel
- Reduced complexity, through the centralized message broker, of many individual connections between components

▶ **Disadvantages**
- In some cases, message transmission is independent of the technology used, while the components know the next processing partner in the chain to which they send the message
- Bottlenecks created by a message broker as all messages must go through a central point
- Increasing complexity of the architecture during use, as additional components have to be managed and monitored

5.2.4 Microkernel Architecture or Plugin Architecture

The *microkernel architecture* is an architectural style in which new functions are added to a core system via extension interfaces. This approach is often referred to as the *plugin architecture pattern*.

An application is structured in such a way that individual function modules, known as *plugins*, can be seamlessly added to or removed from a core application. These plugins are independent components that communicate with the main application via defined interfaces. The *extension points* define these interfaces, as shown in Figure 5.7.

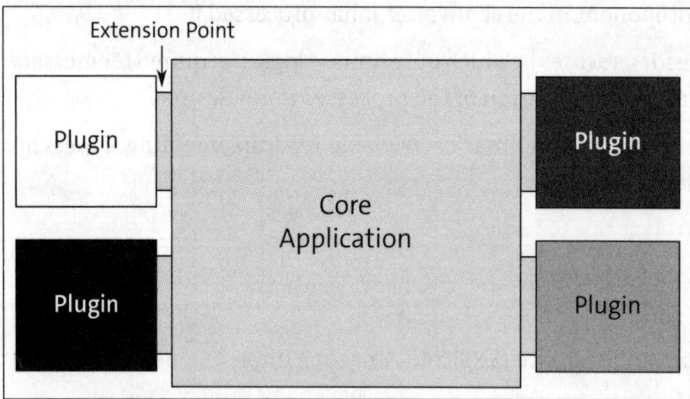

Figure 5.7 Microkernel or Plugin Architecture

With this approach, new functionality can be added without having to adapt the main application. The pattern is often used for developing applications that must be delivered as fully autonomous software, although individual functions can still be developed independently and integrated into a main application. The use of a plugin architecture can also be advantageous when structuring server applications, especially in combination with the concept of *vertical slices*, as shown in Figure 5.8.

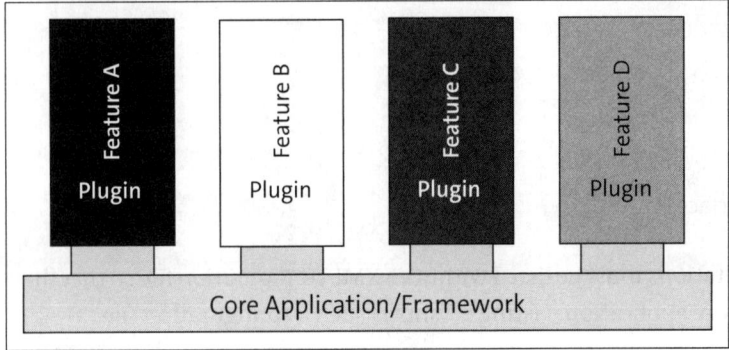

Figure 5.8 Plugin Architecture with Vertical Slices

Vertical slices can create more isolated and more independent modules, with each covering a single business area. These modules, which act as independent verticals, can communicate with each other via a mechanism provided by the core application or by the underlying framework.

Several open-source projects use a microkernel architecture and utilize an extension mechanism, such as the following:

- *Eclipse* plugins
- *Visual Studio Code (VS Code)* extensions
- *WordPress* plugins
- *Blender* plugins

The core principles of the pattern can be summarized in the following way:

- **Modularity and maintainability**
  By dividing the application into independent modules (the plugins), each of which is responsible for a specific functionality, the application is easier to maintain, and the modules do not influence each other.

- **Expandability**
  A cornerstone of this architecture is expandability. Developers can create new functionality without having to adapt the core application. Such a design ensures that the software can be adapted to changing requirements or new technologies without the need for a complete overhaul.

- **Separation of concerns**
  Dividing the application into several modules, each of which takes on a specific functionality, separates their responsibilities (see Chapter 2, Section 2.6).

Our next example illustrates how a microkernel architecture works: An application imports an XML configuration file with the plugin's configuration and executes it according to the plugin interface.

The `Plugin` interface, which must be implemented by all plugins, is shown in Listing 5.1.

```java
package plugins;

public interface Plugin {
    void execute();
}
```

Listing 5.1 Plugin Interface (Java)

All plugin implementations must be listed within an XML configuration file so that they can be used by the core application. Listing 5.2 shows such a configuration file.

```xml
<plugins>
    <plugin>
        <name>GreetingPlugin</name>
        <class>myplugins.GreetingPlugin</class>
    </plugin>
    <plugin>
        <name>AnotherPlugin</name>
        <class>myplugins.AnotherPlugin</class>
    </plugin>
</plugins>
```

Listing 5.2 Plugin Configuration File

The core application shown in Listing 5.3 loads the individual configured plugins via a `PluginLoader` and executes their functionality in a loop.

```java
package app;

import plugins.Plugin;
import java.util.List;

public class MainApplication {
    public static void main(String[] args) {
        String xmlPath = "plugins.xml";
        List<Plugin> plugins = PluginLoader.loadPluginsFromXML(xmlPath);

        System.out.println("Loaded Plugins:");
        for (Plugin plugin : plugins) {
            **plugin.execute();**
```

                }
        }
}

Listing 5.3 Microkernel Application Using Plugins (Java)

The pros and cons of adopting a microkernel include the following:

▶ **Benefits**
- Flexibility and expandability of the application, as changes and extensions are made via plugins. The core application remains unaffected.
- Responsibilities are separated within the architecture.
- That plugins can be developed in isolation from each other can make the development of an application more effective.

▶ **Disadvantages**
- Adaptations to the core application are virtually impossible. Changes to the interface to the plugins would possibly result in a change to all plugins.
- The application can become dependent on plugins if important functionality is not provided by the core application itself, but also by plugins.

5.2.5 Microservices

For a long time, enterprise applications were designed and implemented as large, coherent units in an approach known as the *monolithic architectural style*. All business functionalities were delivered in a single unit and executed at runtime within an (operating system) process.

The internal structuring and division into components of such applications is carried out with the help of the respective programming language, for example, through the use of classes, functions, or namespaces.

Many successful products and projects have been implemented using this approach. However, as the size of an application and its project duration increases, so do problems and thus frustration with the application among users, specialist departments, and developers.

In response, some project teams and companies began to break up their monolithic systems and replace them with smaller, more independent applications or to develop new components as separate applications.

In 2012, James Lewis and Martin Fowler then compiled the similarities between the approaches of these different teams and described the resulting modularization concept. This concept gave rise to the idea of *microservices*. However, no fixed definition exists for microservices. For this reason, the characteristics described by Lewis and Fowler are often to orient the implementation of this architectural style.

Adrian Cockroft, who played a key role transitioning Netflix to a microservice architecture, describes this architectural approach in the following way:

> *"Service-oriented architecture composed of loosely coupled elements that have bounded contexts." —Adrian Cockcroft*

Applications that follow the architectural style of microservices therefore have the following characteristics:

▶ **Components as separate services**
Applications are usually made up of multiple individual components. In monolithic applications, all components are combined and executed as a single unit. With microservices, on the other hand, each component is executed as a separate service or instance.

▶ **Loose coupling between the services**
The dependencies between the individual components are minimal, and communications among the components is *technology-agnostic* (i.e., independent of the technologies of the individual services). Technical changes to one component do not necessarily entail changes to other components.

▶ **Bounded context**
Each service is built around an independent, clearly defined business functionality. Only minimal functional dependencies on other components should exist. In this context, the Unix philosophy is often mentioned: "Do one thing and do it well." The division of components or bounded contexts is often the biggest challenge in a microservice-based application.

> **Smart Endpoints and Dumb Pipes**
>
> In the world of microservices, the *smart endpoints and dumb pipes* approach has become widespread. The idea behind this concept is to equip the individual endpoints (i.e., the services) with everything they need to run independently and at the same time decouple them as much as possible from other services. Communication among services should be as lightweight as possible and not be based on complex protocols or data formats.
>
> A quote often used in this context comes from Ian Robinson: *"Be of the web, not behind the web."* In other words, widespread open standards from the internet should be used wherever possible. HTTP, REST, JavaScript Object Notation (JSON), and XML are examples of these standards.

▶ **Suitable technologies**
Each service is implemented in a technology that is suitable for the corresponding requirement. You don't need a centralized technology stack. Thanks to the clear interfaces using standardized protocols, the services remain interoperable and can continue to work together.

- **Decentralized data storage**
  Each service is responsible for its own data and stores it in its own persistence layer, which can only be accessed via the service's interfaces.
- **Decentralized management**
  In microservice environments, there is no central instance that is responsible for all services and controls or manages them. In a *service-oriented architecture (SOA)*, for example, everything is managed by what is called an *enterprise service bus (ESB)*.

> **Service-Oriented Architecture**
>
> A service-oriented architecture (SOA) is an architectural style in which applications are divided into small, reusable services and centrally controlled by an *enterprise service bus (ESB)*. All services within the SOA use a common data repository, which makes managing the services independently of each other more difficult.
>
> Business processes are mapped or controlled centrally within the ESB, and the failure of one service can result in a failure of the entire business process.

- **One team**
  The responsibility for a service lies with a single team. In contrast to monolithic applications, where several teams share responsibility, in a microservice-based environment, each team is responsible for "its" service. The need for agreement between teams is minimized since releases can be planned independently of each other.

> **The Ideal Team Size for Microservices**
>
> Many statements about the ideal team size can be found. According to Martin Fowler, Amazon Web Service (AWS) uses the term *two-pizza team* to describe team size, for example. The team should be so large enough that two pizzas are enough to satisfy them. (The actual sizes of the pizzas themselves are open to speculation.)
>
> In reality, team sizes of three to twelve people per service have proven to be useful. With more than twelve people, the communication overhead increases again, and with fewer than three people, you can no longer speak of a team.

- **Infrastructure and automation**
  In microservice-based applications, the configuration and setup of infrastructure is automated as much as possible. Technologies such as *Docker* or *Terraform* emerged with the spread of this architectural style and have since become widespread.
- **Be prepared for error situations**
  Splitting an application into multiple separate, independent components increases the likelihood of errors occurring, especially when the components interact with each other. For this reason, in a microservice architecture, each service must be prepared for possible error situations at all times.

> **Criticism of Microservices**
>
> Organizations can benefit from the advantages of the microservice architecture style if they have a complex and highly scalable application landscape. However, you must not neglect the challenges that arise when using this approach, such as the following:
>
> - *Increasing complexity* of the system landscape due to the division of the application into many separate services that must be managed.
> - *Overhead for infrastructure* because additional components, such as central monitoring solutions or service discovery systems, are often required.
> - *Communication overhead* because no more local method calls are made between the services, only remote calls. These calls must be transmitted and secured via the network, for example.
> - *Data consistency and transactions* are more difficult to implement due to the distribution of the application, and data consistency across multiple services may only be established with a time delay.
> - *Testing and debugging* the entire process is difficult.
> - *Organizational challenges* arise because each team acts independently and is responsible for its own service. New teams may need to be set up, or operational teams may need to be divided up into smaller, individual teams. Microservices require clear communication as well as cooperation and coordination between the teams.
>
> Some organizations have split their monolithic application into many smaller services and then found that the cost of implementing and operating the microservices exceeded the benefits that the microservices brought. As with any technology or pattern, its use should be carefully considered, and the pros and cons thoroughly weighed. The challenges that arise would be listed under "Consequences" in a pattern description.

5.3 Styles for Application Organization and Code Structure

As described earlier, an architectural style defines a basic structure for the organization of software components and their relationships with each other. Each style is based on specific principles and procedures to fulfill certain requirements. Examples of requirements include scalability, modularity, or maintainability.

In addition, all applications, regardless of their size, require a clear, unambiguous domain logic that implements the specific rules and the corresponding business processes.

For this reason, the focus of this section is on efficiently organizing the application and creating a structured code base to ensure an effective and maintainable implementation of the business logic.

5.3.1 Domain-Driven Design

In his work, Eric Evans found that many software projects failed or had serious problems due to unclear requirements, due to a lack of communication between the software developers and the business experts, or due to inadequate modeling of business processes. At the same time, he observed how the object-oriented community was creating a better approach for matching software development to real business requirements and to cope better with everyday challenges. He referred to this approach as *domain-driven design (DDD)*.

In 2003, Evans published *Domain-Driven Design: Tackling Complexity in the Heart of Software*, in which he described this philosophy in detail and thus coined the term *domain-driven design*.

> **The Big Blue Book**
>
> Evans's 2003 book is often referred to as the *big blue book* because of its color and size. Ten years later, Vaughn Veron published *Implementing Domain-Driven Design*, which has a red cover and is therefore often referred to as the *big red book* in the community. Veron's book is a little more practice-oriented and focuses on how DDD can be implemented.

Evans makes the following assumptions for business-related and non-technical software:

- The focus of software projects must be on the business domain.
- Complex business logic should be based on a model.

Domain-driven design does not describe any explicit programming principles but instead deals with the modeling of business models and their processes as well as the definition of a vocabulary for modeling a specialist area, called a *domain*. This approach is suitable for developing complex software systems that can evolve over time.

Even if close to object-oriented software development, and often used there in practice for analysis and design, domain-driven design can also be used for other approaches and is not limited to object-oriented programming.

> **Domain-Driven Design: Areas of Use**
>
> Domain-driven design and its approaches have become increasingly widespread and are now regarded as among the most important procedures for organizing and mapping specialist functionalities in complex software systems.
>
> Domain-driven design can, for example, find module boundaries with the concept of *bounded context* and therefore support the division of an application into several components or services. When designing microservice architectures, this concept can help

you define clear service boundaries between the individual microservices and to enable structured communication between them.

One of the most important points is the close collaboration and communication between developers and technical experts, which is why the approach has become particularly widespread in agile, interactive environments.

The core of every domain-driven design consists of modeling a clearly defined domain. In an object-oriented implementation, this domain forms the basis for the definition of domain classes and enables their strict delimitation (or decoupling) from the other functionalities of the system.

Figure 5.9 shows a simple example of a domain model for managing training courses. This diagram is free of technical dependencies and focuses solely on domain logic.

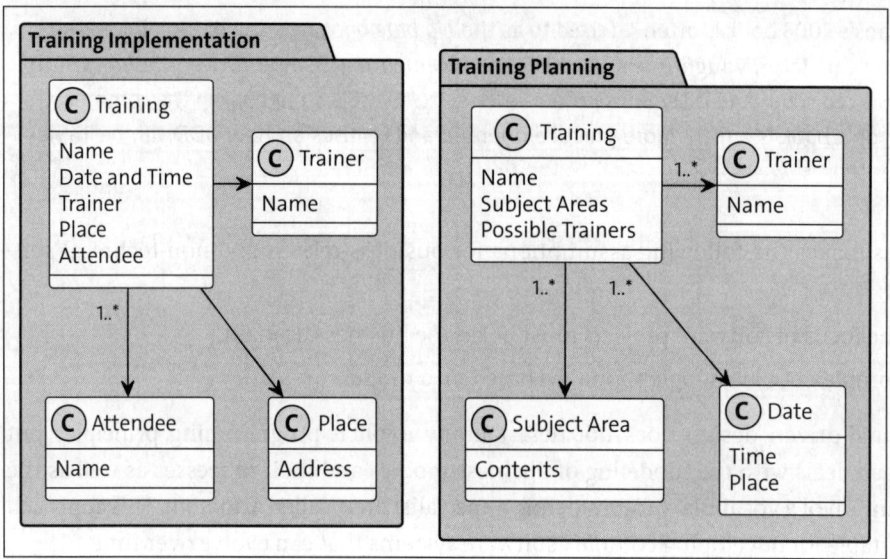

Figure 5.9 Sample Class Diagram

However, let's take this example leads further to two domain-driven design concepts defined within what's called the *strategic design phase*:

- A ubiquitous language
- A self-contained context or a self-contained unit

Ubiquitous Language

In domain-driven design (DDD), the creation of a standardized, ubiquitous language that is binding for all participants is always part of the modeling of a domain logic. This vocabulary is intended to promote communication between business domain experts and software developers and thus ensure that both groups share a common understanding of the business domain. Such *ubiquitous language* is not necessarily generally understandable and often requires in-depth knowledge of the business context.

In projects without standardized language, terms must often be translated or explained between business experts, developers, and other project participants. This translation is sometimes necessary, not only between business experts and developers, but sometimes also among developers themselves. Confusion and misunderstandings caused by the lack of a ubiquitous language inevitably result in increased development effort and ultimately error prone and less stable software.

From the perspective of domain-driven design, therefore, the business-specific terms of your ubiquitous language should remain unchanged in the code and should not be translated.

In the diagram shown in Figure 5.9, for example, the following terms were used:

- Training
- Instructor
- Participants
- Place

A subject matter expert who is deeply rooted in the specific domain could contradict the term "place" in the model presented. In their view, "training location" would be more precise and appropriate. However, these differences in word choice should not be an obstacle; instead, disagreement can become an opportunity for a constructive exchange between experts and developers, so that a common model and a uniform language can develop.

Your solution's ubiquitous language should be recorded in any case and used consistently in documents or diagrams. In the simplest case, you create a glossary with all the terms that occur in the language.

Bounded Context

The example shown in Figure 5.10 illustrates a second challenge. The application consists of two components: training planning and training implementation. Each subarea defines the term "training" differently, and therefore, the term is not clearly defined.

One possible solution is to combine the two models in a joint model. In this context, clear terms such as "training date" and "training definition" can be introduced. Alternatively, a general training term could be defined for use in both sub-areas. Although this approach could be useful, as in this example, you run the risk of a more extensive, more complex, and possibly more confusing model.

If the model is successively expanded in this way, the most likely result is a rather complex, barely maintainable domain model with unclear responsibilities.

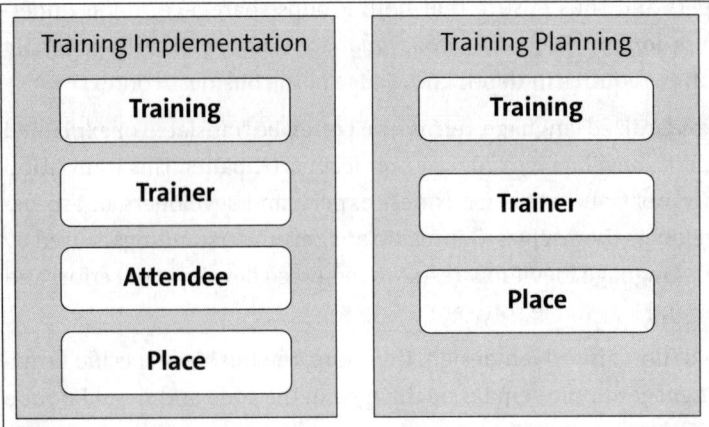

Figure 5.10 Examples of Bounded Contexts

Domain-driven design recognizes the concept of a *bounded context* as a possible solution. This closed and clearly limited context defines an area in which a business-specific model with corresponding rules and a ubiquitous language applies.

A bounded context is characterized by the following points:

- **Clear boundary**
  The context defines which parts of a system are inside or outside. This limit can be of a functional, organizational, or technical nature.

- **Ubiquitous language**
  A ubiquitous language applies within the context so that all terms and concepts are clear and understandable for all team members.

- **Specific models and rules**
  As mentioned in earlier our example, each context has a specific technical model with its own rules and logic. This approach enables an exact separation and a focus on the context-specific business domain to enable precise mapping.

- **Clear responsibilities**
  Not only does a context describe business models; it also describes the associated responsibilities of the models and components it contains. This task should ensure fewer conflicts between the various subsystems. In software development, these responsibilities may also represent clearly defined team responsibilities.

In our training example, this approach can result in two bounded context instances, each with its own business model. Each model has exactly the responsibility for a corresponding area and defines exactly only the dependencies and attributes that are necessary for this area. The example of the Training class shows quite clearly that the two trainings are quite different and that other attributes are of interest in the area of training implementation than in the area of training planning.

Evans clarifies that a bounded context does not necessarily represent a module boundary and that multiple contexts can be defined in parallel in modules. A bounded context is not primarily a technical boundary, even if a boundary is often implemented using technology.

By separating the models and defining two contexts, we hope our example becomes clearer and easier to maintain. Changes in one area have no direct impact on the second area, and further development can take place separately, which would not be possible in a joint model. If the example is divided into two modules, possibly also into services, the two teams can work on it in parallel and deliver their two versions independently of each other.

5.3.2 Strategic and Tactical Designs

Domain-driven design defines several phases the development process goes through. In simple terms, two important phases can be highlighted:

- Strategic design
- Tactical design

Within the strategic design, the business domain is analyzed and modeled. Bounded contexts and their relationships with each other are established. The focus in this phase is clearly on the business domain to be implemented and on a rough structuring of the system.

During the tactical design phase, the domain model is refined and described in more detail using existing components. Domain-driven design defines the following components:

- Services
- Aggregates
- Entities
- Value objects
- Repositories
- Factories

In our training example, so far, we've created a strategic design, and tactical development should follow.

5.3.3 Hexagonal Architecture/Ports and Adapters

The *hexagonal architecture*, also known as a *ports-and-adapters architecture*, was developed by Alistair Cockburn and presented in 2005. With this architectural approach, Cockburn aims to make software systems more flexible, more testable, and easier to maintain.

Cockburn developed this approach after seeing business logic merged with the code for user interface logic in many applications whose source code he had read and used over the years. This resulted in the following problems, among others:

▶ **The applications were difficult to test automatically**
If the logic to be tested is contained directly in the user interface logic, changes to the display, such as a change in size or the position of a button, can result in errors when testing the logic.

▶ **Automated background processing is impossible**
If the logic is merged with the user interface logic, it cannot be executed separately, for example, within a batch process. A call must always be made via the user interface.

▶ **No alternative call paths possible**
For the same reasons, i.e., because the business logic must be called via the user interface, no other clients can be connected. If, for example, another application wants to call the logic, it would also have to use the user interface. An external programming interface is not available.

Many business-critical applications developed at that time followed a logical layer separation and divided the application into multiple logical areas accordingly. The three-layer architecture with user interface, business and data access layers, as shown in Figure 5.11, was widespread.

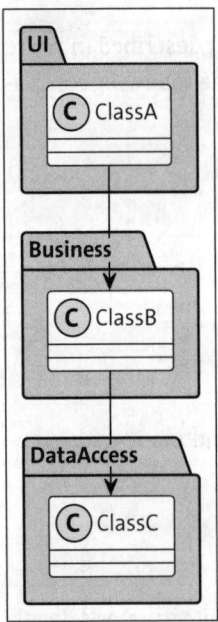

Figure 5.11 Classic Three-Layer Architecture

If violations of the separation of layers were discovered in such applications—for example, due to code that was implemented in a layer not intended for this purpose—attempts were made to move this code to the correct layer. If the layer required for this functionality was missing, it was often introduced without further ado. This approach resulted in an iterative process which, without effective controls, can quickly lead to chaos as more and more new layers are created with tasks that are not clearly defined. Without an effective control over whether the layer structure is adhered to, the use of this model is not practical.

In addition, layer models structured in this way tend not to map dependencies within the application in the way that the business logic would require. As shown in Figure 5.11, for example, a dependency exists between the business layer and the data access layer. As a result, especially in extreme cases, the business logic is dependent on the technical implementation of data storage or on changes in the infrastructure, and thus, technical changes can influence the business logic.

If, for example, you use a database-dependent data type to store object IDs in a database and this database dependency is adopted as an attribute in the business objects, its use is tied to the respective database product. Use without the corresponding database library or access to the database is made considerably more difficult, maybe even impossible. Test execution is impaired, and changes to the database technology used require adjustments to the data model of the business logic.

Every dependency restricts further development and makes an architecture less flexible. Changes in dependencies may force adjustments to the dependent components. In long-lived systems, the longer the software is used, the more frequently infrastructure changes may occur, in contrast to fundamental changes to the business logic.

Such dependency problems always have an impact on the testing of the application. Each dependency in the code must be fulfilled at the time of testing. If the business logic depends on a database, only this database can be used for testing, which makes tests more complex and less flexible.

In classic layer models, dependencies exist from "top to bottom" or "from right to left." The direction depends on whether the model is displayed horizontally or vertically as a graphic.

Cockburn's approach reorganizes these dependencies. The central element is the application logic, which is located "inside." External resources are connected around it, on the "outside," via ports, as shown in Figure 5.12. Things called *adapters* mediate between a port and the business logic. Dependencies only exist from the outside to the inside.

Using the hexagon diagram shown in Figure 5.12, Cockburn wanted to demonstrate that, not only two external dependencies could be connected, such as data storage and display, but that a variety of external resources can be connected. The number six and the name *hexagonal architecture*, which is derived from it, are only symbolic, which is why the approach is often referred to as *ports and adapters*.

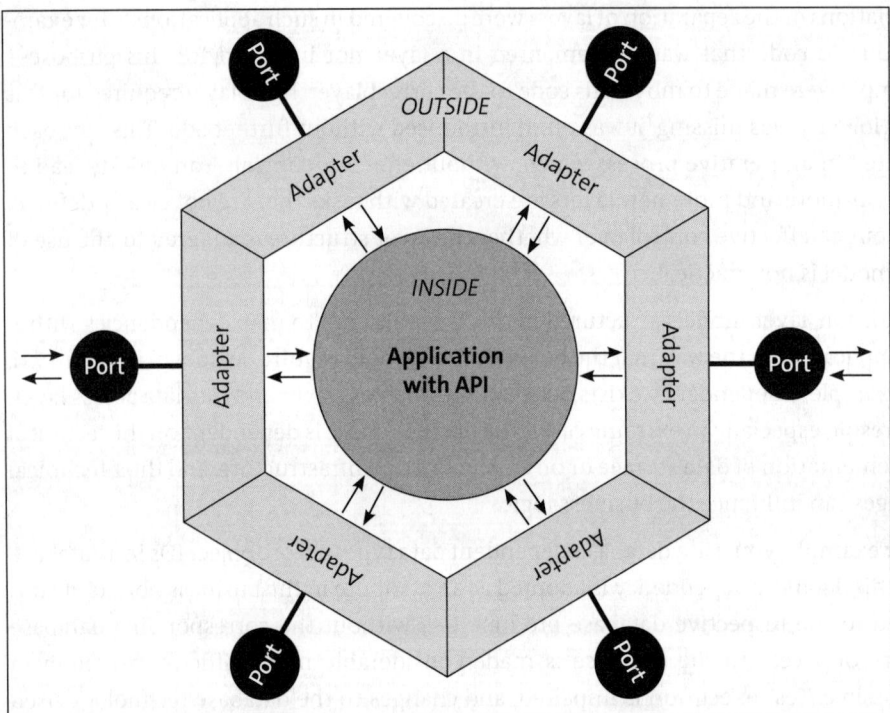

Figure 5.12 Hexagonal Architecture or Ports-and-Adapters Architecture

The term *port* represents the point at which a compatible external component can be connected—similar to an operating system port via which external components that support the corresponding protocol can communicate. The task of the *adapter* is to translate the communication conducted via a port into the interface of the inner core of the application, thus carrying out a protocol conversion.

In the case of a REST call to an application, the HTTP port would receive the external requests, and a corresponding adapter (such as a Java servlet) would evaluate the parameters and pass requests on to the actual business logic in the appropriate format.

The database driver represents the port for connecting a database, while a lean access layer component (the adapter implementation) would perform the translation between the business logic and the database driver.

This type of application structure means you can isolate the core of an application from external changes, thereby increasing flexibility and testability.

To align the dependencies among application parts according to this model, the *inversion of control* principle can be used for implementation at the code level (see Chapter 2, Section 2.5). By defining a business domain-oriented interface within the business layer, as shown in Figure 5.13, the dependency between the two layers can be reversed. An important point to note is that the interface introduced is exclusively business oriented and does not contain any technical details. The *ports-and-adapters* architecture is the logical continuation of this approach.

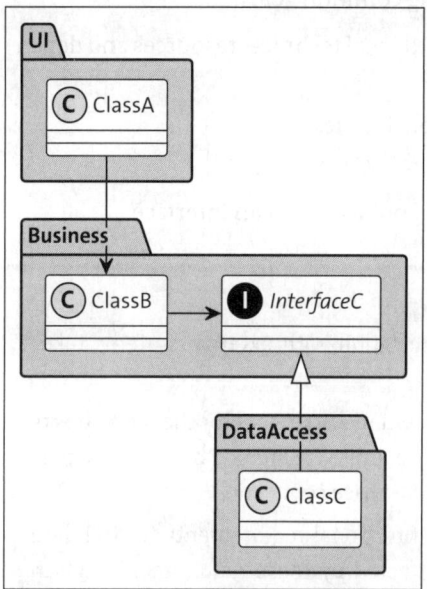

Figure 5.13 Three-Layer Architecture with Inversion of Control

By introducing business interfaces into the business logic component, a proxy object, often called a *mock object*, can easily used as a substitute for the actual implementation for automated testing via a test framework, as shown in Figure 5.14.

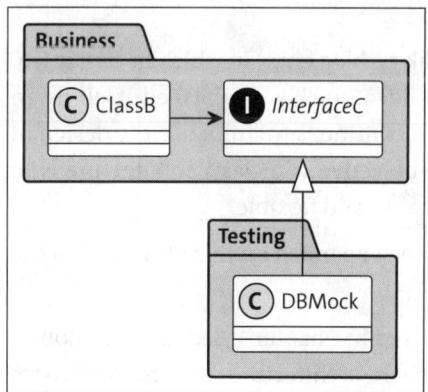

Figure 5.14 Using Mock Objects for Testing

The behavior of that mock object can be adapted individually for each test case. In our example of a database connection, this mock object eliminates the need for time-consuming test data or preconfiguration of the database and can, for example, simulate connection problems.

Alistair Cockburn's basic ideas have been expanded at various points to include additional architectural layers in the form of further rings or hexagons, such as a service layer. However, all approaches have the following common goals:

- Better decoupling of the business logic from external technical resources and dependencies
- Better maintainability thanks to clearer responsibilities
- Improved testability
- Ability to use business logic more flexibly and publish it via an interface

> **Data Mapping in a Hexagonal Architecture**
>
> In the *ports-and-adapters* architecture, communication with external systems takes place via ports.
>
> Usually, each system or technology used has its own requirements for the data objects used. In an HTTP/JSON transfer, for example, only a limited number data types are possible, and mapping information plays a role in a database connection.
>
> In both examples, these requirements are non-functional requirements for the data objects. The business logic should remain unaffected by these requirements, which means that each adapter should contain its own data model and a corresponding mapper for the transformation between the two models if the port requires a special data format. Communication via the interface must always take place with the internal data model.

5.3.4 Clean Architecture

In 2012, Robert C. Martin published a groundbreaking blog entry in which he first introduced the concept of a *clean architecture*. This idea quickly established itself as an important concept in software development and continues to influence the design of software architectures to this day. For him, the objective of a clean architecture is to design software that is well structured, maintainable, and flexible.

In his book of the same name, published in 2017, he expanded on this idea and laid the foundation for a large number of best practices.

In addition to *SOLID* principles, which were covered in Chapter 1, Section 1.4.3, Robert Martin describes the extension of existing software architectures and merged them into a consolidated view, called the clean architecture. You can use this architectural approach to create software systems that fulfill the following basic requirements:

- **Independence from frameworks**
  The software should be independent of libraries or other external software and not only work in interaction with them.

▶ **Testability**
The business logic should be able to be tested independently without relying on external dependencies, such as databases or web servers, at test time.

▶ **Independence of a presentation layer**
The presentation layer should not contain any business logic and should be easy to replace.

▶ **Independence from databases**
The same rules should apply to the persistence layer as to the presentation layer. This layer must not contain any business logic either, and the persistence technologies must be interchangeable.

▶ **Independence from external partners**
The internal business logic must remain independent of the "outside world" and external partners.

Martin used the following concepts as his basis:

▶ The *hexagonal architecture* of Alistair Cockburn
▶ The *onion architecture* of Jeffrey Palermo
▶ His own *screaming architecture*
▶ The *data, context and interaction (DCI)* architectural pattern of James Coplien and Trygve Reenskaug
▶ The *entity-control-boundary* or *boundary-control-entity (BCE)* architectural pattern of Ivar Jacobson

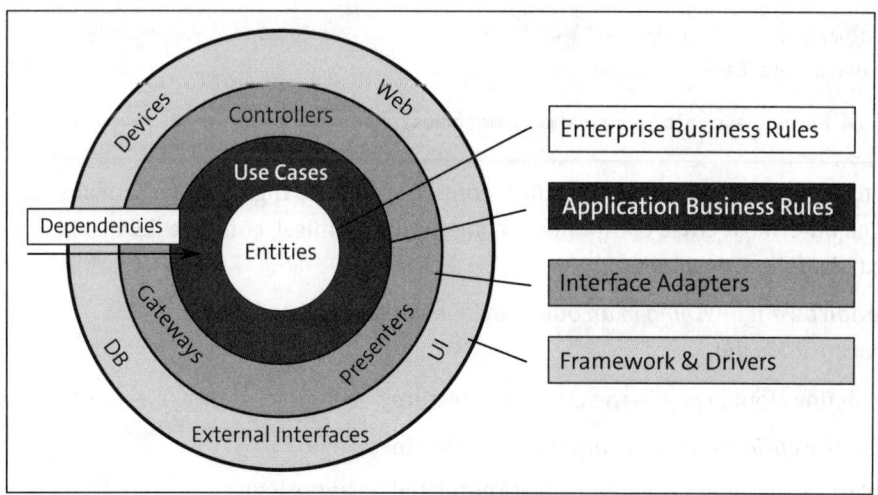

Figure 5.15 Clean Architecture According to Robert C. Martin

Martin's representation of a clean architecture, shown in Figure 5.15, consists of several concentric circles, each representing a specific area of the software. The number of circles is not fixed and can vary from implementation to implementation. However, an important part is that the *dependency rule* is always observed. This rule states that source code dependencies may only be directed from an outer layer to an inner layer, never in the opposite direction.

The inner areas therefore contain business logic that is absolutely free of dependencies on outer areas. This separation applies in particular to dependencies on data formats or structures that are defined in external areas and may be generated or specified by frameworks. In the Java area, for example, Java Persistence API (JPA) entity classes, which are used within the business logic, violate this rule because they have technical dependencies and therefore a dependency on one of the external, technical areas. Listing 5.4 shows an example of such a class.

```java
import jakarta.persistence.Column;
import jakarta.persistence.Entity;
import jakarta.persistence.Id;
import jakarta.persistence.Table;
import java.time.LocalDate;

@Entity
@Table(name = "Attendees")
public class Attendee {

    @Id
    private long id;

    @Column(name = "created_at")
    private LocalDate creationDate;
```

Listing 5.4 Entity Class with Technical Dependencies (Java)

The further an area is inside the circular representation, the higher its level of abstraction. The outermost circle components contain the technical, concrete details. In this context, the following applies:

> *"We don't want anything in an outer circle to impact the inner circles."*
> —Robert C. Martin

Martin defines four layers in the clean architecture:

- *Enterprise business rules* or company-wide business logic
- *Application business rules*, or application-related business logic
- *Interface adapters*
- *Frameworks and drivers*

Let's now take a closer look at each of these layers.

▶ **Entities**
At the core of the application are the *entities*, which comprise the relevant implementations of the company-wide defined business logic and rules. These entities can be objects with corresponding methods or pure data objects and functions.

Changes in external areas or external dependencies should have little impact on the entities in the core area, if any impact at all. Changing the database technology or adapting the presentation of a web application should have no effect on the entities.

For applications, this means that the classes and objects in the inner circle area must not have any external dependencies or references. This can be easily monitored in the source code via the required import statements: If there are import statements for other external packages or areas, the rule is violated.

Dependencies at the Core of the Application

In real life, entities could not exist without dependencies. Even the use of standard data types such as `java.lang.String` from the Java programming language standard library is already a dependency.

From a pragmatic point of view, entities will have dependencies; however, you must ensure that the dependencies you use have stable interfaces so that no breaking changes are expected. Standard libraries (and the classes or types they contain) can generally be regarded and referenced as stable. As explained earlier, technological dependencies should not be included and should therefore not be used at the core of the application.

Logging libraries often require a compromise. If no abstract, technology-independent interfaces can be used, such as those defined by *SLF4J*, you may need to rely on specific libraries and adapt your dependency rules.

▶ **Application business rules or application-related business logic**
Each application has its own business logic. All use cases of the application are encapsulated and implemented in the circle area of the *use cases*, as shown in Figure 5.15. They control the data flow with the entities in the innermost area. Changes in outer layers must not cause any changes in this area of the application. The only permissible reasons for customization are changes to the entities or the business logic on which the application is based.

▶ **Controllers, gateways, and presenters: interface adapters**
The interface adapters layer is responsible for converting the data or data formats from the inner areas to the outer areas. If, for example, a database uses a different data format than the one used in the core of the application, a suitable conversion takes place in this layer. For user interfaces, for example, this layer contains the components required to implement the *model view controller pattern* (see Section 5.4.1).

As with all other layers, no details of this layer should be passed on to the inner layers. For our earlier Java JPA example, the JPA entities would be defined in the interface adapter layer and, if necessary, converted into an inner layer format.

- **Devices, web, databases, UI, and external interfaces: frameworks and drivers**
  The outermost layer of the architecture model contains the framework or library dependencies and the implementation of the technical details. Web frameworks or database drivers are loaded and used here, which are details that are not important in the inner layers.

5.4 Patterns for the Support of Architectural Styles

Structuring an application requires consideration of several levels of abstraction. This section presents design patterns that can be implemented within the application to support the implementation of different architectural styles.

5.4.1 Model View Controller Pattern

Origin: Trygve Reenskaug—Web Presentation Patterns

The *model view controller (MVC) pattern* is one of the best known and most widely used architectural patterns in software development, especially for interactive applications with a user interface. This pattern defines a structured separation of an application into three main components:

- Model: The model contains the data and business logic.
- View: The view visualizes the data for the user.
- Controller: The controller accepts user input and controls the interaction between the model and view.

Problem and Motivation

A central challenge in the development of user interfaces lies in avoiding too close a link between user interaction, data, and business logic. This type of integration often results in source code that is difficult to maintain, difficult to test, and limited in flexibility.

Separating the code for presentation from the code for the business model is a basic principle of good software design for applications with a user interface. This separation is essential for the following reasons, among others:

- **Fundamentally different concerns of view and model**
  The model focuses on the implementation and adherence to business logic and possibly data storage in a data store, such as a database. For the presentation of the data, however, the focus is on user interaction and the best possible display. The combination of these two opposing aspects can often lead to challenges.

- **Different representations of the same data**
  If business models, business logic, and display logic are combined in a single component, you cannot use these elements in different contexts. If, for example, two user interfaces, perhaps an HTML interface and a remote procedure call (RPC) interface, need to use the same data with the same validation logic, it can be difficult to reuse the code.

- **Testability of the business logic**
  The more strongly the display logic is coupled with the business logic, the more difficult testing the actual business logic becomes because the tests require the user interface to be checked at the same time. Thus, test procedures are more complex and may require the use of interface scripting tools.

- **Maintainability**
  Customizations or enhancements to software are made more difficult by a tight coupling of different aspects since developers often must understand and modify large sections of code to implement the desired changes, which increases the risk of errors and delays in development.

Solution

The *model view controller pattern (MVC)* solves the problem by dividing software into three clearly separated components, as shown in Figure 5.16:

- Model
- View
- Controller

The business data and associated logic are represented by the *model*, which can be independently developed, tested, and used in different contexts.

The *view* component is responsible for displaying the model in the user interface. When displaying a customer object, the *view* could, for example, consist of a window with control elements or an HTML page with corresponding display elements.

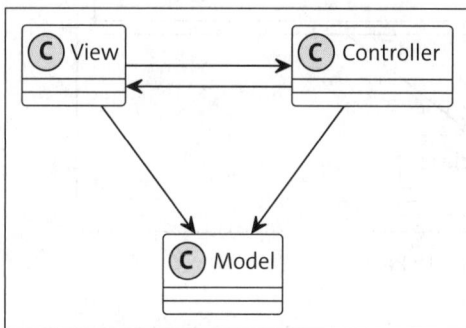

Figure 5.16 Structure of the Model View Controller Pattern

The third component is responsible for controlling the *view* and updating the *model*: the *controller*. It receives the user interactions, processes them, and updates the *view* and the *model* accordingly.

When allocating responsibilities to the individual components, the dependencies between them must be considered and adhered to correctly. The presentation, i.e., the *view*, depends on the *model*, but the *model* must not depend on the *view* in return. This allows the *model* to be used later in other presentations without having to make any adjustments. The same principle applies to the *controller*: It is also dependent on the *model*, but not vice versa.

The user interface of the application therefore consists of the two components: *view* and *controller*.

The structure and participants in this pattern are as follows:

- Model: The *model* contains the data and the business logic.
- View: The *view* contains the code required to display the *model* and accept user input in a special technology.
- Controller: The *controller* evaluates the user input and updates the *model* and the *view* accordingly.

Sample Solution

Ruby on Rails is a framework that supports developers in building server-side, data-driven web applications. This framework is based on the Ruby programming language and the MVC pattern.

In this application, we want to display a list of blog articles on a website. After initializing a new Rails project, a controller, a view, and a model can be created using the `rails` command-line tool. The components and their storage location within the application are shown in Figure 5.17.

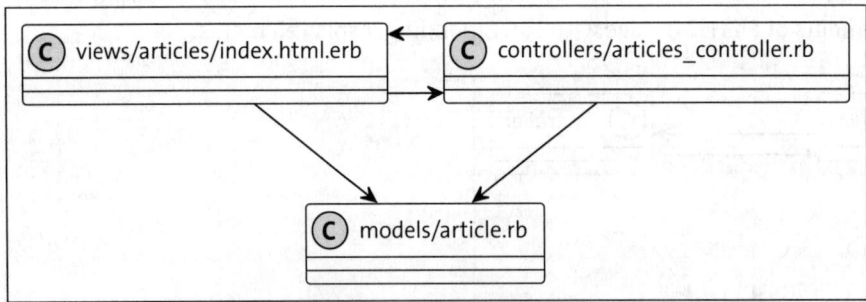

Figure 5.17 Structure of the Sample Application with Rails

The command shown in Listing 5.5 creates a new *model* for the `Article` object containing two attributes: `title` and `body`, with the given data types. The resulting file is saved as `models/article.rb`.

```
./bin/rails generate model Article title:string body:text
```

Listing 5.5 Creating the Article Model

At the same time as the new *model* file is created, a database script is automatically created for updating the corresponding database structure for the `Article` objects. This database script can be applied using the `rails` command line tool, as shown in Listing 5.6.

```
./bin/rails db:migrate
```

Listing 5.6 Rails Command for Updating the Database

Once the two commands have been successfully executed, a *model* object is available in which the business logic can be implemented and which already contains functionality for using the database.

You can use command-line tools to execute a *console* in *Grails*, which enables you to work directly with these objects.

If new database entries need to be created and stored in the database, you can use the highlighted commands shown in Listing 5.7.

```
$ ./bin/rails console
Loading development environment (Rails X.X.X)
mvc-sample(dev)> article = Article.new(title: "Hello Rails", body: "Bla")
=>
#<Article:0x000079effc35dc50
...
mvc-sample(dev)> article.save
  TRANSACTION (0.5ms)  begin transaction
  Article2 Create (11.4ms)  INSERT INTO "article2s" ("title", "body", "desc",
"created_at", "updated_at") VALUES (?, ?, ?, ?, ?) RETURNING "id"  [["title",
"Hello Rails"], ["body", "I am on Rails!"], ["desc", nil], ["created_at",
"2024-12-06 13:55:08.412293"], ["updated_at", "2025-04-05 13:55:08.412293"]]
  TRANSACTION (11.2ms)  commit transaction
=> true
```

Listing 5.7 Commands Used to Create Article Objects in the Database

In this example, note how the business logic of the `Article` object is encapsulated and can already be used via the *console* without an explicit user interface. The *model*, as defined by the pattern, is decoupled from the user interface.

You can use the command shown in Listing 5.8 to create an HTML controller skeleton with an `index` function for the output of `Articles`.

```
./bin/rails generate controller Articles index
```

Listing 5.8 Command for Creating a New Controller Using Rails

This command creates a suitable *controller* and a *view* component, which still needs extension. For data to be displayed, the already generated `Article.all` function must be called within the `index` function of the controller. The entire implementation of the controller is shown in Listing 5.9.

```
class ArticlesController < ApplicationController
  def index
    @articles = Article.all
  end
end
```

Listing 5.9 Controller Implementation in Ruby

The HTML page, which uses *Embedded Ruby* for template control in addition to the HTML markup, takes over the display. Listing 5.10 shows the customized view implementation.

```
<h1>Articles</h1>

<ul>
  <% @articles.each do |article| %>
    <li>
      <%= article.title %>
    </li>
  <% end %>
</ul>
```

Listing 5.10 HTML Page for Displaying the Articles Using Embedded Ruby

An extension for creating new `Article` objects via the HTML interface would be carried out in the following way:

- You create a method within the `ArticleController` that uses the functionality of the `Article` model to save the data to the database.
- You add a button or link to the view component, which calls up the corresponding controller method and performs the action when it is pressed. At the end of the save process, the model would be updated, and the display must be regenerated by the controller by redirecting or reloading the page.

When To Use the Pattern

Some examples of when to use the pattern include the following:

- **Separation between display logic and business logic required**
  With the MVC pattern, the display logic can be separated from the business data and its logic so that the business components can be reused. A separation should always be carried out for non-trivial applications so that the expandability, testability, and so on of the application is not restricted.

- **Complex logic in the business objects**
  Complex business logic within objects should always be separated so that adjustments to the business logic do not affect the display, and the components can be reused.

- **Improvement of testability in the frontend desired**
  Separation of presentation and business logic enables the separate testing of responsibilities.

Consequences

The use of the pattern has the following consequences:

- The application is easier to maintain.
- Tests of the business logic can be created without simultaneous tests of the display logic.
- Business objects and business logic can be used in different contexts, for example, in multiple interfaces, without adjustments.

5.4.2 Model View ViewModel Pattern

The *Model View ViewModel (MVVM) pattern* is a special variation of the *MVC pattern* and is often referred to as the *presentation model pattern*.

The MVC pattern and the MVVM pattern differ in how the logic, data, and user interface interact with each other. While the MVC pattern is often used in desktop applications, the MVVM pattern is mostly used in web applications and web frameworks, for example, in *Angular*.

Problem and Motivation

In many applications, status information supplements the business data and logic in order to display the user interface correctly or more clearly. This information is used, for example, to dynamically control the detail view of data records or to validate the text entered in an input field and store it in a model.

Integrating this status information into the business objects or the *model* in the MVC pattern unnecessarily increases the coupling between the model and the view and mixes up the actually separate responsibilities.

Solution

The *Model View ViewModel pattern* uses special models for the view components, which are called *ViewModel* objects. These objects are linked to the view by means of *value binding*. Thus, the view components monitor changes to the ViewModels and display them directly. The individual fields in the ViewModel are "bound" to the display components in the view accordingly.

In contrast to the *model view controller pattern*, in which the controller has a dependency on the view and updates it, in the *MVVM pattern* the view monitors the ViewModel, and the value binding performs the update. Thus, ViewModel is more independent and can be used for different views. You can also easily combine multiple ViewModels in a new view.

The business logic and the data of the model objects are referenced and used within the ViewModel. If actions must be executed from the view, they fall back on the functionality that is implemented in the ViewModels. Figure 5.18 shows the structure.

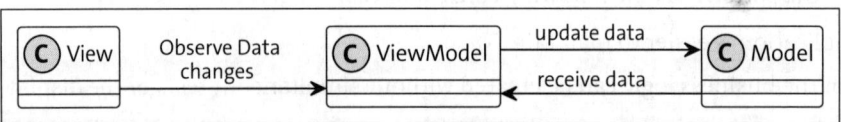

Figure 5.18 Structure of the Model View ViewModel Pattern

The structure and participants in this pattern are as follows:

- Model: The *model* contains the data and the business logic.
- View: The *view* is the display component that accesses the *ViewModel* via *data binding* to display data. The view also executes actions via the *ViewModel*.
- ViewModel: The *ViewModel* is a *model* developed specifically for the display, which can also use multiple business models and is integrated into *views* via *value binding*. In addition to the data, the ViewModel provides the view with actions that can be executed.

Sample Solution

Angular is a client-side web framework developed by Google that is based on *TypeScript* and can be used to create dynamic *single page applications (SPAs)*. Although *Angular* is not a strict MVVM framework, it integrates the concepts of the MVVM pattern into its architecture.

In Angular, data and business logic are bundled in *services*. These services take on tasks such as API calls or validations and form the *model* for the implementation of the MVVM pattern. The *view* is represented in Angular via templates. These templates are HTML pages with Angular-specific code for value bindings and other elements for logic control, such as `if` statements.

The *ViewModel* is implemented using *component classes*. In addition to all the data to be displayed in the view, these classes contain the logic required for the display.

In the sample application, messages that are retrieved from a server should be displayed in a view in the browser. Within that view, you can use a simple button to control whether the message should be displayed (as shown in Figure 5.19) or not (as shown in Figure 5.20).

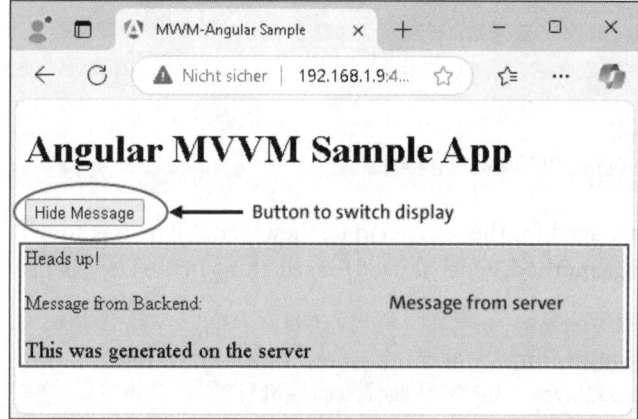

Figure 5.19 Server Message Displayed

Figure 5.20 Message Not Displayed

The application uses two views that are nested inside each other. The message from the server is displayed in the inner view (`MessageView`); the outer view (`AppView`) contains the logic for showing and hiding the inner view. Listing 5.11 shows the implementation of the outer `AppView` with the logic for showing and hiding. The `app-message` HTML tag references the inner `MessageView` and is only included or displayed if the `showMessage` variable contains the `true` value. The value can be changed by clicking on the button.

```
<main class="main">
  <h1>{{title}}</h1>
  @if (!showMessage) {
  <button (click)="toggleMessage()">Show Message</button>
  } @else {
  <button (click)="toggleMessage()">Hide Message</button>
  }
  @if (showMessage) {
  <app-message></app-message>
  }
</main>
```

Listing 5.11 Outer View: App Component View (Angular)

The `showMessage` variable is managed for the `AppView` in its ViewModel, which is shown in Listing 5.12. The `toggleMessage` method, which is used for switching in the view, is also defined.

```
import {Component} from '@angular/core';
import {MessageComponent} from './message/message.component';

@Component({
  selector: 'app-root',
  imports: [MessageComponent],
  templateUrl: './app.component.html',
  styleUrl: './app.component.css'
})
export class AppComponent {
  title = 'Angular MVVM Sample App';
  showMessage = false

  constructor() {
  }

  toggleMessage() {
    this.showMessage = !this.showMessage;
  }
}
```

Listing 5.12 ViewModel for AppView (Angular)

The implementation of the message component consists of the following individual parts:

5.4 Patterns for the Support of Architectural Styles

- Service (`BackendService`) for accessing the backend (model), as shown in Listing 5.13
- ViewModel (`MessageComponent`) for the logic or for communication with the service, as shown in Listing 5.14
- View for displaying the message, as shown in Listing 5.15

```
import {HttpClient} from '@angular/common/http';
import {Injectable} from '@angular/core';
import {Observable} from 'rxjs/internal/Observable';
import {BackendMessage} from './backendMessage';

@Injectable({
  providedIn: 'root'
})
export class BackendService {

  constructor(private http: HttpClient) {
  }

  getMessage(): Observable<BackendMessage> {
    return this.http.get<BackendMessage>("http://...")
  }
}
```

Listing 5.13 Service Implementation (Model)

```
import {Component} from '@angular/core';
import {BackendService} from '../backend.service';

@Component({
  selector: 'app-message',
  imports: [],
  templateUrl: './message.component.html',
  standalone: true,
  styleUrl: './message.component.css'
})
export class MessageComponent {

  message: string | undefined

  //DependencyInjection
  constructor(private service: BackendService) {
    service.getMessage().subscribe(body => {
      this.message = body.message;
```

      });
   }
}

Listing 5.14 Message Component as ViewModel

```
<div class="message">
  <p>Heads up!</p>
  <p>Message from Backend:</p>
  <p class="backend">{{message}}</p>
</div>
```

Listing 5.15 HTML View for Displaying the Message

An overview of all components shown in Figure 5.21.

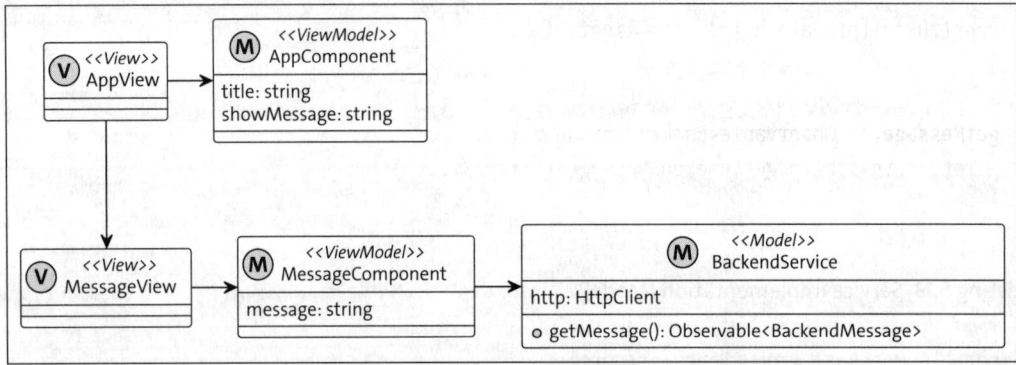

Figure 5.21 Overview of Angular Components

When To Use the Pattern

Some examples of when to use the pattern include the following:

- **State management for the display**
  If the display is not only responsible for displaying data from business models, but the state of the display itself also needs to be managed, these states can be saved in the ViewModel and not included in the actual business model.

- **Reactive user interfaces**
  The pattern and the corresponding frameworks are particularly suitable for the creation of reactive user interfaces if changes need to be displayed or updated almost in real time. The data binding between view and ViewModel makes synchronization much easier.

- **Reusable user interfaces**
  As the view has no dependencies on the ViewModel, both can be reused more easily and independently of each other. Most MVVM frameworks take advantage of this

independence and provide extensive libraries for common user interface components. This can simplify and thereby accelerate the development of your own applications.

▶ **Complex user interactions required**
The MVVM pattern is ideal for complex applications with many user interactions and complex processing steps. The ViewModel takes over the logical processing of these actions, separating it cleanly from the user interface.

Real-life examples include the validation of user information and the dynamic adjustment of input fields.

Consequences

The use of the pattern has the following consequences:

▶ A clear separation of responsibilities makes it easier to maintain the application.
▶ The individual components can be tested more easily since there are fewer dependencies.
▶ The reusability of the components is increased.
▶ The *two-way binding* allows changes in the ViewModel to be displayed directly in the view, and vice versa, ultimately resulting in more responsive and user-friendly applications.
▶ The complexity of the application increases.
▶ Without the use of a framework, the effort is high, and by using a framework, you bind yourself to it.

5.4.3 Data Transfer Objects

Origin: Fowler: Patterns of Enterprise Applications—Distribution Patterns

You can use the *data transfer object (DTO) pattern* to decouple and optimize the data structure that is exchanged between different layers of an application or via network interfaces from the underlying business logic. This simplifies the transmission and serialization in common formats such as JSON or XML.

Problem and Motivation

Interfaces should be customized according to the *interface segregation principle*, described in detail in Chapter 2, Section 2.3.4, and only provide data and information that is relevant for the respective client according to the *information hiding* principle. For invoicing software, for example, data on the size of an item is probably of secondary importance, while this data will certainly be of greater significance for logistics software.

As a result, not all the attributes of a business object need to be transferred. In some cases, a complete transfer is even impossible since business objects form complex, interwoven dependencies that cannot simply be serialized in their entirety.

The direct use of business objects for data transfer would also mean that their class definition would also need to be available on the client. The result would be a tight coupling between the clients and the internal business logic of the server, which should be avoided as well.

In addition, remote calls can result in a longer response time due to the transmission and processing times on the client and server sides. If a client requires multiple data records, multiple separate calls often have to be made to the server, which leads to increased resource requirements and a potentially noticeable delay for the user.

Solution

When you use the *DTO pattern*, only lean and flat data structures are exchanged, instead of complex objects, and these data structures contain only the information required by the client. These objects are exchanged between the individual layers of an application.

The objects do not contain any logic but merely serve as data containers with fields and associated getter and setter methods, which simplifies the serialization in formats like JSON, XML, or Protobuf (Protocol Buffers) and optimizes data transfers.

Because clients cannot access the internal data models of the called layer directly, they cannot manipulate its data directly either. The DTO pattern thus protects the internal data models from unauthorized access and changes by external clients, which increases the maintainability and flexibility of the business logic and the application.

Transfer objects are often created using code generators. In Java, for example, you can use the *Project Lombok* library to quickly create DTOs. An annotation on a class can be used to automatically generate access methods, as shown in Listing 5.16.

```
import lombok.Setter;
import lombok.Data;
import lombok.ToString;

@Data
public class DataExample {
  private final String name;
  private double score;
  private String[] tags;

  @ToString(includeFieldNames=true)
  @Data(staticConstructor="of")
  public static class Exercise<T> {
    private final String name;
    private final T value;
```

        }
    }

Listing 5.16 Using Project Lombok to Create a DTO (Java)

Since Java 14, you can also use what are called *Java records*, as shown in Listing 5.17. This special type of class can define unchangeable data objects in a simple and concise way. They reduce the *boilerplate code*, but as an integrated feature of the language, and can replace Lombok in this area.

```
public record DataExample(String name, double score, String[] tags) {
}
```

Listing 5.17 Sample Java Record

Sample Solution

In a *TypeScript*-based application, the internal application data model consists of a user object that contains many fields that should not be transferred directly to the client. These fields include, for example, the change timestamps and the user's password hash, as shown in Listing 5.18, as well as references to the address and the orders placed.

```typescript
// UserEntity.ts
export class UserEntity {
  id: number;
  username: string;
  email: string;
  passwordHash: string; // Sensitive data
  createdAt: Date;
  updatedAt: Date;
  orders: Order[];
  address: Address;
}
```

Listing 5.18 Class of a User with Internal Information (TypeScript)

An extract from this data, which is stored in a database, can be made available to external clients via a REST interface. For this purpose, a DTO is created that only contains the fields that are actually to be sent to the clients. Listing 5.19 shows such a DTO that only contains the user name and email address.

```typescript
// UserDTO.ts
export class UserDTO {
  username: string;
  email: string;

  constructor(username: string, email: string) {
    this.username = username;
```

```
    this.email = email;
  }
}
```

Listing 5.19 DTO for Transfer to the REST Interface

Within a frontend controller, the data is loaded from the database via the service and copied into the DTO object so that it can be delivered to the clients. Listing 5.20 shows the corresponding implementation.

```
// UserController.ts
import { UserService } from './UserService';

export class UserController {
  private userService = new UserService();

  getUserResponse(id: number) {
    const user = this.userService.getUserById(id);
    const userDTO = new UserDTO(user.username, user.email);
    return {
      status: 'success',
      data: userDTO,
    };
  }
}
```

Listing 5.20 Frontend Controller for Delivering the Data

Accordingly, the client receives an object with the fields relevant to it, as shown in Listing 5.21.

```
{
  "status": "success",
  "data": {
    "username": "john_doe",
    "email": "john@example.com"
  }
}
```

Listing 5.21 Results Document Transferred to the Client

When To Use the Pattern

Some examples of when to use the pattern include the following:

- **Distributed systems with various interfaces**
  In distributed applications, data transfers and the serialization of business objects

needs to be enabled and optimized. The DTO pattern enables each service to transfer only the required data to the next service and prevents unnecessary or sensitive data from being passed on.

- **Larger systems**
  In smaller applications, the DTO pattern is often regarded as a developmental and operational overhead since additional work is required to manage the DTOs and since objects must be copied into each other at runtime. However, DTO makes sense in larger applications if the business logic can be decoupled from the transfer and the business objects need to offer more complex functions than simple read and write operations. However, even smaller applications can benefit from the improved separation of data models and remain expandable and maintainable.

Consequences

The use of the pattern has the following consequences:

- The business logic is decoupled from the data transfer.
- Data transmission in a distributed environment can be optimized by combining multiple calls into one overall call.
- The serialization of objects is simpler since no complex business data structure needs to be transferred.
- *Information hiding* and the interface segregation principle are adhered to.
- Code may need to be maintained twice in DTOs and business objects because both have similar attribute and field definitions.
- The conversion between objects can be complex and error prone.
- Performance issues due to the conversion between the business and transfer objects are possible, which can be a real problem, especially with large amounts of data.

5.4.4 Remote Facade Pattern

Origin: Fowler: Patterns of Enterprise Applications—Distribution Patterns

The *remote facade pattern* is an architectural pattern that abstracts and simplifies the communication between an application and a remote system.

Problem and Motivation

When domain logic is implemented within an application, fine-grained objects are often created that are put together by composition and communicate intensively with each other. Dividing the application into multiple components in this way and the associated decoupling of responsibilities provides the advantage of improved maintainability and flexibility. The components are easier to understand, maintain, and modify. Thus, an application can be adapted more quickly to changing requirements without having to revise large parts of the code. However, if these objects are used across a process boundary, intensive communication via an external interface can have a negative impact on the performance of an application. Each individual request between the processes leads to longer response times and therefore to lower overall performance.

Figure 5.22 shows a sequence diagram of an application that communicates with a domain object via a remote interface. Each field of the object is read or updated individually via a remote call.

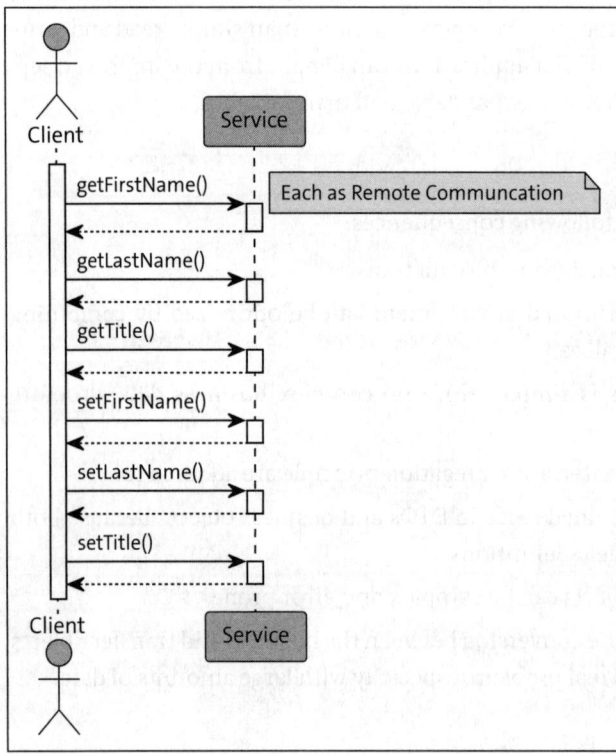

Figure 5.22 Remote Communication Using Individual Calls

Solution

The definition of a coarse-grained interface for object calls from remote clients reduces the number of necessary calls and would thus solve the problem of the accumulating execution times of remote requests. However, using this interface would affect the method calls as well as affect the called objects themselves: Instead of individual methods for retrieving and setting values, these objects must provide a customized interface, which would dissolve the finer structure of the objects and result in code that is difficult to maintain.

The *remote facade pattern* introduces a *remote facade*, which you can use to translate a coarse-grained interface into an underlying fine-grained object structure. It does not contain any additional logic.

In this case, the *separation of concerns* principle comes into play in that different responsibilities are separated: Complex business logic is implemented in the domain objects, while the code that enables efficient remote access is encapsulated in the facade.

Our earlier example changes with the introduction of a *remote facade*, as shown in Figure 5.23. The client now only uses the two coarse-grained methods: getData and setData. When these *bulk accessor methods* are called, the facade collects the corresponding data from the domain objects and delivers them collectively or writes them back to the object. This process can be achieved using a *DTO*, as described earlier in Section 5.4.3, for example.

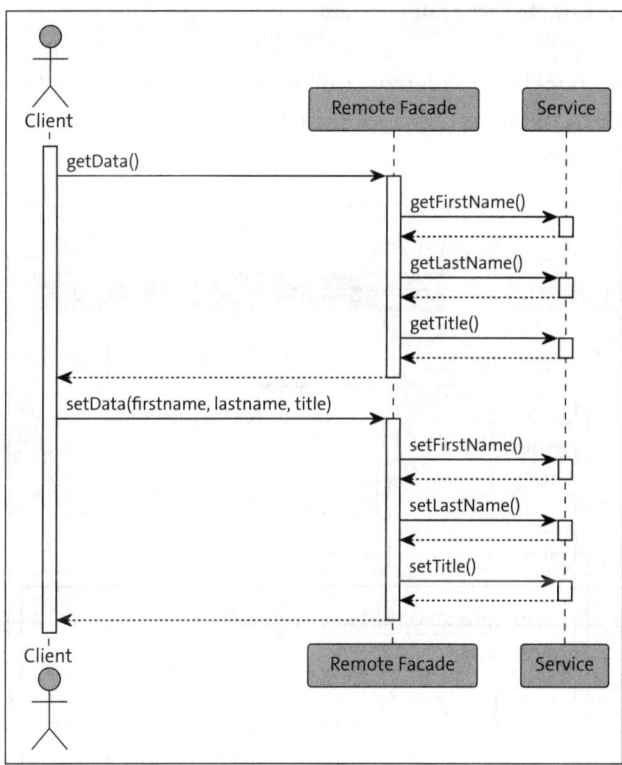

Figure 5.23 Introducing a Remote Facade

Although the remote facade interfaces for communication with the client could be designed to be stateful, a stateless design is recommended to enable a more scalable and fault-tolerant implementation on the server (see Chapter 7, Section 7.5).

Sample Solution

In this final example, let's say a tour operator uses software involving multiple services for different tasks within a booking process. Now, a client must use the individual services via the *remote facade* to make a hotel booking.

The class diagram of the application is shown in Figure 5.24. The application comprises of the `BookingFacade` class and three interfaces for the services used, namely, `UserService`, `HotelService`, and `PaymentService`.

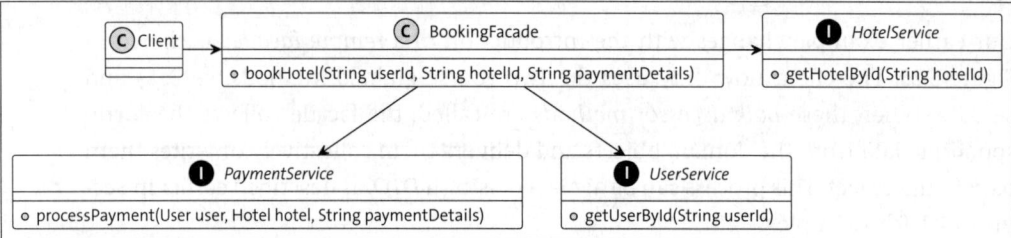

Figure 5.24 Class Diagram for Our Remote Facade Application

A client should be able to make a booking via a remote call, as shown in Figure 5.25. The `BookingFacade` therefore provides a rough interface for the client. This interface can be accessed via gRPC or a REST call, among others.

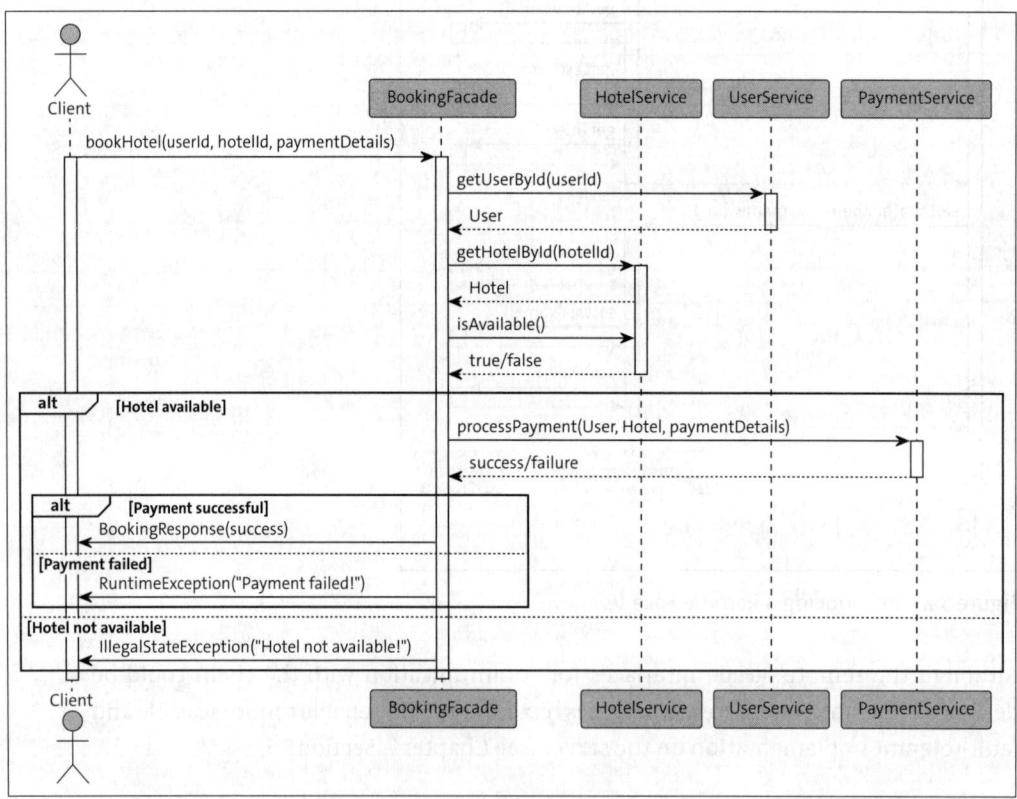

Figure 5.25 Sequence Diagram for Our Remote Facade Application

The facade can be implemented in the same way, as shown in Listing 5.22.

```java
public class BookingFacade {

    private final HotelService hotelService;
    private final UserService userService;
    private final PaymentService paymentService;

    public BookingFacade(HotelService hotelService,
                         UserService userService,
                         PaymentService paymentService) {

        this.hotelService = hotelService;
        this.userService = userService;
        this.paymentService = paymentService;
    }

    // Method for booking a hotel
    public BookingResponse bookHotel(String userId,
                                     String hotelId,
                                     String paymentDetails) {
        // 1: Check user
        var user = userService.getUserById(userId);
        if (user == null) {
            throw new IllegalArgumentException("User not found!");
        }

        // 2: Check hotel availability
        var hotel = hotelService.getHotelById(hotelId);
        if (!hotel.isAvailable()) {
            throw new IllegalStateException("Hotel is not available!");
        }

        // 3: Make payment
        var paymentSuccess =
                paymentService.processPayment(user, hotel, paymentDetails);
        if (!paymentSuccess) {
            throw new RuntimeException("Payment failed!");
        }

        // 4: Confirm booking
        var response = new BookingResponse();
        response.setUser(user);
        response.setHotel(hotel);
        response.setStatus("Booking successful!");
```

```
        return response;
    }
}
```

Listing 5.22 Implementation of the Facade

The client code uses the *remote facade* to execute the action, as shown in Listing 5.23.

```
public class BookingClient {

    private BookingFacade bookingFacade;

    public void book() {

        // Hotel booking via facade
        String userId = "123";
        String hotelId = "456";
        String paymentDetails = "Credit card: 1111-2222-3333-4444";

        try {
            BookingResponse response =
                bookingFacade.bookHotel(userId, hotelId, paymentDetails);
            System.out.println(response.getStatus());
        } catch (Exception e) {
            System.err.println("Error: " + e.getMessage());
        }
    }
}
```

Listing 5.23 Client Implementation

When To Use the Pattern

Some examples of when to use the pattern include the following:

▶ **Avoidance of complex and extensive communication**
  The pattern is suitable for simplifying complex and extensive communication between a client and a service by providing a coarse-grained interface for the interactions.

▶ **Minimization of data transmission**
  Each remote call increases the response time and the volume of data to be transmitted by a system. The pattern can be used to optimize transfers and make them more efficient.

- **Summarizing common functionalities**
  Many remote interfaces share overarching requirements such as logging or security checks. These *cross-cutting concerns* can be centrally managed and consistently provided by introducing a *remote facade*.

Consequences

The use of the pattern has the following consequences:

- The development of a client is easier because the client can access a customized interface and does not need to know all the details about the business objects, just the most relevant details for that client.
- Changes to business objects are separate from the remote interface and can be made independently.
- All complex processes can be brought together in a single place and managed jointly.
- Data transfers and interactions between systems are optimized.
- Extensive facades can result in load problems since the individual calls can no longer be routed to multiple remote service instances.

Chapter 6
Communication Between Services

The exchange of information is essential in distributed systems. Only with communication between different applications can more complex use cases be mapped. This chapter explores different communication styles between systems and describes proven patterns that you can use to increase communication resilience and minimize sources of error.

Very few software applications are executed completely independently and in isolation. Almost every application needs to exchange data with other systems for a variety of reasons. For example, an online store must query information from the inventory system to display current availability, or a software delivery platform must be supplied with the latest software versions developed.

Communication between different applications or systems makes it possible to implement complex use cases whose individual steps extend across multiple systems and which could not be fully implemented in a single system.

With the internet, especially since the emergence of the microservices architecture style and the use of cloud-based solutions, communication between different systems and applications has become much more intensive and complex compared to the monolithic application structures that were common before and deployed on premise.

Microservices rely on a large number of small, independent services that communicate with each other and exchange information via defined *application programming interfaces (APIs)*. Such a distributed architecture requires increasingly efficient and reliable mechanisms for data transfer and for synchronization between the individual components. In general, this architectural style leads to a constantly growing number of interactions.

Every interaction or integration of a system usually takes place via a network and always faces the same challenges, such as the following:

- **Networks are unreliable.**
  Communication via a network has a much higher error potential than local data transmission, as the data must be redirected via multiple components (such as switches, firewalls, or proxies). Each component involved can cause a delay or interruption in transmission and thus make the entire communication unreliable. The more components are involved in the data exchange, the less reliable the transmission becomes.

- **Networks are slow.**
  Compared to local method calls or inter-process communication between processes on a computer, network calls are extremely slow. This slowness is not only due to slow transmission speeds through the network, but usually also due to the necessary serialization or deserialization of the data to be transmitted. If distributed applications are designed in the same way as locally communicating applications, poor processing speeds might plague an application.

- **Applications and environments are different.**
  Various technologies are usually used in distributed environments, such as different programming languages, operating systems, or data formats. Even when in-house software development is based on a homogeneous technology stack, in most cases, external systems must be connected within or outside the company boundaries. Every integration or communication must therefore be adapted to or designed for different technologies.

- **Change is unavoidable.**
  Changes to business or technical requirements often require the adaptation of a deployed component. This adaptation may be a new or updated technical function or an update of the framework used, for example. These internal changes can in turn affect the external communication mechanisms of an application and result in adjustments there too. For this reason, the dependencies between the individual systems should be minimized as far as possible. The goal when setting up communication connections should be to couple them as loosely as possible so that adapting an application does not trigger a chain reaction of changes in multiple applications.

You can explore various communication styles or solution patterns to overcome challenges in data transmission and integration. In addition to traditional integration methods such as file transfer or the use of a shared database, direct synchronous calls or messaging are often used today.

Regardless of the technology or communication style, data transmission between different services or components should always be designed in such a way that it functions reliably and stably. In addition, the services should be as loosely coupled as possible to ensure a high degree of flexibility and adaptability in the system.

Accordingly, this chapter looks at different interaction models and their strengths and weaknesses. Furthermore, I'll present established patterns for ensuring stable communication to overcome the challenges mentioned earlier.

This chapter is divided into the following four sections:

- Styles of application communication (Section 6.1)
- Resilience patterns (Section 6.2)
- Messaging patterns (Section 6.3)
- Interface versioning patterns (Section 6.4)

In the following sections, I will introduce you to various patterns and solutions for each area.

6.1 Styles of Application Communication

Various communication styles, also known as *protocols* or *interfaces*, enable data exchanges and interactions between software applications. The choice of the appropriate style depends on the specific requirements and circumstances of the systems involved and should be selected carefully.

The following criteria can be decisive in this context:

- **Coupling between applications**
  Tightly coupled systems are based on common *assumptions* that are necessary for data transmission. These assumptions may, for example, relate to data formats, the structure of the data, or the protocols used. These assumptions and the resulting direct dependencies between the systems can result in simple, high-performance data transfers. However, a tight coupling of the systems reduces flexibility, and changes in one system can result in further changes in the dependent system. For this reason, changes in a system can, in the worst case, lead to an interruption or failure of the data transmission. In most cases, direct and strong dependencies between systems should be avoided and replaced by a loose coupling. Interfaces should be defined as specifically as necessary, but also as generally as possible.

- **Data format**
  The communicating applications must agree on a suitable data format. Ideally, this format should be a common, standardized data format that can be processed in both systems. In some systems, however, only a few data formats are supported, and thus the interfaces that can be used are also limited. Although XML or JavaScript Object Notation (JSON) would be desirable, some host applications can only input and output comma-separated values (CSV) text files, for example.

- **Communication speed**
  Data should always be exchanged between different systems as quickly as possible to ensure consistent data. The more regularly or more promptly data is exchanged, the lower the risk. Systems might immediately exchange many small data packets, for example, through many individual HTTP queries, which can cause inefficient data transmissions and negatively impact the entire environment.

- **Data or functions**
  Either data or functions can be shared or used between systems. Data is transferred explicitly, while certain tasks are carried out in another system when functions are shared. The use of functions often provides a higher level of abstraction and leads to less dependency between the systems, as they hide details and provide only one interface for the call.

▶ **Scalability and availability**
In corporate environments, data is often not only required in one target system but must be distributed to multiple systems. Information that a new item will be sold in a store, for example, can be relevant to the store itself, to accounting, or to warehouse management. If the data is required in multiple systems, the communication styles used should be sufficiently flexible to accommodate additional receivers. If an error occurs in the system, communication styles could potentially also ensure that the data is still available for the various systems.

The following sections describe different communication styles and highlight their strengths and weaknesses.

6.1.1 Synchronous Communication

Synchronous communication is often used to connect multiple systems with each other. It is easy to implement and offers the advantage of providing an immediate response to a request. In a synchronous communication scenario, a client calls a server or service and waits for its response. In the meantime, the client blocks its further processing.

As shown in Figure 6.1, for example, the client calls a service synchronously and waits until the service has completed its work by another synchronous call of the Backend component. The user receives direct feedback as to whether the call was executed with or without errors.

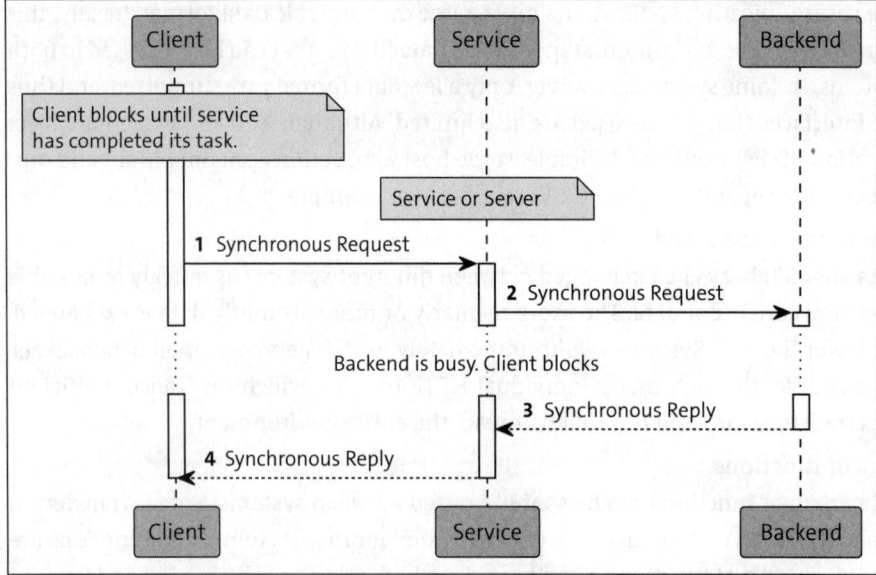

Figure 6.1 Synchronous Communication

The use of synchronous communication lends itself to applications with user interactions or API calls that expect a direct and immediate response.

Some benefits include the following:

- Direct, immediate feedback on the success or failure of the call
- Simple interaction

Some disadvantages include the following:

- Blocking calls force the client to wait.
- With long call chains across multiple components, the possible sources of error are multiplied.

Some examples include the following:

- HTTP/REST
- gRPC
- SOAP

6.1.2 Asynchronous Communication and Messaging

In contrast to synchronous communication, in asynchronous communication, a client sends a message and continues processing without waiting for an immediate response to its message. Thus, the client can perform other tasks while the message is being processed by the receiver (even with a time delay). A possible response from the receiver can also be created and transmitted back to the sender.

A special type of asynchronous communication is the concept of *messaging*, as shown in Figure 6.2, in which a central intermediary can take over the receipt and delivery of messages. In modern architectures, this central intermediary is a *messaging system*. Other possible alternative implementations of messaging-based communication are presented in Section 6.3.

When using a messaging system, for the client, the call ends with the successful transmission of the message. It can then complete other tasks without having to wait for a response. The messaging system then delivers the message to the receiver (in our example, to the Service component), which then starts processing it. Figure 6.2 shows this process.

> **Messaging and Messaging Systems**
> *Messaging* refers to a special type of asynchronous communication in which a central intermediary—a *messaging system* or a *message-oriented middleware (MOM)*—takes over the transmission of messages.

6   Communication Between Services

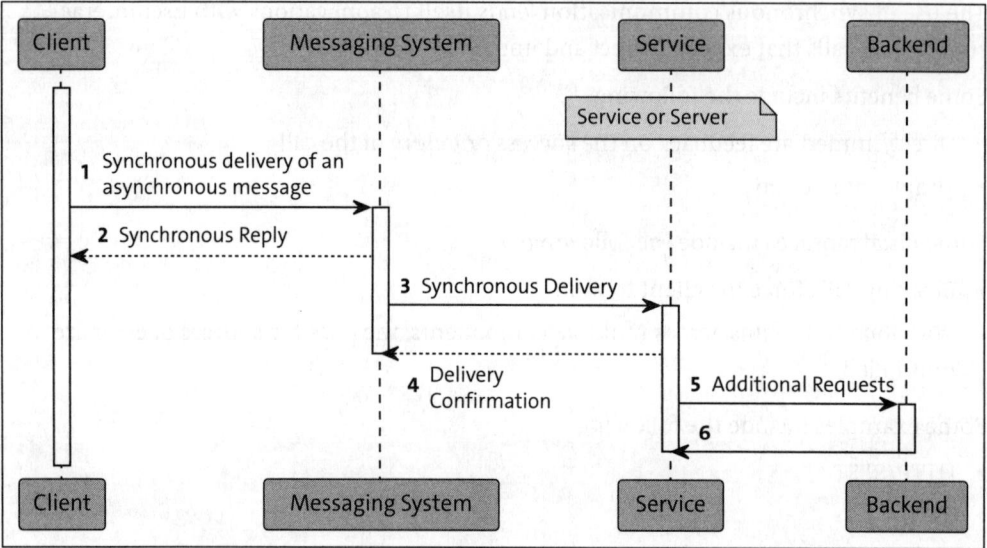

Figure 6.2 Asynchronous Communication

> **Responses in Asynchronous Communication**
>
> *Asynchronous communication* is characterized by the fact that the sender does not receive an immediate response to its message. This scenario doesn't always make sense, however, and in most cases, a response is expected from the called system—especially if the message is a request, a question, or a task.
>
> With the *request-reply mechanism* and a correlation identifier (for more information on this topic, see Section 6.3.3), this type of communication can be established with a response, and asynchronous communication can be extended.

Asynchronous communication is suitable for the exchange of information or messages between systems in which a time delay in processing is acceptable or in which the processing of messages must be decoupled with respect to time.

Examples of asynchronous communication include batch processing or APIs that perform large database queries in the background and do not send the result to the client until processing is complete.

You can use messaging solutions based on asynchronous communication to make the exchange of messages more robust and flexible. For example, messages can be delivered to multiple receivers or temporarily stored by the messaging system so that processes can be repeated or, if a communication partner is not available, reliably carried out.

Some benefits include the following:

- There are no blocking calls within the client.
- The communication partners messaging each other don't need to be available at the same time and don't need to be directly connected to each other. A decoupling takes place.
- Messaging solutions often provide mechanisms for caching messages so that they are not lost even if a system fails. Messages can therefore be processed later, as soon as the system is available again.
- The buffering and caching of messages during messaging can also reduce loads on processing systems by processing messages with a time delay and sequentially. For synchronous calls, these messages must be processed directly. In asynchronous scenarios, for example, the receiver can determine the processing speed and adopt parallelism.
- Messages can be delivered flexibly to multiple receivers using a messaging solution.

Some disadvantages include the following:

- Asynchronous communication is generally more complex and more difficult to implement, as additional concepts must be considered and implemented, such as parallel processing, time decoupling, and the sequences to be adhered to in processing or repetition logic.
- When using a messaging solution, the complexity of the infrastructure used increases. New components must be integrated and managed.
- In asynchronous communication, errors may only occur with a time delay and must be reported back to the caller. In addition, mechanisms for repeating failed processing operations must be implemented.
- The messages are not processed immediately, and a consistent state of the overall system is therefore only guaranteed at a later point in time.
- The sequence of message processing can be a challenge with parallel processing.

Some examples include the following:

- Event-driven actions
- Messaging
- Batch processing
- Webhooks

6.1.3 Streaming

Streaming refers to the continuous transmission of data between systems. In contrast to request-response-based communication, in which the entire data volume is transmitted at once, streaming transmits data packets continuously in smaller units. This approach is particularly useful if large amounts of data or real-time data must be transferred or if the latency between the request and the first processing of the data is to be reduced. In streaming-based communication, the client rarely knows the final amount of data that will be transmitted by the server.

The example shown in Figure 6.3 is a sequence diagram in which a client starts communication via streaming and is then continuously supplied with new data packets by the server. The client can process the data immediately and doesn't need to wait until all the data is available on the server or has been transferred from the server to the client.

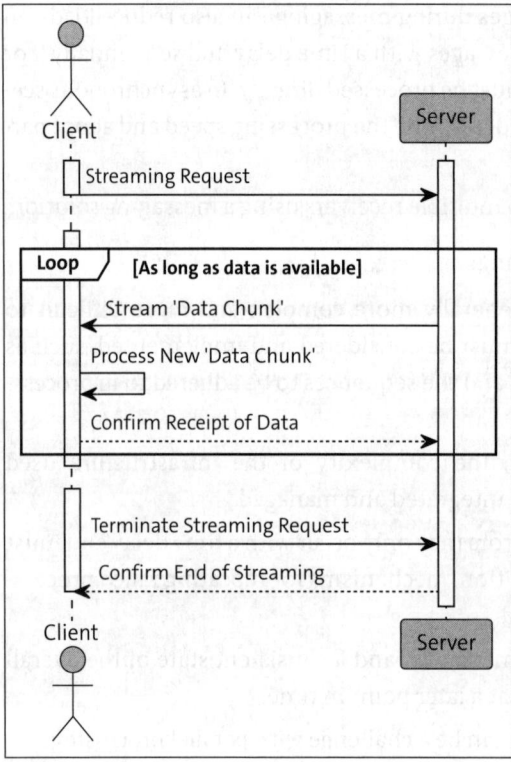

Figure 6.3 Sequence Diagram for Streaming

Communication via streaming is useful when large amounts of data must be transferred efficiently and with low latency in real time. Classic use cases for streaming include video telephony or stock market data, which must always be updated in real time in the browser.

Streaming is also often used for communication between microservices due to its low latency and more efficient use of resources. On both the server side and the client side, the complete data no longer needs to be cached in the memory for requests in order to build a complete return data structure (e.g., a JSON document). In addition, clients can start processing as soon as the first data packet is received, thus achieving short response times.

With streaming, transmissions take place, not only from the server to the client, but also in the opposite direction (i.e., from the client to the server). Depending on the technology used, bidirectional streaming is also possible.

> **HTTP Range Request and Streaming**
>
> Request for comment (RFC) 7233 specifies *HTTP range requests* to allow an HTTP client to request only specific data blocks of a larger file from a server instead of downloading it completely. This technology is also often used with microservices to transfer large amounts of data.
>
> In the first HTTP request, the server informs the client of the entire size of the file, and in subsequent requests, the client can query specific parts of the file using a special HTTP header. In the following request, for example, the client requests bytes 1000 to 5000 from the server:
>
> ```
> GET /video.mp4 HTTP/1.1
> Host: example.com
> Range: bytes=1000-5000
> ```
>
> Strictly speaking, this type of block-based transmission is *not* streaming-based communication because the client is already informed of the final data volume at the start. Further, the transmission is not carried out continuously, but at the explicit request of the client.
>
> HTTP range requests are useful for resuming interrupted downloads, for example, but they are not real streaming.

Some benefits include the following:

- Low latency since the data is transferred continuously in smaller portions and a client can start processing without having to wait for the complete transfer
- Efficient use of resources since the entire data volume of a request doesn't need to be buffered in the memory but instead can be processed successively
- Scalability since streaming technologies are designed for large amounts of data and distributed systems

Some disadvantages include the following:

- The complexity of the environment and the application increases.

Some examples include the following:

- WebSockets
- Event streaming with products such as Apache Kafka or NATS
- gRPC with streaming
- Voice over Internet Protocol (VoIP)

6.2 Resilience Patterns

Errors can occur during any data transfer between systems, regardless of the communication style used. These can be caused, for example, by technical faults, faulty software, or human intervention.

If a system calls another system, a total of 10,000 errors can occur with 10,000 calls, for example. The probability of one of these possible 10,000 errors occurring increases with the number of systems or components involved in the call. The longer a call chain of the components involved is, the higher the actual probability of an error occurring.

The example shown in Figure 6.4 is an example of a call chain for a simple service that uses a central database to answer client queries. In this call chain, however, not only are the aforementioned application components—client, server, and database—involved, but also other infrastructure components that might fail or cause errors. The call chain in our example already contains eight components.

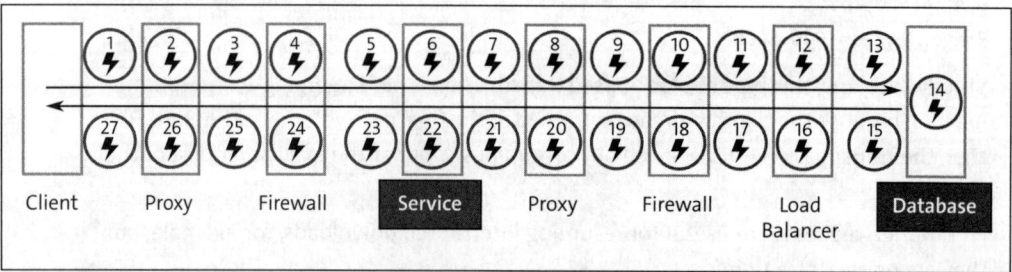

Figure 6.4 The 27 Possible Errors When Calling a Database

If we only count the *integration points* (i.e., the points at which a system queries another system or receives a response), we already have 14 points at which errors can occur (7 for one route, for a total of 14 for the roundtrip). If we also assume that errors can also occur within the components, this small example already includes 27 possible sources of error that influence the probability of errors occurring.

One question to ask when operating distributed systems and applications that communicate across their own system boundaries is therefore not *whether* an error will occur, but rather *when* it will occur.

As distributed applications and the microservices style of architecture gained popularity in the early 2000s, so did the bugs in these systems and their communication links. As Michael T. Nygard noted in 2007, software design up to that point was not complete:

> "Software design as taught today is terribly incomplete. It talks only about what systems should do. It doesn't address the converse—things systems should not do."
> —Michael T. Nygard

Nygard then published *Release IT! Design and Deploy Production-Ready Software*, which laid the foundation for what are called *resilience patterns*. These patterns describe solutions that you can use to make systems more resistant to fault and error.

Up to that point, applications and systems were designed with maximum redundancy. In other words, for each system, you had at least one other system that could take over the tasks if the first system failed. Often, one of the primary objectives was to ensure absolute reliability to keep the application available.

However, a central principle of this redundancy-based approach is the assumption that programs usually crash due to internal errors such as bugs. In modern programming languages with improved memory security, however, complete crashes have become rare and are hardly a relevant problem nowadays.

In increasingly distributed systems, on the other hand, problems often arise at the *integration points* since different components are connected to each other at these points and can result in issues within the applications themselves.

> "Integration points are the number-one killer of systems." —Michael T. Nygard

Resilience patterns, on the other hand, focus on accepting errors as an unavoidable element of processing. The goal of these patterns is no longer to make the environment absolutely fail-safe, but instead to create a more fault-tolerant architecture that can continue to function and react appropriately even when problems occur. This approach aligns with the concept of failing gracefully, where systems are designed to handle failures in a way that minimizes impact, provides informative feedback, and maintains as much functionality as possible rather than crashing outright.

Errors should be accepted as part of reality. Thus, by planning ahead and designing mechanisms to cope with possible errors, systems can become robust against disruptions and build up a certain resilience.

We want users to still be able to carry out and successfully complete their use cases (called *units of work* or *business transactions*). Ideally, in the event of an error, the user won't notice any differences to a problem-free execution.

6.2.1 Error Propagation

A high risk when dealing with complex systems is that seemingly harmless local problems can spread like chain reactions across an entire system and ultimately lead to a complete system collapse. Such a complete failure often begins with a small fault that initially seems insignificant, such as a slow-responding database connection.

This delay can be caused, for example, by the fact that there are many simultaneous requests and as a result, the available connection pool is quickly filled. New connections can then no longer be established quickly enough or at all. This results in blocking threads that wait for new connections from the database connection pool but block further resources or requests themselves in the meantime.

In the case of a microservice, for example, remote connections might no longer be answered, which could affect other systems.

An initially insignificant delay or malfunction can gradually develop, similar how a tiny chip in your car's windshield can result in a giant crack—that is, to further widespread problems or failures—until the entire system fails. In this context, Nygard speaks also of *cracks* that spread throughout a system.

For this reason, an important task is to prevent the spread of errors to protect the overall system. Analogous to the automotive industry, which adopted crumple zones that absorb an impact of a collision and deliberately collapse to protect passengers, defined weak points or isolators can also be built into software systems, which absorb any errors that occur and take care of them, thus protecting the overall system.

These disconnectors, also called *crackstoppers*, can be indispensable components in a system to protect against overload.

> *"At each step in the chain of failure, the crack can be accelerated, slowed, or stopped. High levels of complexity provide more directions for the cracks to propagate in."* —Michael T. Nygard

The explicit processing of errors ensures transparency: Dealing with them is predictable and thus prevents unforeseen consequences.

6.2.2 Subdivision of the Resilience Patterns

Resilience patterns can be subdivided into different categories depending on their basic task:

- Error avoidance
- Error detection
- Error correction
- Restore
- Elasticity and scaling patterns

The following sections present selected patterns for these purposes.

Patterns for Error Avoidance

Patterns in this category aim to proactively avoid errors. Their goal is to ensure the stability of a system right from the design stage and during implementation and to minimize potential weak points and thus possible causes of errors.

The pattern is important for the following reasons, among others:

- **Avoiding overly complex structures in the application design**
  The more complex the design of an application, the higher the number of potential sources of error. If data must be loaded from external services or databases, for

example, the number of integration points increases, as I have already described earlier, and so does the risk of malfunctions. However, an application design that is too complex internally can also lead to faulty code and result in failures.

- **Redundancy**
  System redundancy continues to play a major role in distributed systems, and the use of multiple instances for critical components is useful to mitigate failures.

- **Efficient timeout control and time limits**
  Carefully chosen timeouts can be used to cancel blocking calls and thus free up resources. This approach can prevent or minimize uncontrollable error propagation. Appropriate timeout settings are the absolute foundation for resilient applications. This topic is examined in more detail in Section 6.2.3.

Patterns for Error Detection

As described earlier, a central goal of a robust and resilient architecture is to prevent the propagation of errors in the entire system. The earlier errors are identified and isolated, the better the necessary measures can be initiated.

The task of the patterns in this category is the early detection of errors that could potentially jeopardize the overall system. The patterns in this category include, among others:

- **Heartbeats and health checks**
  Heartbeats and *health checks* are recurring, regular checks of components to determine whether the respective component is still active and functional.

- **Circuit breakers**
  A *circuit breaker* interrupts communication with a faulty component, similar to an electrical fuse, to isolate external issues and contain or prevent the spreading of errors. This pattern is examined in more detail in Section 6.2.5.

- **Monitoring and alerting**
  You can use these measures and patterns to implement monitoring mechanisms to identify and report anomalies at an early stage.

Patterns for Error Correction

This category includes patterns that are intended to minimize the effects of errors and keep the system functional despite partial failures.

This category includes the following patterns:

- **Retry/backoff**
  With a *retry* mechanism, previously executed requests to a system are repeated to bridge short-term errors. If the repeated call also fails, a *backoff* value controls the successive increase in delays between additional retries.

▶ **Graceful degradation**
Graceful degradation is the approach whereby an application remains available even if parts of the system fail or are no longer available. For example, an online store application can continue to operate without the comment function. Another example is when orders are not forwarded directly to another system but are temporarily stored in the database of the online store application. Instead of failing completely, the functionality of the application is gradually reduced to a still acceptable minimum.

▶ **Load shedding**
Load shedding is a strategy for relieving overloaded systems. Certain queries or processes are specifically rejected or executed with a time delay in order to avoid overloading the system. By reducing the load, important processes or requests can continue to be processed, and the system can be kept stable.

Patterns for Restore or Recovery

The patterns in this category focus on the most efficient possible recovery process for a system after a failure. This approach minimizes downtimes and reduces the negative impact on users through faster system availability.

This category includes the following patterns:

▶ **Autorecovery**
The term *autorecovery* refers to the automatic recovery or automatic restart of a service or instance after an error.

▶ **Failover**
If a service or instance fails, a *failover* redirects the data traffic to another, redundant backup system or an alternative instance.

▶ **Snapshot and rollback**
After a failure, the system can be restored to a previous state using a *snapshot* created. A snapshot is a saved copy of the system status at a defined point in time.

Elasticity and Scaling Patterns

Patterns in this category deal with the ability to recognize fluctuations in the current workload and to react dynamically to these changes. Automatic adaptation to the current situation is intended to avoid overloading individual systems and thus protect the entire system.

This category includes the following patterns:

▶ **Autoscaling**
The actual resource requirements are determined at application runtime and, if necessary, new resources (e.g., additional servers running the application) are made available, or resources that are no longer required are released again.

6.2 Resilience Patterns

- **Load balancing**
  Any load can be distributed across different resources (e.g., multiple instances of an application) to avoid overutilized areas (called *hot spots*). Their excessive use could lead to bottlenecks elsewhere, which could have a negative impact on the overall performance of the system.

- **Partitioning**
  Systems, services, or data are divided into multiple logical units to improve scalability and isolation.

6.2.3 Timeout Pattern

One of the most basic and effective resilience patterns for increasing the stability, robustness, and responsiveness of a system is the *timeout pattern*. Carefully selected timeouts can isolate errors and prevent unexpected events from spreading uncontrollably within a system.

Timeouts terminate processes that are no longer expected to respond after a certain time frame has been exceeded.

To illustrate this concept, Figure 6.5 shows a process for an internet store in which a customer places an order via the OrderService and waits for a synchronous response. The OrderService in turn uses the PaymentService and ShippingService services to process the request. In this example, both services are designed as independent microservices and are accessed via the network.

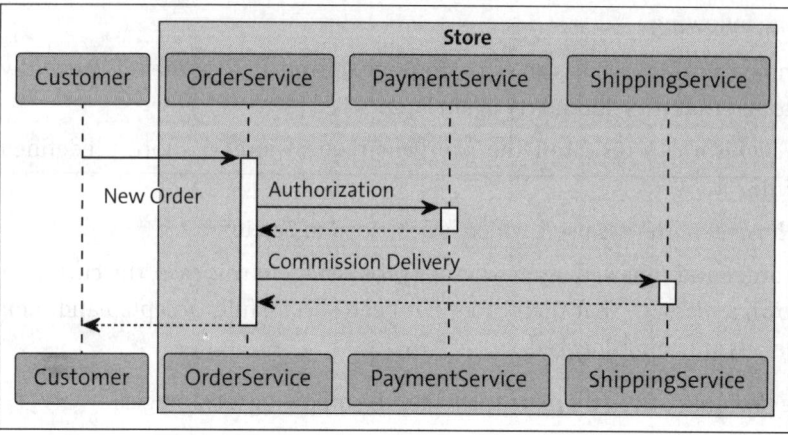

Figure 6.5 Sample Order Process

Without a timeout control, the process sequence could look as follows:

- The customer completes the order in the store and creates a new order in the OrderService accordingly.

- The OrderService calls the PaymentServive to verify the customer's payment information and authorize the payment.
- The OrderService waits for a response from the PaymentService before informing the ShippingService.
- However, the PaymentService has problems answering the request at this time due to an overload or a network failure. Its response is either delayed or does not come at all. As a result, the entire processing of the process is blocked and gets stuck.

In this example, the customer is either forced to wait a long time for a response, or the process ultimately fails. This problem impairs the user experience, and the system may fail to deliver further data at all. If, for example, multiple customer orders are blocked, the HTTP connection pool for incoming requests could also be fully utilized, which results in further incoming requests (from other customers) being blocked.

This problem can be solved by using the timeout pattern. Then, the process could proceed as follows:

- The customer completes the order in the store and creates a new order in the OrderService accordingly.
- The OrderService calls the PaymentServive with a defined timeout of, for example, two seconds to verify the customer's payment information and authorize the payment.
- If the PaymentService does not provide a response within the configured two seconds, the OrderService cancels the request.

After the timeout, you have several alternatives in OrderService to deal with the situation, such as the following:

- The customer promptly receives an error message stating that payment is currently not possible and that they should try again later.
- A *retry mechanism* can resubmit the payment request after a short, predefined period of time.
- The *circuit breaker pattern* is used, which is described in Section 6.2.5.
- Saving the order and renewed asynchronous processing: In this case, the customer may be shown a message that their order has been successfully accepted and their payment details are being checked.

The use of the *timeout pattern* therefore provides the following advantages:

- Resources are not blocked and are available for other processes.
- The user experience can be improved through early feedback on the process.
- Errors are isolated so that problems with a resource or an external service do not affect the entire system.

To quote Michael T. Nygard:

"The Timeouts pattern is useful when you need to protect your system from someone else's Failure." —Michael T. Nygard

However, timeouts should not only be configured for external connections, but also for all potentially blocking resource access, for example, when using thread pools. As mentioned previously, blocking threads can paralyze entire systems and result in unpredictable chain reactions.

Configuring timeouts is not only useful at a low abstraction or application level, such as in network configurations. Higher levels of abstraction can also benefit from an appropriate configuration. However, these values should always be coordinated, which is not trivial. If changes are made to the system (e.g., perhaps introducing a new backend system that is called during a process), these values must be checked again.

In our example, the OrderService could be configured with a timeout of five seconds, for example, so that the client has a maximum wait time of this duration. Each included call could be configured with a shorter time span than the total timeout, as shown in Figure 6.6.

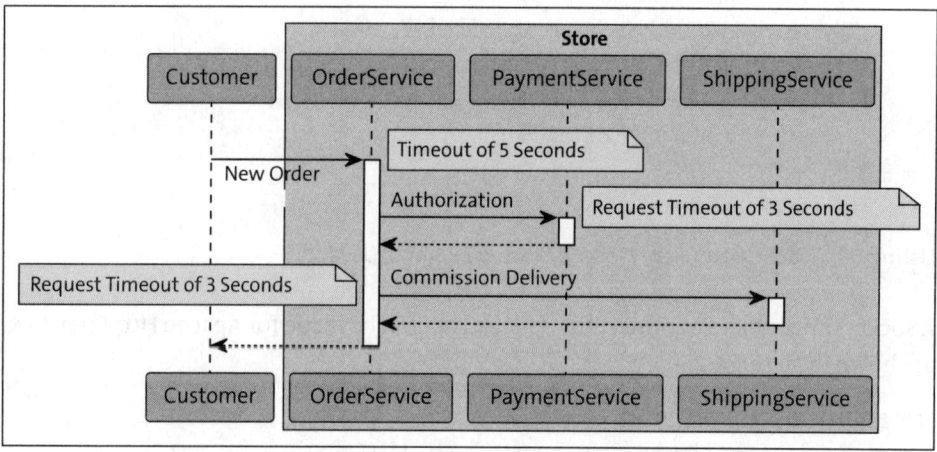

Figure 6.6 Timeout Configuration in the Sample Ordering Process

---

Timeouts for Asynchronous Processes and Messaging

If you use asynchronous processes and messaging, a message can be provided with *time-to-live (TTL) information*. This information defines either an expiration date or a maximum lifespan for the message, for example, to prevent the message from remaining in a queue indefinitely or to prevent obsolete data from being processed.

Before the actual processing starts, each component or the messaging system checks whether the message has already expired. If yes, processing is aborted or rejected, and the message is not processed.

Most communication libraries, such as database drivers or HTTP clients, provide the option of configuring various timeout values. However, the specified values should always be checked and adapted to your own use case. The unreflective use of the default values should be avoided at all costs.

In many HTTP client libraries, no timeouts are configured at all. This setup is useful for streaming applications where open connections should not be closed. However, having no timeout is a problem in all other cases and can jeopardize the stability of an entire environment.

Listing 6.1 shows the "Getting Started" example for the *Apache HTTP Components* library, version 5, where the default settings include a socket timeout of three minutes. In other words, the client blocks processing for a maximum of three minutes if the remote station does not deliver any data, which can be too long in some situations.

```java
try (CloseableHttpClient httpclient = HttpClients.createDefault()) {
    ClassicHttpRequest httpGet = ClassicRequestBuilder.get("…")
            .build();
    httpclient.execute(httpGet, response -> {
        System.out.println(response.getCode());
        final HttpEntity entity1 = response.getEntity();
        EntityUtils.consume(entity1);
        return null;
    });
}
```

Listing 6.1 Apache HttpClient Getting Started Example (Java)

A socket timeout and a connect timeout setting can be made for Apache HttpClient 5.x, as shown in Listing 6.2.

```java
ConnectionConfig connConfig = ConnectionConfig.custom()
        .setConnectTimeout(200, TimeUnit.MILLISECONDS)
        .setSocketTimeout(200, TimeUnit.MILLISECONDS)
        .build();

BasicHttpClientConnectionManager conMan = new BasicHttpClientConnectionManager();
conMan.setConnectionConfig(connConfig);
...
HttpClients.custom().setConnectionManager(conMan).build();
```

Listing 6.2 Apache HttpClient Timeout Setting (Java)

You should always research the configuration options available in your client libraries and configure them according to your requirements. You can use tools like *Toxiproxy* to test the values used and ensure their actual use.

Toxiproxy for Testing TCP Connections

Toxiproxy is an open-source framework to simulate poor network connections and is an excellent way to check the timeout settings of a client library. Toxiproxy is available at *https://github.com/Shopify/toxiproxy*.

Within a central Toxiproxy instance, *proxies* can be configured for individual TCP connections through which application clients can communicate with their actual target server, referred to as an *upstream server*. Each of these connections can be configured with one or more misbehaviors, called *toxics*, to disrupt communication between the individual partners.

Figure 6.7 shows a Toxiproxy instance for which a proxy has been configured that receives requests from a client on port 9999 and forwards request to an upstream server on port 80. This proxy can now be equipped with misbehavior via a management API.

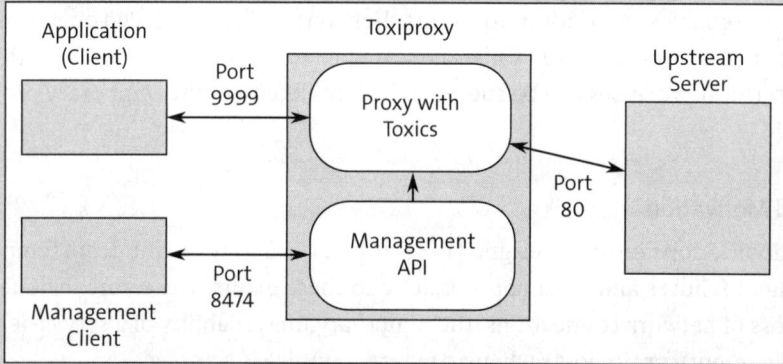

Figure 6.7 Toxiproxy Structure

The Toxiproxy distribution contains the actual Toxiproxy instance as well as management clients as programming libraries in various languages and a command-line client with which the proxies can be configured.

The following toxics can be configured:

- latency: This toxic can be used to include a time delay in the data transfer through the proxy.
- down: Proxies can be completely activated or deactivated to simulate accessibility problems.
- bandwidth: Data connections through the proxies can be limited with this toxic and configured to a maximum value per second.
- slow_close: At the end of a data transfer, when the connection should be closed again, the closing of the connection can be delayed.
- timeout: Using the *timeout* toxic, connections can be closed after a defined time or blocked indefinitely.
- reset_peer: This toxic simulates a *TCP RESET (connection reset by peer)*, which can also be provided with a time delay.

> - `slicer`: Larger data packets can be *sliced* into smaller portions during transmission with this toxic and optionally equipped with a time offset.
> - `limit_data`: Closes the connection after a defined data limit.
>
> For example, you can use the command-line client to create and configure a new proxy for a Redis communication with the following commands:
>
> ```
> toxiproxy-cli create -l localhost:26379 -u localhost:6379 redis
> toxiproxy-cli toxic add -t latency -a latency=1000 redis
> ```
>
> In this case, if a client then connects to the proxy and not directly to the upstream Redis server, communication is delayed, and the client configuration can be checked.

6.2.4 Retry Pattern

The *retry pattern* enables applications to handle short-term failures or failed operations in a way that is transparent to the caller. Repeated execution of the relevant action ensures that critical processes can be successfully completed despite temporary problems.

Problem and Motivation

Distributed applications must be designed in such a way that they can intercept temporary component failures and react appropriately to these errors. The errors include a temporary loss of network connections; the temporary unavailability of a service (e.g., due to a deployment); or timeouts when an external service is accessed.

Unavailability problems when accessing an external service often resolve themselves after a short time, and a repeated call can be made successfully after a short wait.

If the availability problems are caused by a deployment, for example, this deployment might be complete after a certain time, and the service is available again. Using what's called a *throttling strategy*, in which only a certain number of requests are processed in the event of an overload, may also be implemented for a resource.

Solution

Every application in a cloud environment should be able to deal with short-term component failures and handle them as transparently as possible for the business process without any noticeable impact. One simple option is to introduce a *retry mechanism* that repeats the respective action until it has either been performed successfully or until a maximum number of repetitions has been reached.

Figure 6.8 shows a sequence diagram for a retry mechanism for an HTTP call. In the event of incorrect calls, the action is repeated multiple times until the data can finally be retrieved or until the maximum number of repetitions has been reached. In this case, the application must process the error accordingly.

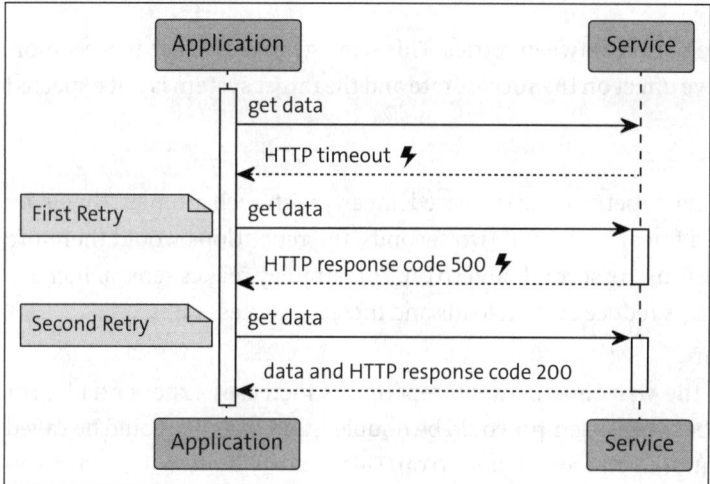

Figure 6.8 Sequence Diagram for a Retry Mechanism

Sample Solution

A retry mechanism can be implemented using custom source code or alternatively with the help of implementations that are already built into many libraries.

In both cases, an important decision to make within the implementation depends on the error which determines how an error situation should be dealt with. Several strategies are possible in this context, such as the following:

- **Direct abort with an error, no repetition**
  If the error is such that a repetition is highly unlikely to be successful, the action should be aborted directly. For example, if a server reports back that the user was not found or input data could not be validated during a login attempt, a repeat attempt will also be unsuccessful.

- **Immediate retry**
  If the cause of the error is most likely only temporary and short-lived, an immediate retry may be useful, for example if it is due to packet loss in network transmission or rarely occurring *race conditions* or resource restrictions.

- **Retry after wait time**
  In most cases, a certain amount of time is required for failed services or network components to become available again. Calls with errors that indicate a temporary, short-term failure, but where immediate repetition of the action doesn't make sense, should be executed with a time delay to minimize the likelihood of another error and a possible overload of the target system.

The wait time between the repetition of failed calls is determined through a *backoff strategy*. Common variants, which can also be used in combination, include the following:

- **Fixed backoff**
  Use a constant time period between retries. This strategy makes sense if even short delays have a positive effect on the success rate and the target system is not expected to be overloaded.

- **Linear backoff**
  The wait time between repetitions is increased linearly with each attempt, that is, by a fixed value. With a linear backoff of two seconds, the repetitions would therefore take place after two, four, six seconds and so on. This strategy makes sense if increasing the wait time helps reduce system loads and increase success rates.

- **Exponential backoff**
  With this strategy, the wait time between repetitions increases exponentially. For example, the time between attempts could be doubled, and a service could be called after two, four, eight, sixteen seconds and so on. This strategy is often used for heavily frequented systems in order to reduce the load on the external system. Setting an upper time limit makes sense to limit the exponential growth.

- **Exponential backoff with jitter**
  This strategy is based on the exponential backoff, but uses a random deviation, a so-called *jitter*, between the calls in addition to the calculated values. A random component is particularly useful for many simultaneous requests from different clients in order to avoid an overload caused by too many repetitions.

  "Fast retries are very likely to fail again." —Michael T. Nygard

In Java or Kotlin applications, for example, you can use the *Resilience4j* library to implement a retry mechanism. This library already contains implementations of various strategies for error determination and backoff strategies and also includes extensions for various frameworks and environments, such as the Spring Framework.

The example shown in Listing 6.3 demonstrates the implementation of a retry mechanism using the *Resilience4j* library, in which an exponential backoff is used. A distinction is made as to whether a call to the HelloWorldService should be repeated made on the basis of the string return value of the called method (response equals "error") or on the basis of a generated exception (IOException and TimeoutException).

```
// Backoff interval configuration
IntervalFunction intervalWithExponentialBackoff = IntervalFunction
        .ofExponentialBackoff(IntervalFunction.DEFAULT_INITIAL_INTERVAL);

// Configuration for the retry mechanism
RetryConfig config = RetryConfig.<String>custom()
        // maximum number of repetitions
        .maxAttempts(10)
        // Backoff configuration
        .intervalFunction(intervalWithExponentialBackoff)
```

```
            // Repetition for certain results
            .retryOnResult(response -> response.equals("error"))
            // Exceptions that should lead to a retry
            .retryExceptions(IOException.class, TimeoutException.class)
            // Exceptions that are to be ignored for a retry
            .ignoreExceptions(BusinessException.class)
            .build();

Retry retry = Retry.of("helloworld", config);

// Service to be called
HelloWorldService helloWorldService = new HelloWorldService();

// Create a decorated service with retry functionality
CheckedSupplier<String> supplier = Retry
        .decorateCheckedSupplier(retry, helloWorldService::sayHelloWorld);

// Call the service via the decorator
String result = supplier.get();
```

Listing 6.3 Resilience4j Retry Example (Java)

You should consider the following points when implementing or using libraries.

- **Runtime behavior**

  If incorrect calls are repeated, this inevitably leads to a longer execution time for the entire process. For this reason, the retry strategy used must always be compared with the requirements of the application. For interactive applications where a user is waiting for results, a good choice is to repeat the failed calls only a few times and only briefly so that a user isn't forced to wait for a long time. If an error occurs, the user can be prompted to perform the action again later. In batch processes, on the other hand, multiple and longer retries may be useful to ensure that the corresponding processes can be carried out successfully.

  Note that a retry strategy with many short repetitions can result in an unwanted increase in load on the target system and could therefore cause that system to crash completely if already working at its load limit.

- **Idempotence**

  Idempotent operations can be called once or called multiple times without changing the result of the action beyond the first execution. If the call of an order function returns an error to the caller, problems may have occurred after the data has been saved, for example. It is possible that only the response data could not be transferred to the client. In this case, repeated execution would be an issue if the order function is not designed to be idempotent since the order would be executed a second time.

In the case of idempotent actions, it's not the actual response that is decisive, but the internal state of the called system. If, for example, a customer data record is deleted via an HTTP DELETE method, a positive response can be given on the first call, and error information can be returned on subsequent calls since the customer no longer exists. The status of the system does not change with additional calls.

> **HTTP Methods and Idempotency**
>
> *HTTP methods*, also known as *HTTP verbs*, are commands transmitted to the server in HTTP requests to tell it what action it should perform with a particular resource. For example, the GET method is used to call a web page, and the POST method is used to send a form.
>
> Idempotency is an important concept for HTTP methods that improves the reliability, fault tolerance, and performance of web applications. RFC 7231 contains a list of the different HTTP methods and their defined idempotence. Unfortunately, not all APIs adhere to this definition.

Method	Idempotence
CONNECT	No
DELETE	Yes
GET	Yes
HEAD	Yes
OPTIONS	Yes
POST	No
PUT	Yes
TRACE	Yes
PATCH	Yes and No (see RFC 5789)—with corresponding header

Table 6.1 HTTP Methods/HTTP Verbs

- **Considering error types**

  Calling an external service can fail for a variety of reasons. The specific error should always be considered within the retry mechanism and handled accordingly. Some errors should lead to repetitions, while for other errors repeating the action does not make sense. Errors can be better defined and processed with typified error descriptions such as RFC 7807, available at *https://datatracker.ietf.org/doc/html/rfc7807*. The RFC defines a standardized, machine-readable format (typically JSON) for conveying error details in HTTP API responses.

> **HTTP Status Codes for Error Classification**
>
> The HTTP protocol defines status codes with which a server can provide information about the outcome of a request. The client can use these status codes to select an appropriate retry strategy. Table 6.2 contains an overview for assigning the error code to a retry action.
>
> | Status Code | Retry Action |
> |---|---|
> | 2xx | Success; no retry. |
> | 3xx | Redirects; no retry. |
> | 4xx | Client error; a retry is usually unsuccessful because the input data does not meet the expectations. Example:
`Status 405 Method not allowed` |
> | 5xx | Server-side error; a retry can be useful here. |
>
> **Table 6.2** List of Retry Actions for HTTP Status Codes

▶ **Transactions**

Each retry of an operation within a running transaction can result in problems with the consistency of the overall transaction. If, for example, a process terminates due to a faulty call, previously successful remote calls may have to be rolled back. In these situations, it may make sense to complete the entire process with retrieving of individual steps and accept a time delay.

When To Use the Pattern

The pattern should always be used if short-term errors can occur when calling external resources and a repetition of the call will usually be successful.

The use of the pattern is problematic when for errors that persist over a longer period of time, and for which retries are highly unlikely to be successful. Also, errors in the business logic sometimes cannot be meaningfully repeated because certain preconditions or input data may already be incorrect.

Too many frequent retries of calls due to load situations can also indicate the need for scaling the target system.

Consequences

With the *retry pattern*, external calls can be made more robust and fail-safe so that overall processes can be carried out despite temporary failures or problems.

6.2.5 Circuit Breaker Pattern

You can use the *circuit breaker pattern* to ensure that a system remains robust against failures or issues in other systems. At the same time, this pattern can protect a called resource from being overloaded by too many call attempts.

Problem and Motivation

Distributed applications should not only be resistant to short-term component failures; they should also react effectively to unforeseen, longer-lasting error situations. If, for example, a service fails completely and the restoration of availability takes a long time, repeated calls (as provided for by the retry pattern) are wasteful because these requests are also highly unlikely to be successful.

Let's explore a scenario to illustrate the problem: In a web store application, an external service provider is connected via an HTTP connection for payment processing. As soon as a customer place an order, the payment information entered is verified with this service provider, and the corresponding invoice amount is requested. The *timeout pattern* (described in Section 6.2.3) is implemented for the corresponding HTTP client used for communication, and each request is aborted after a specified time. The *retry pattern* (described in Section 6.2.4) is also adopted to prepare for short-term failures.

If, for example, hardware issues on the payment service provider's side result in a partial failure of the external service and the requests sent to it are answered, but only slowly, the two patterns used (*timeout* and *retry*) cause all calls by the HTTP client to be aborted after a certain wait time and then executed again. However, the repeated requests are also aborted again by a timeout since the hardware issues cannot be solved immediately.

Important resources are blocked within the calling service during these timeout wait times. Each calling and blocked customer may have at least one TCP connection to the service, which is managed in a connection pool for incoming connections. The more customer requests are blocked because of the payment service provider, the higher the probability that this connection pool will be exhausted, and the service will no longer be able to accept new connections.

In this example, the failure of an external service can lead to a complete unavailability of the entire web store.

Solution

The introduction of the *circuit breaker pattern* can isolate the longer-term failures of external systems and protect your own system from cascading errors.

The basic principle of the pattern corresponds to that of a fuse in an electric circuit: If a consumer overloads, the fuse will trip, and the consumer will be disconnected from the main. This mechanism prevents an overload from causing further damage to the electrical circuit, such as a fire due to overheated cables.

Similar to this safety mechanism, the *circuit breaker pattern* interrupts the communication with a called resource if too many errors are detected during calls within a defined time interval. The pattern also has a mechanism that automatically checks whether the resource is ready for use again and can be called again. For this task, the pattern defines three states for its work, as shown in Figure 6.9, between which the circuit breaker switches:

- Closed: The addressed service functions normally. All requests are forwarded directly to the service department.
- Open: No requests are forwarded to the target system.
- Half-open: In this state, individual requests are forwarded to the target system for testing.

If the addressed external resource works without errors and no problems are detected when it is called, the circuit breaker is in the closed state and all calls are forwarded directly to the target system.

However, if too many faulty calls are detected within a defined time interval, the circuit breaker switches to the open state and rejects all new requests to the target resource directly (i.e., without passing them on to the target system, with an error).

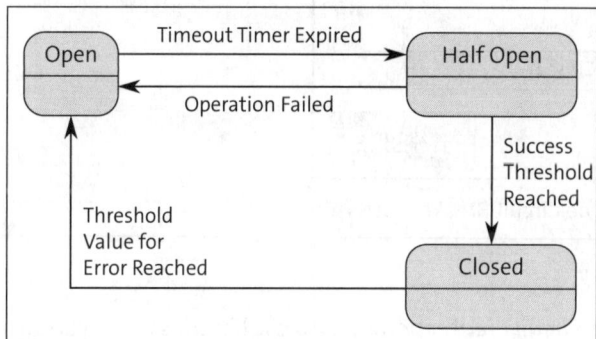

Figure 6.9 State Diagram of the Circuit Breaker Pattern

After a defined wait time, the circuit breaker finally switches to the half-open state and forwards individual calls to the target resource to check whether it is available again and working correctly. If these requests are answered successfully, the circuit breaker switches back to the closed state, and the external system is called again normally. However, if these test calls fail, the circuit breaker switches back to the open state, and the wait time starts again. Figure 6.10 shows a corresponding sequence diagram.

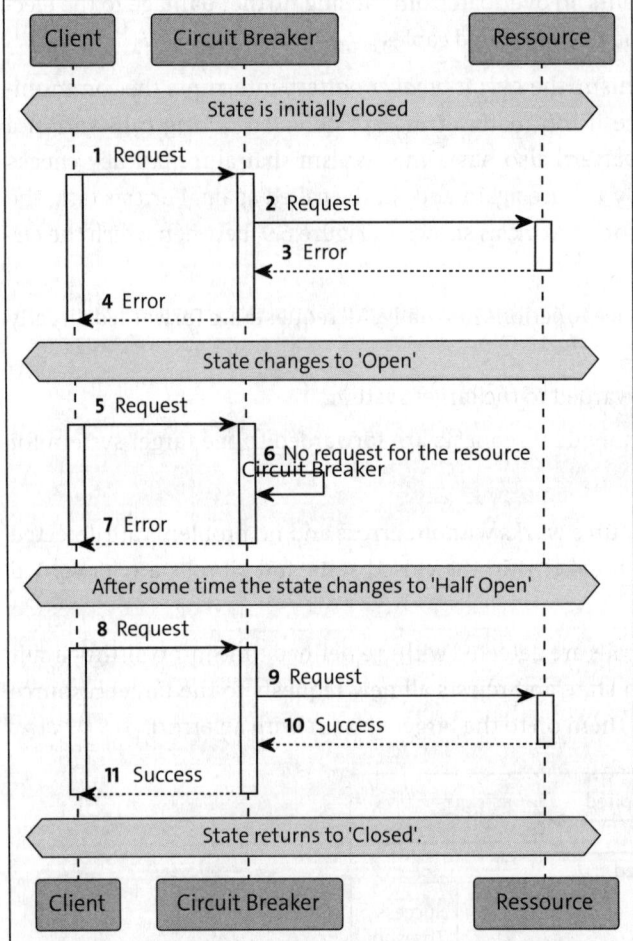

Figure 6.10 Sequence Diagram of the Circuit Breaker Pattern

Sample Solution

The easiest way to implement the *circuit breaker pattern* is to use libraries in the various programming languages. In Java, for example, you can use the *Resilience4j* library mentioned earlier. For Go, for example, Sony has published the *github.com/sony/gobreaker* library that you can also use for this purpose.

All implementations are based on a similar principle and use the *decorator pattern* in which the decorated object represents the actual resource. The decorator, which encapsulates access to the resource, usually uses a circuit breaker implementation in which, among other things, the current state is stored.

Figure 6.11 shows a class diagram (even if no classes exist in Go) for a sample Go implementation for our example.

6.2 Resilience Patterns

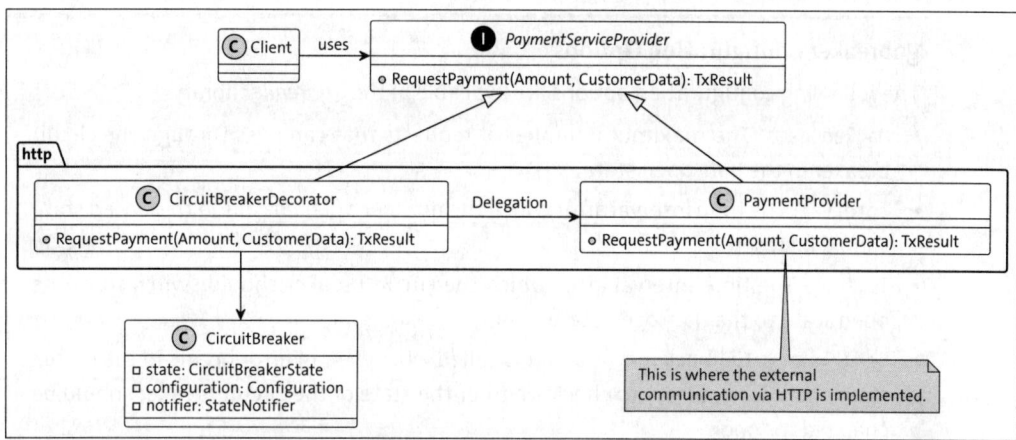

Figure 6.11 Structure of the Circuit Breaker Pattern

For the sample solution, the `PaymentServiceProvider` interface is defined using the `RequestPayment` method, which is used to validate the payment information and request the invoice amount, as shown in Listing 6.4.

```go
type PaymentServiceProvider interface {
    RequestPayment(amount float64, data CustomerData) (TxResult, error)
}
```
Listing 6.4 The PaymentServiceProvider Interface (Go)

Corresponding to Figure 6.11, two implementations are available for this interface:

- `http.PaymentProvider`: Where the actual HTTP connection to the external service is located.
- `http.CircuitBreakerDecorator`: The circuit breaker functionality is implemented in the decorator. If necessary, the decorator forwards the requests to the actual provider.

The implementation of the decorator uses the *gobreaker* library, which defines the typable `CircuitBreaker` data type. This type can be configured via various fields and has the `Execute` method with which a function or method passed to it can be executed according to the configured values. The signature of the method is shown in Listing 6.5.

```go
func (cb *CircuitBreaker[T]) Execute(req func() (T, error)) (T, error)
```
Listing 6.5 gobreaker Library Execute Method (Go)

> **gobreaker Configuration Options**
>
> The following configuration options are available in the *gobreaker* library:
>
> - `MaxRequests`: The maximum number of requests that can pass through the circuit breaker in the half-open state
> - `Interval`: The time interval after which the number of failed calls in the closed state should be reset
> - `Timeout`: The time interval after which the circuit breaker should switch from the open state to the half-open state
> - `ReadyToTrip`: Callback method that is called whenever an error occurs in the closed state and it is necessary to check whether the state of the circuit breaker should be changed to "open"
> - `OnStateChange`: Callback method that is always called when the circuit breaker changes its state
> - `IsSuccessful`: Callback method that is called by the circuit breaker to check whether a problem that has occurred should be evaluated as an error

Within the `NewCircuitBreakerDecorator` factory method, a `CircuitBreaker` instance is configured and created for the sample implementation, as shown in Listing 6.6.

```go
func NewCircuitBreakerDecorator(
    wrapped internal.PaymentServiceProvider) *CircuitBreakerDecorator {

    settings := gobreaker.Settings{
        Name:        "",
        MaxRequests: 1,
        Interval:    30 * time.Second,
        Timeout:     15 * time.Second,
        ReadyToTrip: func(counts gobreaker.Counts) bool {
            return true
        },
        OnStateChange: func(name string,
            from gobreaker.State,
            to gobreaker.State) {

            log.Println("state changed", name, from, to)
        },
        IsSuccessful: nil,
    }
    cb := gobreaker.NewCircuitBreaker[internal.TxResult](settings)
    return &CircuitBreakerDecorator{wrapped: wrapped, circuitBreaker: cb}
}
```

Listing 6.6 gobreaker Configuration and Creation (Go)

Listing 6.7 shows the remaining Go-based implementation of the decorator and also shows how the Execute method uses the RequestPayment method of the decorated object. This call only takes place if the CircuitBreaker is in the correct state.

```go
type CircuitBreakerDecorator struct {
    wrapped         internal.PaymentServiceProvider
    circuitBreaker *gobreaker.CircuitBreaker[internal.TxResult]
}

func (c *CircuitBreakerDecorator) RequestPayment(
    amount float64,
    data internal.CustomerData) (internal.TxResult, error) {

    transactionResult, err := c.circuitBreaker.Execute(
        func() (internal.TxResult, error) {

            result, err := c.wrapped.RequestPayment(amount, data)
            return result, err
        })

    return transactionResult, err
}
```

Listing 6.7 Implementing the CircuitBreakerDecorator (Go)

When To Use the Pattern

This pattern should always be used if you expect long-lasting errors when calling external resources and when simple repetition using a retry mechanism does not make sense or may lead to additional cascading errors.

By using the pattern, you can prevent repeated (most likely unsuccessful) calls and thus make the external system and your own system more stable and robust against overloading.

Not only do circuit breakers stabilize the system, but they also make it possible to return alternative responses instead of errors in error situations, thus contributing to an improved user experience and better application response times.

The automatic recovery of the system also enables a return from faulty operation to normal operation without manual intervention, which minimizes downtimes and also improves the user experience.

The *circuit breaker pattern* is not a general protective shield against all possible failure scenarios. You should use it specifically where it makes the most sense to improve the stability and reliability of an application. The pattern is less suitable for local access to private resources, such as memory access, since the additional complexity cannot justify the low added value.

Consequences

The integration of the *circuit breaker pattern* into the architecture of an application can lead to an increase in complexity because additional components and more detailed logic for error handling are introduced.

You'll need a way to evaluate all errors that occur and sort them into different error categories. Not every error must necessarily lead to a status change of the circuit breaker. For example, if a service delivers an error for a customer that has not been found, this error can be neglected within the scope of the circuit breaker since the service is still operational in this case.

In general, the effectiveness of the pattern is determined by the careful use of the *timeout pattern*. If you have no timeout settings, no error may be detected for blocking calls, and the circuit breaker will then have no effect because no status is changed due to an error. Timeouts are thus a basic requirement for the use of a circuit breaker.

In the event of interruptions caused by the circuit breaker, you must define a fallback strategy and think through its consequences: If calls are aborted immediately without waiting, this can lead to problems or overload in calling systems, as can slow response times. The calling logic may be implemented in such a way that a short wait time is assumed, and errors occur in concurrent calls in the event of fast responses. Predefined responses in the event of an error are also conceivable; however, they must make sense technically.

Every use of a circuit breaker should be accompanied by a suitable monitoring strategy so that error situations are made transparent and can be resolved accordingly. In this way, you can analyze any issues that occur later or, if necessary, stop processing in other system components for the duration of an open circuit breaker.

6.2.6 Bulkhead Pattern

You can use the *bulkhead pattern* to limit the effects of faults or failures of individual system components and thus protect the entire system from unpredictable chain reactions or failures.

In shipbuilding, from which the term *bulkhead* was adopted, a bulkhead is a closed partition within a ship that limits flooding to areas that have sprung a leak and thus prevents the chaotic spread of water ingress. Bulkheads are supposed to prevent ships from taking on water too fast and sinking.

Problem and Motivation

In distributed applications, complex tasks are often accomplished through the interaction of multiple services. If the data or the functions of an external service are required, these resources are accessed via the network. This access then requires internal resources within a service, such as threads or connections from a connection pool.

If all resources within a service are managed and used jointly, the following scenario is conceivable, for example: In a web store application, orders and item data are stored in a database. When a customer triggers an order, an order is created in the database, and payment is ordered from a payment service provider. The entire checkout process takes place within a database transaction, which is completed after successful confirmation by the payment service provider. The database connection used for this transaction is returned to the pool.

If problems or time delays occur during the payment confirmation, the corresponding open transactions, as shown in Figure 6.12, cannot be completed, and the database connections used for this transaction cannot be returned to the connection pool. As a result, you have many connections open in parallel, blocking orders, the connection pool to the database can be exhausted, and further connections to the database are not possible.

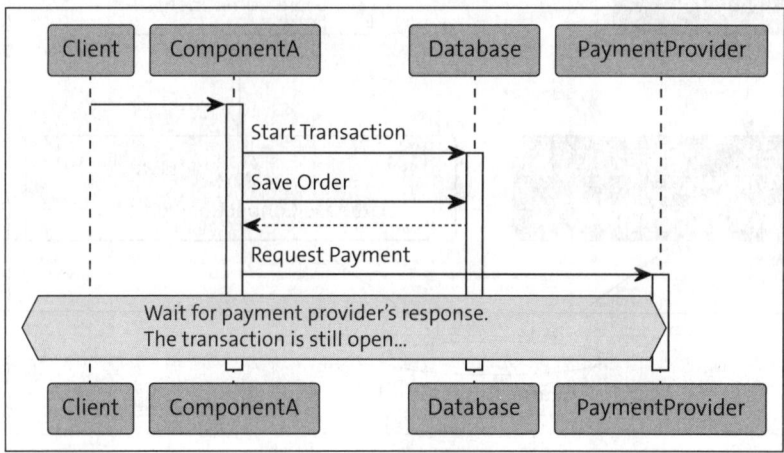

Figure 6.12 A Call Blocking the Database from Further Connections

Because item data might also be stored in the database and the delivery of the HTML interface of the store application is dependent on this database access, item pages can no longer be delivered due to the exhausted connection pool. In this case, even customers who have not yet started an order process will be affected by the consequences of the misconduct.

Solution

The *bulkhead pattern* solves this problem by subdividing the resources used or subdividing the resource access requests within the system into different groups or independent areas based on the individual load and availability requirements. Faults that occur in these independent areas are isolated within their area, and further propagation is prevented. Unaffected and separate parts of the system can still be used in the event of a fault.

In our example, we'll set up two connection pools to the database, as shown in Figure 6.13. In this case, the order and store components each use a separate connection pool, and the problems in one category cannot impact the other components.

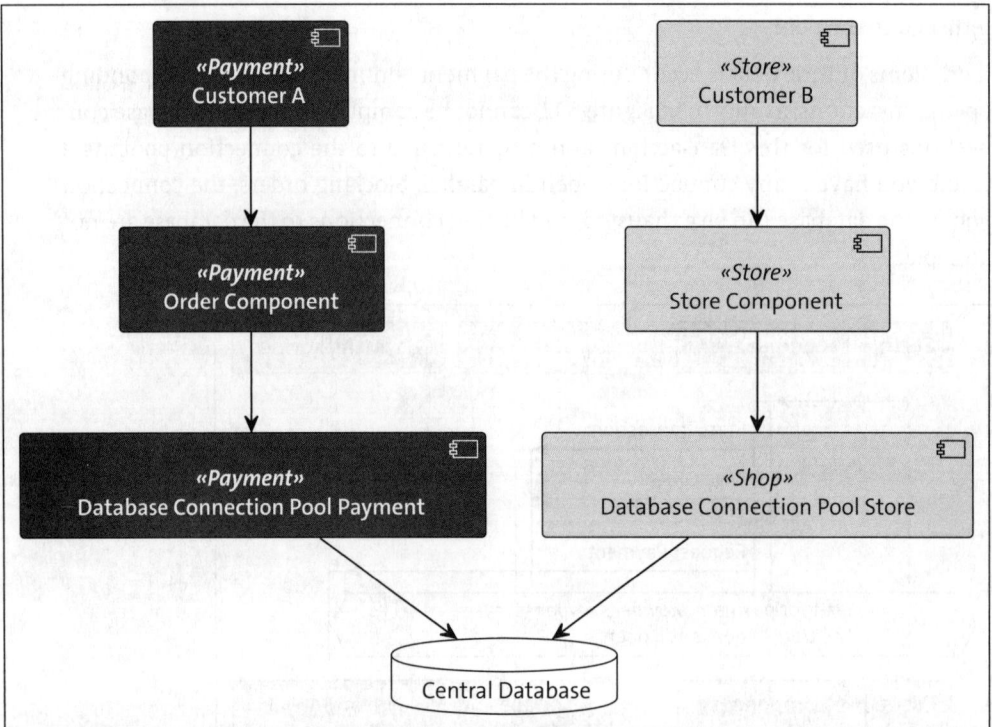

Figure 6.13 Allocating Resource Access

Alternatively, the system can also be subdivided at the service instance level, as shown in Figure 6.14. In this approach, different clients of the system are each assigned a complete, dedicated instance with which they can work. In our example, two store instances could be operated, one of which takes care of the delivery of the item data, while the second instance is used exclusively for the ordering process. The two use cases would be separate and would not influence each other.

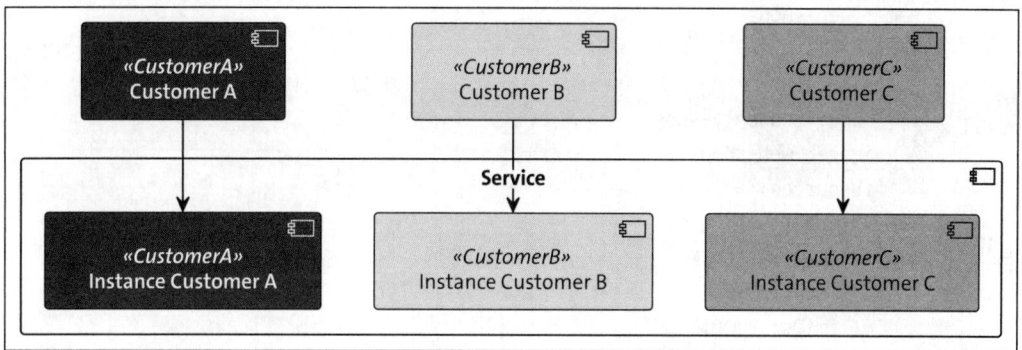

Figure 6.14 Multiple Instances of a Shared Resource

A subdivision of systems and the use of the *bulkhead pattern* have several advantages, such as the following:

- The subdivision of a system into several areas ensures good isolation and localization of errors so that errors in one area cannot spread to other areas. This isolation prevents the entire system from failing.
- Despite errors occurring, a system can still provide functionality.
- By subdividing a system, several areas with different priorities or quality of service requirements can also be operated.

You can implement the *bulkhead pattern* in a variety of ways at different levels of the architecture. Some possible areas of use include the following:

- Thread pool bulkhead for the use of different thread pools for the execution of tasks
- Service bulkhead for the instantiation of several similar services for different clients
- Database bulkhead for the separation of connection pools or partitions, for instance, possible division by write and read operations
- Infrastructure bulkhead for the subdivision of resources, such as the implementation of a management network or execution of a service on different physical machines

Sample Solution

As described earlier, the *bulkhead pattern* can be used in many different ways. At the infrastructure level, for example, two store containers could be started within a pod in Kubernetes, as shown in Listing 6.8. Even if not ideal for a Kubernetes configuration, this example should illustrate that two instances can be operational at the same time, for different purposes, without any problems.

```
apiVersion: v1
kind: Pod
metadata:
```

```yaml
  name: web-shops
spec:
  containers:
    - name: readonly-shop
      image: web-shop
      resources:
        requests:
          memory: "64Mi"
          cpu: "250m"
        limits:
          memory: "128Mi"
          cpu: "1"
    - name: order-shop
      image: web-shop
      resources:
        requests:
          memory: "64Mi"
          cpu: "250m"
        limits:
          memory: "128Mi"
          cpu: "1"
```

Listing 6.8 Kubernetes Configuration for Two Identical Containers

Frameworks such as *Resilience4j* (for Java) or *Polly* (for .NET) can be used to implement the pattern efficiently in your own code. In addition, almost all resource drivers or HTTP libraries make it possible to define separate connection pools for previously defined areas. You should always make use of this option first.

When To Use the Pattern

You should always use the *bulkhead pattern* when any of the following statements are true:

- A service should still provide functionality if a connected external resource fails—even if only in part.
- A complete isolation of multiple clients is desired.
- A system is to be protected against cascading errors.

Consequences

The implementation of the *bulkhead pattern* increases the complexity of the system since it requires multiple independent areas, each of which must be defined and configured. One consequence might be a more inefficient use of resources because, for example, connection pools require separate configurations with capacity reserves for each area, instead of being optimally managed together, thus increasing the overall consumption of resources.

A meaningful division of areas requires not only the consideration of technical dependencies, but also a clear definition of functional relationships and responsibilities. This approach is the only way to recognize which areas can operate independently without causing new problems through technical dependencies.

The *bulkhead pattern* alone is not enough to reliably guarantee the performance of a system. However, this pattern is efficient in combination with other patterns such as the *retry*, *timeout* or *circuit breaker patterns*.

6.2.7 Steady State Pattern

The aim of the *steady state pattern* is to keep systems in a stable, predictable operating state without manual intervention.

Problem and Motivation

In traditional IT environments without cloud components, major system failures are often triggered by manual interventions in production environments. Administrators potentially need to log directly into the production systems and use command-line commands to perform complex data cleansing, as fully utilizing the available storage space can lead to instability and performance issues. The data to be deleted might include, for example, old log files or cache files that are no longer required.

A single incorrect command during such a cleanup run can result in irreversible data loss or the crash of an entire system.

A manual deletion of data no longer seems necessary in the cloud environment since external resources are seemingly infinitely available. But in this context, too, the accumulation of data can cause problems:

> "It sometimes seems that you'll be lucky if the system ever runs at all in the real world. The notion that it will run long enough to accumulate too much data to handle seems like a 'high-class problem'-the kind of problem you'd love to have."
> —Michael T. Nygard

For the following reasons, you should always clean up data that is no longer required:

- **Manual actions jeopardize systems**
  As I have already mentioned, manual interventions in systems are always associated with risks of errors and can affect the stability of systems.
- **Storage limitations and costs**
  Storage space is limited on local computers, and in the cloud environment, large amounts of data can result in considerable costs.

▶ **Performance and scalability**
Processing large amounts of data requires a lot of computing power and therefore time, longer query times in databases, for example. Backups and their restoration also require significantly more time when the amount of data increases.

▶ **Data protection and compliance**
Data may often only be stored for a limited period of time. The *General Data Protection Regulation (GDPR)* in the European Union (EU) or the *California Consumer Privacy Act (CCA)*, for example, stipulate retention periods for personal data. Unnecessary storage can lead to legal consequences.

▶ **Relevance and quality**
Outdated data can distort current evaluation results and negatively influence important business decisions.

▶ **Security risks**
The more data is stored in a system, the greater the risk of misuse or a data leak. Let's say, for example, confidential data is temporarily stored in a cache directory on the server before a download by an authorized user and this directory is not automatically cleaned up. In this case, unauthorized individuals may be able to view this information.

Solution

You must define and implement a strategy for cleansing all data as early as possible in the development process. This strategy should define clear guidelines for handling different types of data, such as deleting irrelevant or no longer required information, compressing large datasets, or archiving historical data. The definition of such a strategy depends heavily on the business or legal requirements for the storage period of the data.

Ideally, the corresponding strategies are implemented and put into operation directly with the delivery or installation of the software, with the aim of a permanently running solution without manual intervention.

In addition to data cleansing, applications must also be prepared for processing large volumes of data. In most cases, only limited test data is available at the time of development, and data access does not have to be restricted in these cases. However, as soon as an application runs in a production environment, the existing or resulting data volumes often cause load and stability issues.

If, for example, all orders are displayed in the software, this works well in the development environment with five data records; however, as soon as there are more than 50,000 data records in a production system, some the software reaches its limits in load and memory usage.

For this reason, all data queries should be limited to a specific section of the data from the outset. In databases, for example, a *paging mechanism* can be implemented in which only a certain section of the result set is requested and returned by the query.

In some applications, only the first hits of a result set are often interesting and therefore sufficient. Loading further data is superfluous in such a case. A good example in this context is Google search, where few users bother navigating to page 2 of the search results.

Sample Solution

In this sample application, training participants (attendees) are to be read from an existing database and displayed in the command line. The implementation takes place in Java using the *Spring Framework* and the *Spring Data JPA*, the latter being an extension of the Spring Data project that allows you to easily create repository implementations for the Java Persistence API (JPA).

Spring Data combines various libraries based on the *Spring Framework* that significantly facilitate the access to different persistence technologies (e.g., databases). By automatically generating access codes and abstracting complex configuration steps, *Spring Data* significantly reduces the workload for application developers.

To access a relational database via SQL with *Spring Data JPA*, for example, one entity and one interface definition is sufficient. The corresponding access code is generated automatically by the framework at runtime.

As shown in Listing 6.9, the Java class of the JPA entity is listed without any additional access methods for the attributes.

```
@Entity
public class Attendee {
    @Id
    private long id;
    private Date CreationDate;
    private String Salutation;
    private String FirstName;
    private String LastName;
    private String EMail;
    ...
}
```

Listing 6.9 JPA Entity for Attendees (Java)

As I have already mentioned, only the definition of an interface is required for the database connection, which in turn extends certain interfaces of *Spring Data*. Figure 6.15

shows the class diagram for our example with the application-specific `AttendeeRepository` interface, which implements the `PagingAndSortingRepository` and `CrudRepository` interfaces of Spring Data.

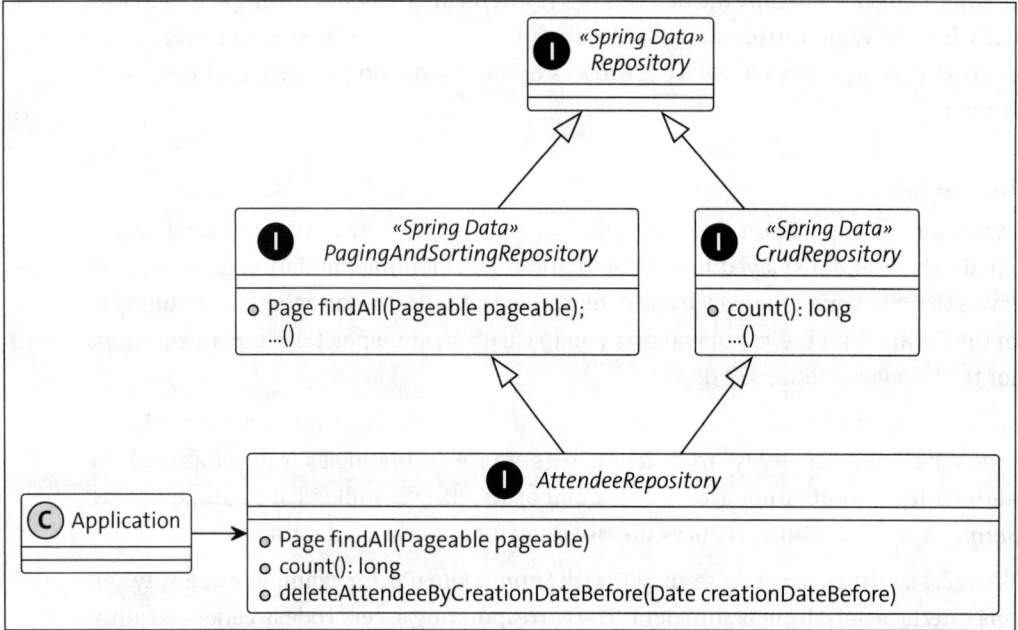

Figure 6.15 Class Diagram with Spring Data

> **CRUD: A Crude Implementation?**
>
> The four basic operations that can be applied to data are summed up by the acronym *CRUD*, which stands for create, read, update, and delete.
>
> These operations usually form the basis for database interactions and are used in many applications to manage data. Some applications are also referred to as *CRUD applications* if they only manipulate data in a database and lack more extensive application logic.
>
> Interfaces often have the CRUD suffix to indicate that they provide precisely this functionality.

By extending the two interfaces, the application-specific repository already has methods for block-by-block reading and for counting the data records in the database. The `deleteAttendeeByCreationDateBefore` method name also corresponds to a predefined method name pattern, which is used to define the functionality for deleting old data records. Listing 6.10 shows the complete interface.

```java
import org.springframework.data.jpa.repository.JpaRepository;
import org.springframework.data.repository.PagingAndSortingRepository;

public interface AttendeeRepository
        extends PagingAndSortingRepository<Attendee, Long>,
        CrudRepository<Attendee, Long> {

    void deleteAttendeeByCreationDateBefore(Date creationDateBefore);
}
```
Listing 6.10 Repository Implementation Using Spring Data JPA (Java)

The relevant code to implement the command-line client is shown in Listing 6.11, which illustrates how the `findAll` method of the repository can be used with the help of the `Pageable` interface so that not all data is read at once. The application-specific interface automatically contains an implementation for a paging mechanism to prevent excessive resource utilization.

```java
@Bean
public CommandLineRunner printAttendees(AttendeeRepository repository) {
    return (args) -> {
        int elementCount = (int) repository.count();
        int pageSize = 10;
        for (int pageNumber = 0;
             pageNumber <= elementCount/pageSize;
             pageNumber++) {

            Pageable request = PageRequest.of(pageNumber, pageSize);
            repository.findAll(request).forEach(attendee -> {
                System.out.println(attendee.toString());
            });
        }
        System.out.println("end");
    };
}
```
Listing 6.11 Use of the Repository for a Command Line Output (Java)

The implementation of the `deleteAttendeeByCreationDateBefore` method, which is also generated by the framework, can be used periodically to delete old data. Listing 6.12 shows an implementation that can be created without much effort. With the `@Scheduled` annotation, the code is automatically executed every 12 hours.

```java
@Scheduled(fixedRate = 12, timeUnit = TimeUnit.HOURS)
public void cleanDB() {
    System.out.println("cleanDB");
```

```
    LocalDate date = LocalDate.now().minusDays(30);
    repository.deleteAttendeeByCreationDateBefore(date);
}
```
Listing 6.12 Periodic Deletion Job with the Spring Framework (Java)

When To Use the Pattern

The *steady state pattern* should always be used when applications generate or save data. The aim of its use should be that an application can be executed endlessly without a manual interaction.

Consequences

Systems that implement the *steady state pattern* can be operated endlessly without any manual interaction. The applications themselves ensure that only the necessary data is accumulated and that large amounts of data do not cause system instability. However, the use of this pattern also increases the complexity of the solution since additional automated cleanup steps must be defined and implemented.

6.3 Messaging Patterns

As described in Section 6.1.2, *messaging* is an *asynchronous communication style* in which messages are exchanged between the individual communication partners involved. In contrast to synchronous communication, the applications don't need to be active at the same time, which makes the exchange between them more robust and less dependent. A central *messaging system* takes over the task of reliable transmission and delivery of messages, which simplifies application development. Your applications can focus on the content of the transmitted information rather than on how the information is exchanged.

This section introduces various patterns for setting up messaging-based communication.

6.3.1 Messaging Concepts

Messaging comprises certain basic concepts, such as the following:

- **Channel**
  Applications that use messaging transfer their data via *message channels*. This channel is a logical connection between a *sender* (also referred to as a *producer*) and a *receiver* (or *consumer*), Channels are not a concrete technical implementation, but a concept that abstracts the details of physical communication and defines a clear interface for the exchange of messages.

- **Message**
A *message* contains the actual information that is exchanged between applications. The sending application packs up the desired data, including instructions, into one or more messages to send to the receiving application. The receivers then extract the relevant information from these messages for processing. During communication, a central messaging system assumes responsibility for the reliable transport of messages until they are successfully received by the receivers.

- **Pipes and filters**
In the simplest use cases, a messaging system can transmit a message sent by a sender directly to a receiver. However, additional actions often must be carried out before the messages can be processed by the actual receiver. In such cases, what are called *pipes and filters* perform validations or format conversions so that the receiver of the message can process it correctly.

- **Routing**
Multiple applications, channels, pipes, and filters can be present in messaging-based systems. *Routing* describes the process by which a messaging system determines where a message should be sent to and therefore processed. This decision can be based on specific rules, criteria, or attributes of messages.

  Particularly with more complex logics, the sender of the message often does not know the receiver and the messages are sent by a *message router* either directly to a receiver, another channel, or another message router.

- **Transformation**
The way messages should be formatted in a messaging-based environment may vary from application to application and require *transformation* between the different expected formats. This task is performed by special filters, called *message transformers*, which convert the messages into formats that are identical in content and understandable for a target application.

- **Endpoints**
Endpoints are the entry and exit points of messages through which senders and receivers communicate with the messaging system. At these interfaces (the adapters), messages are either generated (produced) and transferred to channels or received and processed (consumed) by channels via APIs that depend on the respective messaging system.

  These *message endpoints* abstract and encapsulate the code that is dependent on the message system, thus decoupling it from the rest of the application.

6.3.2 Messaging Channel Patterns

Message channels form the basis for communication in messaging-based systems, and various decisions must be made for their appropriate use, including the following:

- Are there one or more receivers?
- Which data should be transferred?
- What should happen to invalid or undeliverable messages?
- How should messages be handled in the event of an error?

The following sections describe some important patterns so that you can choose the right pattern for your purposes based on these questions.

Point-to-Point Channel

Synchronous *remote procedure calls (RPCs)* have an advantage in that they are received and processed by exactly one server-side process. Because the call only reaches the process once, the action is only executed once.

In contrast, with messaging, a message that has been transmitted to a channel could be discovered by multiple consumers in the channel and then processed multiple times.

One possible solution is to restrict access to the channel through the messaging system so that only one consumer can read messages from this channel. However, doing so would prevent the parallel processing of different messages from different consumers.

Alternatively, multiple competing consumers could agree among themselves who is responsible for which message and which consumer should process which message. The result would be a complex logic within the consumers, requiring closer coupling between them since, in this case, all these consumers must know each other and communicate with each other.

A *point-to-point channel*, as shown in Figure 6.16, a feature provided by most messaging systems, can solve the problem as well. This kind of channel ensures that a message transferred to a channel is only processed by one of the *competing consumers*. The synchronization of which consumer should process the message takes place in the messaging system and not through the communication of the consumers.

This pattern enables scalable message processing since it ensures that each message is only processed once and the number of possible consumers can be varied at runtime.

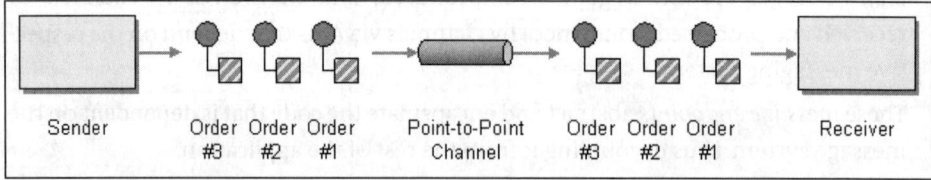

Figure 6.16 Point-to-Point Channel (Source: https://enterpriseintegrationpatterns.com/)

Some examples include the following:

- Queues in general
- *Java messaging system (JMS)* queues

- *RabbitMQ* queues
- *Amazon Simple Queue Service (SQS)*
- *Apache Kafka* topics with a configured *consumer group*

Publish-Subscribe Channel

In some use cases, messages must to be sent to multiple receivers via a *broadcast mechanism* and only processed once by each of them. This scenario is important for financial transactions or for event logging so as to avoid duplicate actions or duplicate data changes, for example. The *observer pattern* defined by the Gang of Four (GoF) provides a description of how objects (*observers*) can receive updates from other objects (*subjects*) and still remain loosely coupled.

Even if the *observer pattern* does not guarantee a one-time processing of a message, it provides a basic framework for managing registrations and notifications. The *publish-subscribe channel* extends this basic functionality of the pattern to include messaging technology.

The pattern works according to a simple principle: The producer generates its messages and transfers them to an input channel. A separate output channel is created for each consumer, from which only they can read and process messages. If a new message arrives in the input channel, it will be copied to all the consumers' existing output channels. This setup ensures that each consumer only processes the message once and that these processed messages can also be deleted from the corresponding channels. Figure 6.17 shows the process schematically.

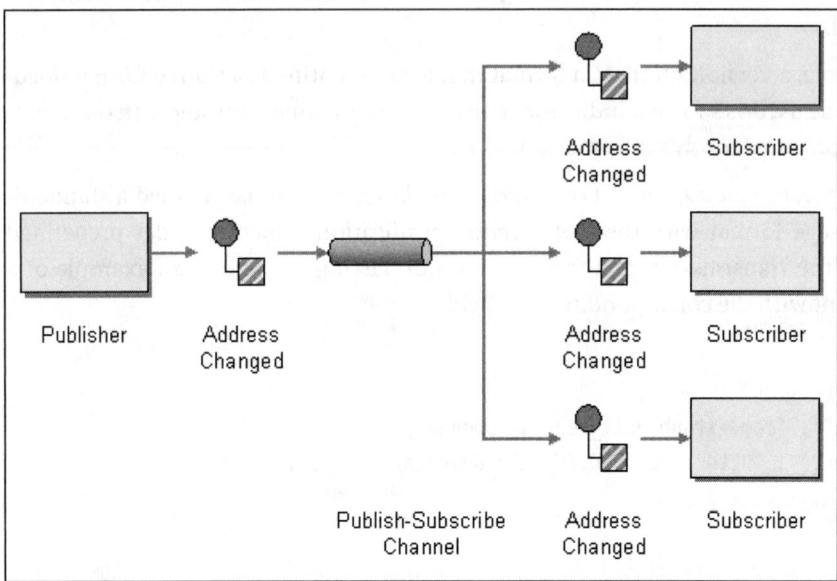

Figure 6.17 Publish-Subscribe Channel (Source: https://enterpriseintegrationpatterns.com/)

In this pattern, consumers are also referred to as *subscribers*, as they register for certain channels or subscribe to these channels. In addition, messaging solutions often provide the option for consumers to log in to multiple channels at the same time using different types of wildcards.

Some examples include the following:

- *Apache Kafka* topics
- *Java Messaging System (JMS)* topics
- *MQTT* protocol
- *Fanout-Exchange* from RabbitMQ

Format Indicator and Datatype Channel

In a messaging-based environment, there's usually an exchange of multiple messages. For the data to be processed successfully, the producer and the receiver must agree on the transmitted user data format.

For example, if a producer transmits multiple messages (i.e., orders, item updates, and evaluations of items), on the consumer side, it must be possible to decide which logic should be used to process the messages. If only one message channel is used for all messages, differentiating between the individual messages is difficult if different data formats are used without any additional information.

This type of information on the transmitted data format can, for example, be provided as an additional metadata field of the transmitted message using what's called a *format indicator*, and the consumer can use logic to decide in what way a message should be processed.

For example, a version number, a format name, or an entire descriptive format document can be used as a format indicator. In these cases, an important step is that the consumer is prepared for this information.

For example, the *Cloud Native Computing Foundation (CNCF)* has defined a standardized message format with the *CloudEvents* specification, which provides predefined fields for the transmission of a format indicator. Listing 6.13 shows an example of a CloudEvent with the corresponding type field.

```
{
    "specversion" : "1.0",
    "type" : "com.github.pull_request.opened",
    "source" : "https://github.com/cloudevents/spec/pull",
    "subject" : "123",
    "id" : "A234-1234-1234",
    "time" : "2018-04-05T17:31",
    "comexampleextension1" : "value",
    "comexampleothervalue" : 5,
```

```
    "datacontenttype" : "text/xml",
    "data" : "<much wow=\"xml\"/>"
}
```

Listing 6.13 CloudEvent Format Indicator Example

CloudEvents

The *Cloud Native Computing Foundation (CNCF)* has combined an open standard and several related specifications under the name *CloudEvents*, which are used to describe and transmit events in cloud-based environments.

The objective of CloudEvents is to create interoperable communication between different services that is independent of the technology used.

As part of CloudEvents, client APIs are also provided, with which the CloudEvents specifications can be used rather easily.

Some major cloud platforms, such as Amazon Web Services (AWS), Google Cloud, or Microsoft Azure use this standard for the communication between their components.

As an alternative to the format indicator, dedicated channels called *datatype channels* can be used for each data format. Ultimately, each data format gets its own channel. In our example, we have a channel for orders, a channel for item updates, and a channel for rating items.

Some advantages of this approach are that a producer can choose the right channel based on the message type and that all messages of a channel can be processed in the same way on the consumer side.

The two patterns can also be used in combination so that multiple channels are used for different formats, but a version specification for the transmitted format is provided via an additional specification.

The topic of versioning interfaces is described in more detail in Section 6.4.

Some examples include the following:

- *CloudEvents* with type specification
- *Apache AVRO object container file* (schema as part of the message)
- *MIME (Multipurpose Internet Mail Extensions) type* of the message in the header
- Different topics or queues for different data formats

Invalid Message Channel and Dead Letter Channel

Errors can also occur in messaging-based systems during the processing of messages. A message may be prevented from processing due to a format error or due to incorrect data. A message might not be transferred to the receiving channel for various other reasons.

If, for example, an incorrect message is detected in a channel, the consumer must react and can either return it to the channel or discard the message. Neither option is ideal because, in the first case, the message would be processed again and would most likely be sorted out again with the same error. In the second case, any serious errors that occur would not be visible and would be ignored, which is the worst possible case. Losing messages should be avoided in almost all cases.

In addition to messages that cannot be processed, the delivery of messages in a channel can also fail. A channel may have been accidentally deleted in the meantime, or the configuration of the producer may be incorrect. Regardless of the cause of the problem, these errors should not be ignored either since they may indicate major problems in the environment or configuration.

Two relatively similar messaging patterns describe solutions to this problem: The *invalid message channel pattern* deals with messages that were received by a consumer but could not be read correctly there, while the *dead letter channel pattern* describes a solution in the event that messages could not be delivered to the correct receiver by the messaging system.

Both patterns use a separate channel in which the messages that cannot be processed are transferred in the event of an error and later analyzed or further processed. The structures of these patterns are shown in Figure 6.18 and Figure 6.19.

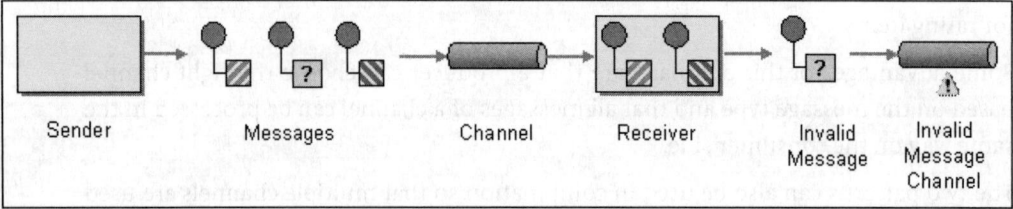

Figure 6.18 Invalid Message Channel Pattern (Source: https://enterpriseintegrationpatterns.com/)

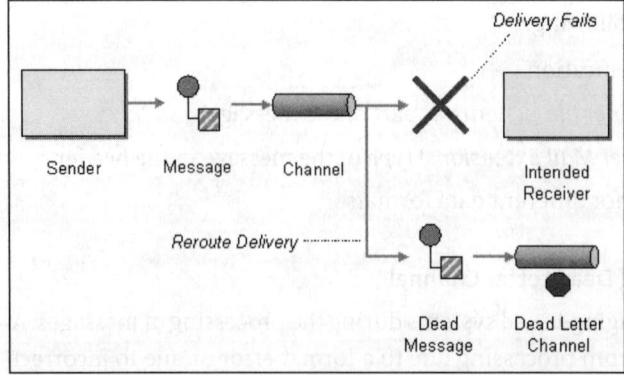

Figure 6.19 Dead Letter Channel Pattern (Source: https://enterpriseintegrationpatterns.com/)

For the invalid message channel pattern, the most important task is that only incorrectly formatted and therefore invalid messages are transferred to the channel. Application-specific errors—such as an invalid triggered RPC call—must still be processed and logged within the consumer.

> **Standardized Definition of Valid and Invalid Messages**
>
> If channels are processed by multiple consumers in parallel, all consumers should have a standardized definition to differentiate between valid and invalid messages. If a message is invalid for one consumer of the channel, this lack of validity should also apply to all other consumers of the channel.
>
> If this rule is not observed, you cannot use the *invalid message channel pattern*, for example, since messages that are classified as invalid by one consumer and removed from the channel can no longer be processed by other consumers.

Messages in the invalid message channel indicate integration problems in the messaging environment: Obviously, a consumer produces messages that cannot be processed by consumers. In a few cases, an automated process can eliminate these format errors. Manual intervention, such as the adaptation of software, is then necessary.

As I mentioned earlier, the *dead letter channel pattern* also uses a separate channel for messages that could not be delivered through the messaging system. In contrast to the invalid channel pattern, these messages are not necessarily invalid, and an automated retry of the delivery process is possible under certain circumstances.

Both patterns can either be developed by the mechanisms provided by the messaging solution used or by mechanisms developed in-house.

Some examples include the following:

- *Dead Letter Exchange* with RabbitMQ
- *Error Channel* in the *Spring Integration Framework*
- *Dead-letter queues* in IBM MQ

6.3.3 Message Construction Patterns

The structure and content of transmitted messages play a decisive role in messaging-based environments, and a basic distinction can be made between three message types:

- **Command message**
  This type of message contains specific instructions about which actions should be called by the consumer in the form of function or method calls.
- **Document message**
  These messages are used to transfer data structures from the producer to the consumer, but without direct instructions on what needs to be done with this data. Consumers can derive corresponding actions for themselves.

- **Event message**
  In this case, the producer provides information about events that have occurred. The message contains information about this event, but no instructions on how to respond to it. The response is the responsibility of the consumer.

The type of message to be used in a communication depends on the type of interaction and the intention of the message transmission. Let's take a closer look at the three message types.

Command Message

As I have explained previously, synchronous calls are executed immediately, and the executing process remains blocked until the server sends its response. This scenario is particularly advantageous if the result is required immediately and needs to be processed straight away.

However, if errors occur during the call and the called service cannot be reached, for example, the supposed advantage becomes a disadvantage, and processing must be aborted.

With an asynchronous call, however, the actual processing in the calling program could be continued in this case, and the incorrect call could be repeated in the background until it can be executed successfully.

To make a call asynchronous and executable via messaging, it must be packaged in a message. The Gang of Four's *command pattern* provides a solution for local method calls. The *command message pattern* transfers this approach to message transmission in messaging and makes it possible to transfer calls as messages in channels.

No specific message format is defined for the implementation of the *command message pattern*. The only requirement is that the method name and all parameters can be integrated and saved in the message.

The *SOAP-RPC* transmission style is an example of how command objects can be serialized as messages. Although these messages are often exchanged synchronously, an asynchronous execution is also possible.

However, proprietary formats are also possible, such as the JSON document shown in Listing 6.14, which contains a command.

```
{
  "command": "WAKE_UP",
  "device_id": "12345",
  "timestamp": "2024-11-21T10:30"
}
```

Listing 6.14 A JSON-Based Command Message

The *command message pattern* is usually used in conjunction with a point-to-point channel since the commands should only be executed once.

Document Message

In distributed applications, the same data is often required in several systems and must therefore be transferred between them.

Traditional integration approaches, such as data exchanges via file transfers or the use of a shared database, can represent a solution. However, these methods also have weaknesses compared to messaging-based techniques in terms of the robustness, reliability, and freshness of the data.

If, for example, data is updated in a shared database for a data transfer, the users of the data must be informed of the update. This update may fail under certain circumstances or only be possible after a time delay. The same applies similarly to file transfers.

Messaging, on the other hand, provides a more robust, reliable, and flexible way of transmitting data. The individual data records can be packaged as what are called *document messages* and transmitted in dedicated channels. The messages only contain the user data and no information about how the data is to be processed. Large volumes of data can also be split into many smaller data packets or messages if required.

For integration solutions, *point-to-point channels* can be used to transfer data from one application to another. If there are multiple consumers who are interested in the data, the messages can alternatively be transferred to a *publish-subscribe channel* and all interested consumers can read the channel data if required. However, if multiple receivers are expected, you must ensure that the transmitted data is not changed by individual consumers, as otherwise you run the risk of inconsistencies.

Similar to the command messages, no specifications exist for the format of the document messages. Listing 6.15 shows an example of the contents of a document message as an XML document: It does not contain any information about how the data is to be processed.

```
<ProductDataDocument>
    <GeneratedAt>2024-11-21T14:30:00Z</GeneratedAt>
    <Products>
        <Product>
            <ProductID>101</ProductID>
            <Name>Notebook</Name>
            <Category>Electronics</Category>
            <Price>999.99</Price>
            <InStock>true</InStock>
        </Product>
```

```
        <Product>
            <ProductID>102</ProductID>
            <Name>Desk Chair</Name>
            <Category>Furniture</Category>
            <Price>149.50</Price>
            <InStock>false</InStock>
        </Product>
    </Products>
</ProductDataDocument>
```

Listing 6.15 Sample Document Message

Document messages can be used not only for integration tasks, but also for workflows that provide for document processing across multiple stages. A document is forwarded to various instances (the *pipes* and *filters*) until it finally reaches the final receiver.

Event Message

In software systems, components often need to be informed about events that have occurred in other components. If the components are located within an application, a corresponding solution can be set up using the *observer pattern* from the Gang of Four.

The same basic concepts of the pattern can also be used in distributed environments when the components extend across multiple systems. However, for reasons of robustness and flexibility, synchronous communication should be replaced by an asynchronous mechanism in this case.

On the one hand, synchronous communication would mean that the subject (i.e., the informing component) would have to know all the components that need to be informed and call them directly. Second, the subject should also be able to deal with incorrect calls and possibly define alternative actions.

If, on the other hand, messaging-based asynchronous communication is used via a *publish-subscribe channel*, the receiving components can be loosely connected to the sending subject. If a receiver is not available at the time of event transmission, the messages can also be consumed by them from the channel at a later time.

The events that occur can then be packaged and sent in messages as so-called *event messages*. In this case, too, no fixed formatting is specified by the pattern, and the messages are similar to *document messages*. In addition, they do not provide any instructions on how to handle the information they contain. The difference between messages lies in the importance of timing and content: With event messages, the timing is usually more important than the content. In some cases, even event messages without content can be completely sufficient to inform about an event.

If other systems are to be informed of events by messages, the corresponding user data is also of interest. Two transmission models for *event messages* are easy to use in practice:

- **Push model**
  In this case, the transmitted message is a combination of an event message and a document message. The message contains the corresponding data or the data of the new status directly and the receiver can use it. The model always makes sense if the transmitted data is small, i.e., can be packed into the message, and all receivers are interested in this data.

- **Pull model**
  In this model, the event does not contain any user data, and the individual observers must obtain the data separately if required. For example, the message can contain a URL or the address of a channel from which the data can be requested. The model is always suitable if the event messages are to be kept as small as possible since they may be generated at a high frequency.

Request Reply

Messaging always takes place in one direction and without a reply. The sender sends a message to a channel and the receiver receives the message from the channel. On the one hand, this asynchronous *one-way communication* enables robust message transmission, and on the other hand, the message creation processes are decoupled from message processing.

Even if this mechanism has some advantages, many components expect a response for executed actions. With synchronous calls, a method call provides the result or at least the information as to whether an execution successful or not.

In asynchronous messaging-based environments, you can use the *request-reply pattern* to implement a *two-way communication*, that is, a call that expects a response.

In this context, two participants are defined:

- Requestor: The *requestor* generates a request as a message and waits for a corresponding response.
- Replier: The *replier* receives the request message, processes it, and sends a reply message.

As shown in Figure 6.20, two channels are used for communication: one channel for the request and another for the reply. Typically, the request channel is a *point-to-point channel*, but *publish-subscribe channels* are also possible if multiple *repliers* need to be addressed and multiple replies are to be processed. The reply channel is always a point-to-point channel since the reply should always be sent to the *requestor* and broadcasting would not make sense. No other receiver can process the reply for a special requestor call in a meaningful way.

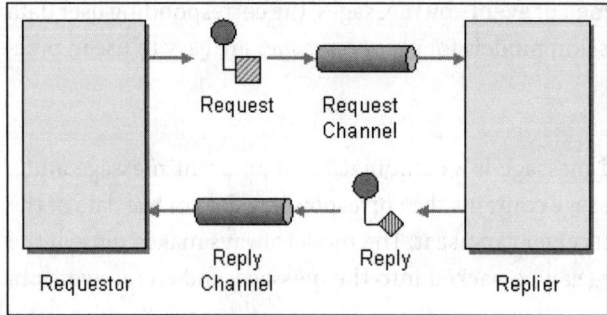

Figure 6.20 Structure of the Request-Reply Pattern
(Source: https://enterpriseintegrationpatterns.com/)

By using asynchronous messaging, the requestor can either wait synchronously for a reply and block the calling thread in the meantime or can continue asynchronously with other tasks and receive and process the reply via a callback mechanism.

If a synchronous, blocking call is used in the requestor, the call logic remains simple, and the advantages of robust message transmission are utilized. However, if, for whatever reason, a replier does not send a response, the requestor must be prepared for this problem and react accordingly. The request-reply mechanism can be used for three different types of request:

- **Messaging RPC**
  In this variant, an RPC is implemented using messaging. *Command messages* are sent to a replier, which executes these commands and then sends back a *document message* in response.

- **Messaging query**
  With this type of query, the requestor sends a *command message* with a search query to the replier, which responds with one or more *document messages* containing the individual result data records. In this case, an important capability is that the requestor can assign the response documents to a request again, which can take place via a sequence number contained in the messages, for example.

- **Notify/acknowledgment**
  Often, only a reply that states that the message has been successfully received and processed is expected. In this case, the communication takes place via *event messages*.

Correlation Identifier

In asynchronous environments with multiple senders and receivers, the processing sequence of messages can be difficult to predict. Even if the messages are read sequentially from a channel, a message that was written to the channel later may have been processed earlier than a message that was transferred to the channel earlier due to a different processing speed within different receivers. Messages can therefore "overtake" each other.

If, for example, a requestor sends multiple messages in quick succession to a number of repliers in a request-reply mechanism, the order of the incoming replies may not correspond to the query sequence.

For this reason, integrating a unique request ID, called a *correlation identifier*, into these messages makes sense. This ID is created by the requestor when the outbound message is generated and transferred to the replier as part of the message. This *correlation identifier* is saved so that later the corresponding response can be linked to right request.

However, a correlation ID is also useful for tracking messages, as it allows a message to be uniquely identified. The ID can, for example, be included in log entries for the message processing steps in order to track how the message was processed.

When the message is processed by the replier, the integrated *correlation identifier* is read and sent back to the requestor as part of the response message. This information can then create the appropriate link to the request for proper processing.

The *correlation identifier* can be transmitted as part of the payload within a message, as shown in Listing 6.16, or as metainformation about the message stored in the message header.

```
{
  "messageId": "abc123",
  "correlationId": "order-98765",
  "event": "OrderShipped",
  "timestamp": "2024-11-21T15:00",
  "data": {
    "orderId": "98765",
    "shipmentId": "56789",
    "carrier": "DHL",
    "trackingNumber": "1Z999AA10123456784"
  }
}
```

Listing 6.16 Inclusion of the Correlation Identifier in the User Data

If using the aforementioned `CloudEvents` specification to format messages, you can also write the data in the message header. Listing 6.17 shows an example in which the Go software development kit (SDK) is used.

```
correlationId := uuid.New()
event := cloudevents.NewEvent()
event.SetID(uuid.New().String())
event.SetSource("source-fellows.com/training-fellow/training-administrator")
```

```
event.SetType("type")
event.SetData(cloudevents.ApplicationJSON, bytes)
event.SetExtension("correlationid", uuid.New())
```

Listing 6.17 Including a Correlation Identifier for CloudEvents (Go)

Some examples include the following:

- *X-Request-ID* or *X-Trace-ID* in HTTP headers
- *Trace Context* specification from the World Wide Web Consortium (W3C) or *distributed tracing*

6.3.4 Messaging Endpoint Pattern

Messaging endpoints abstract and encapsulate the code dependent on the messaging system and thus assume responsibility for receiving and sending messages.

As a rule, the adapter code uses an API provided by the messaging system, for example, a *Java Messaging System (JMS)* or *RabbitMQ* library, to ensure communication with the messaging system. For communication with the application, an application-dependent API is provided for sending or for receiving.

Ideally, a messaging endpoint abstracts the communication with the messaging system in such a way that the application components that use the code don't need to know that the connection is based on messaging.

Based on the following patterns, you'll see that you can also implement endpoints without using messaging system-dependent libraries at all.

Asynchronous Request-Reply

The *asynchronous request-reply pattern* enables the sender to make a request in a non-messaging system-based communication without having to wait for the result of the processing and receive the reply later.

Problem and Motivation

In addition to a server component, many modern applications have an HTML-based interface that is connected via synchronous HTTP calls with REST semantics. Such a connection via synchronous calls is easy to implement and usually benefits in fast server response times.

However, if the response times on the server increase or if asynchronous processes are triggered in the server, these synchronously blocking clients could become a problem. The client code is forced to wait, and the user experience suffers as a result.

Depending on the technological environment, you can use messaging-based solutions to implement an asynchronous call. For clients that are executed in the browser as

HTML/JavaScript, a direct connection of message channels via special network ports, network protocols, or libraries is not usually possible.

Solution

The *asynchronous request-reply pattern* describes a solution based on *HTTP polling*, in which a client sends multiple requests to the server until the result is available.

The sequence of calls, shown in Figure 6.21, follows this order:

❶ The client makes its request, which takes longer to process on the server side.

❷ The server replies with HTTP status code 202 (ACCEPTED) and signals to the client that the request has been accepted and is now being processed. As part of the reply, the client receives a reference or a URL where it can query the status of the started job. Before executing the process, the server should validate the data transferred by the client and, if necessary, reply to the request with a corresponding HTTP status code, for instance, status code 400 (BAD REQUEST).

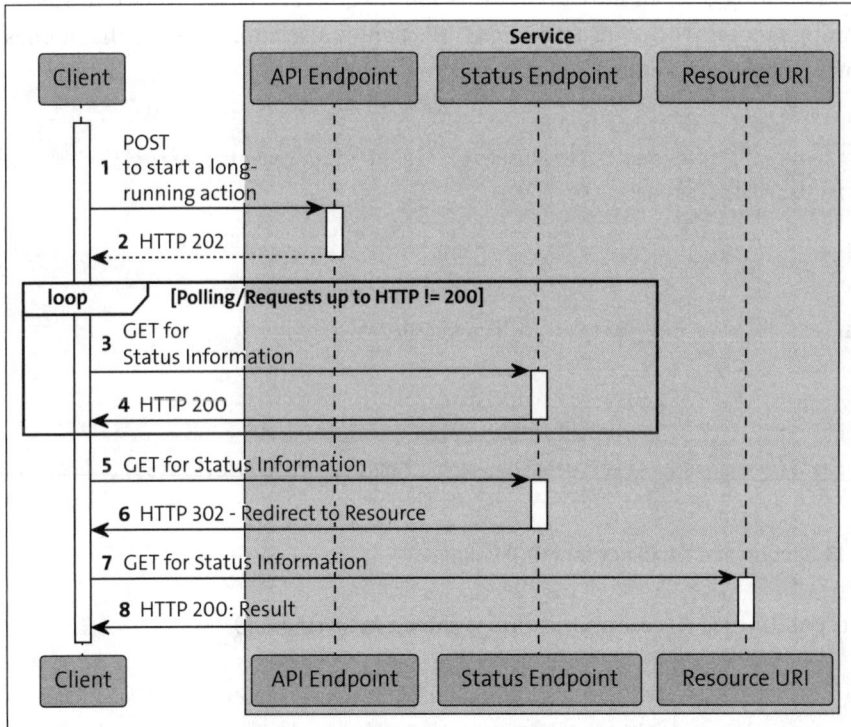

Figure 6.21 Asynchronous Request-Reply with HTTP Polling

❸ After a certain time, the client queries the status of the initiated process via the returned URL or at a dedicated status endpoint with the process reference that the first call returned.

❹ If the process continues to run, the server returns an HTTP response with status code 200 to signal that everything is still OK and that it should make another request after a certain time.

❺ The status request is repeated until the server no longer returns a 200 HTTP status code.

❻ Once the background process has finished, the server returns HTTP status code 302 (REDIRECT) with a URL from which the client can retrieve the result of the processing.

❼ The client retrieves the result.

❽ The client receives the corresponding reply.

Sample Solution

HTTP polling can also be used for communications with clients that are connected to a server through firewalls or other restricted connections. Our next example shows a Go client and the associated HTTP communication.

As shown in Listing 6.18, an initial request is sent to an HTTP-based service to start a long-running process. The corresponding HTTP client is also configured so that it does not follow an automatic redirect.

```
http.DefaultClient = &http.Client{
    CheckRedirect: func(req *http.Request, via []*http.Request) error {
        return http.ErrUseLastResponse
    },
}

response, err := http.Get("http://localhost:8082/start")
if err != nil {...}

if response.StatusCode != http.StatusAccepted {
    log.Fatal("could not start process. got status ", response.Status)
}
```

Listing 6.18 Section of a Go Client for HTTP Polling

The corresponding HTTP communication is shown in Listing 6.19.

```
### request

POST http://localhost:8082/start
Content-Type: application/json

...

### response

HTTP/1.1 202 Accepted
Location: http://localhost:8082/status123
```

```
Date: Sat, 23 Nov 2024 17:25:25 GMT
Content-Length: 0
```

Listing 6.19 HTTP Communication to Start a Process (Request and Response)

After the first call, the client starts a loop to check the status of the process. The code shown in Listing 6.20 illustrates how the client uses the returned URL in the `Location` header for additional requests.

```
resultStatus := http.StatusOK
for resultStatus == http.StatusOK {
    waitTime := response.Header.Get("Retry-After")
    if waitTime != "" {
        sleep, err := strconv.Atoi(waitTime)
        if err != nil {...}
        time.Sleep(time.Duration(sleep) * time.Second)
    }

    newUrl := response.Header.Get("Location")
    response, err = http.Get(newUrl)
    if err != nil {...}
    resultStatus = response.StatusCode
    log.Println("got status", resultStatus)
}
```

Listing 6.20 Loop for Retrieving the Status (Go)

This reply to a status request, with its `Location` and `Retry-After` headers, is shown in Listing 6.21. The two HTTP headers give the client information on how long it should wait until the next request and which URL it should use for that.

```
HTTP/1.1 200 OK
Location: http://localhost:8082/status123
Retry-After: 1
Date: Sat, 23 Nov 2024 17:33:37 GMT
Content-Length: 0
```

Listing 6.21 HTTP Response from Server when Process Is Still Running

When the long-running process is complete, the server responds with HTTP status code 302 and redirects the client to the results page. For this reason, the automatic redirect was deactivated in the client at the very beginning.

```
HTTP/1.1 302 Found
Location: http://localhost:8082/result123
```

```
Date: Sat, 23 Nov 2024 17:39:08 GMT
Content-Length: 0
```

Listing 6.22 HTTP Response from Server when Process is Ready: Redirect to the Result

Listing 6.23 shows the code of the last part of client interaction to retrieve the result from the long running process and the output in the client.

```
newUrl := response.Header.Get("Location")
response, err = http.Get(newUrl)
if err != nil {
    log.Fatal(err)
}
defer response.Body.Close()
bytes, err := io.ReadAll(response.Body)
if err != nil {
    log.Fatal(err)
}

log.Println("result is:", string(bytes))
```

Listing 6.23 Retrieving the Result (Go)

When To Use the Pattern

The *asynchronous request-reply pattern* can always be used in the following situations:

- It is very difficult or impossible to provide callback methods for the notification of asynchronous replies in the client environment, such as in a browser.
- The client can only be connected via HTTP because connection restrictions, such as a firewall, prevent other protocols.
- The server can only be connected via HTTP and does not support current technologies such as WebSockets.

However, its use is not recommended if the following statements are true:

- No time delay can be tolerated when receiving the data due to the wait time between status queries.
- Newer technologies (such as WebSockets, gRPC streaming, etc.) can be used with all communication partners.
- Messaging-based implementations or corresponding libraries can be used directly since there are no connection restrictions.

Consequences

Although technical restrictions exist, clients can benefit from asynchronous communication with a service and thus improve the user experience.

Long Polling

You can use the *long polling pattern* between a frontend component and a backend component to transfer real-time data from the backend to the frontend or vice versa.

Problem and Motivation

Some client applications require data to be exchanged between a server and a client component with as little delay as possible. Good examples include stock market data or chat messages.

With newer technologies like WebSockets or unrestricted network transmissions, data can also be transmitted in web browsers in near real time. However, as soon as transmissions are restricted by firewalls or proxies, for example, browsers might find it difficult to receive data from a service without long delays and without generating load peaks themselves with constant and possibly superfluous requests. Because the client does not know the update time for the new data, it must constantly make new requests (polling) and thus waste resources on the server side.

Solution

Long polling is an efficient way of informing clients about new data in almost real time without using a traditional polling mechanism, which could generate a large number of requests.

With the HTTP protocol, a single request is used to return multiple data records. The connection is kept open, and the server writes its data to the HTTP response body as soon as this information is available, which is also read in portions by the client.

The long polling process, shown in Figure 6.22, involves the following steps:

❶ The client makes its initial request, perhaps also transmitting user data that is important for data retrieval.

❷ The server may answer the request directly and send some initial data back to the client. Importantly, the server does not close the HTTP request connection and thus retains a reference to the HTTP body of the response so that it can continue to write data to the body.

❸ The server itself waits for data received from the messaging system that is to be delivered and leaves the request from the client open.

❹ New data arrives from the messaging system that is to be transferred to the client.

❺ The server writes further data into the already opened response body of the HTTP request.

❻ The client processes the data and waits for more.

❼ The server keeps the connection open and waits for data.

❽ Further data arrives on the server.

❾ The server writes the data back into the HTTP response body, which is still open.

❿ The client can also process the retransmitted data.

⓫ If the connection times out, the client makes another request and the process starts again.

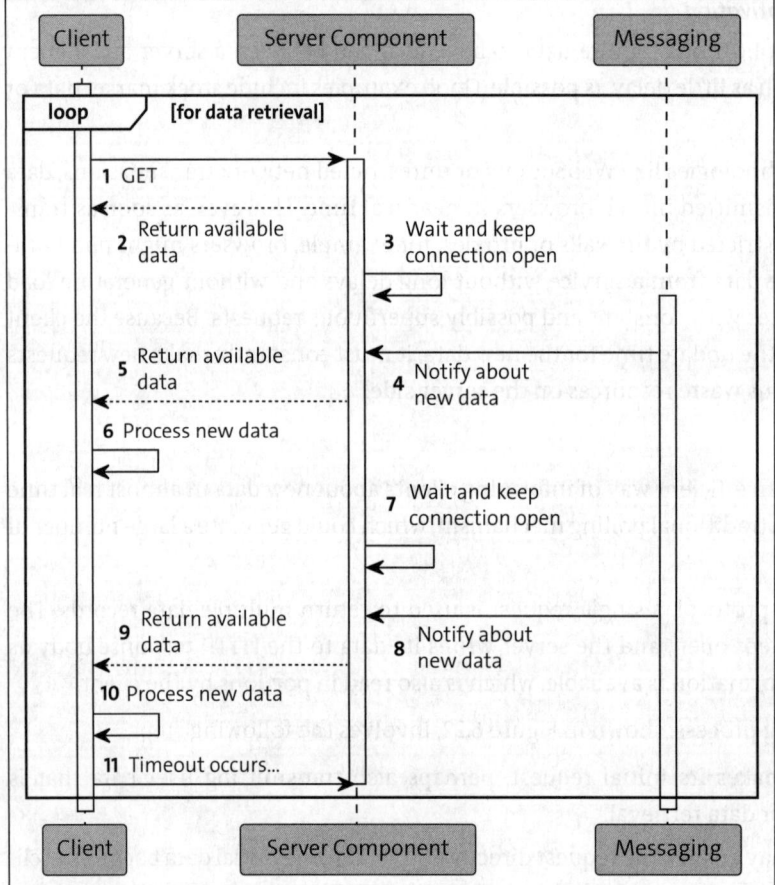

Figure 6.22 Sequence Diagram for Long Polling

Sample Solution

The pattern can be implemented quite easily within modern browsers using pure JavaScript.

Listing 6.24 shows the `callEndless` function with which a corresponding `Reader` is determined for a transferred HTTP response promise and passed on for evaluation. If the HTTP call or the passed promise has a result or has been terminated, the method will be restarted. The result is an endless loop with repetitive calls.

```
/**
 * @param {Promise<Response>} httpRequest to use
 */
```

```
function callEndless(httpRequest) {
    httpRequest.then(response => {
        const reader = response.body.getReader();
        return readDataFromStream(reader);
    }).catch((error) => {
        console.log("error", error);
    }).finally(() => {
        callEndless(fetch("http://localhost:8080"));
    });
}
```

Listing 6.24 Long Polling Implementation with Pure JavaScript

Listing 6.25 shows the evaluation logic of the reader, which always evaluates a new data packet from the server or reads from the transferred reader. In our example, a list in the frontend is updated with some JSON data transferred from the server.

```
let decoder = new TextDecoder();

/**
 *
 * @param {ReadableStreamGenericReader<Uint8Array>} reader
 * @returns {ReadableStreamGenericReader<Uint8Array> | undefined}
 */
function readDataFromStream(reader) {
    return reader.read().then(({ done, value }) => {
        // If no more data is available, close the stream
        if (done) {
            return;
        }
        let jsonString = decoder.decode(value);
        let jsonData = JSON.parse(jsonString);
        addEntryToList(jsonData.message)
        // Process next data
        return readDataFromStream(reader);
    });
}
```

Listing 6.25 Evaluation of the Data in the JS Client

The last step is to start the endless loop to retrieve the data with the following code:

```
callEndless(fetch("http://localhost:8080"));
```

> **Sample Server for Long Polling**
>
> To test the client code shown, a Go server can be implemented that writes several JSON documents in the response body of the HTTP request. Note that no buffering for the `ResponseWriter` may be used for this purpose; instead, the handler function must send the data "directly" to the client.
>
> The following lines of code show the important part of the corresponding handler implementation, in which the buffer of the data is manually emptied or sent to the client by means of a `flush` call.
>
> ```go
> func longPolling(response http.ResponseWriter, request *http.Request) {
>     flusher, _ := response.(http.Flusher)
>     for range 3 {
>         data := []byte(`{"message": "Long Polling"}`)
>         response.Write(data)
>         flusher.Flush()
>         time.Sleep(1 * time.Second)
>     }
>     flusher.Flush()
> }
> func main() {
>     log.Println("Starting server")
>     log.Fatal(http.ListenAndServe(":8080",
>         http.HandlerFunc(longPolling)))
> }
> ```
>
> **Listing 6.26** Sample Server to Test Long Polling: Imports Missing (Go)

When To Use the Pattern

You can always use this pattern when a client cannot be connected via more modern technologies (such as WebSockets or a messaging system API), but you still require data in near real time.

If newer communication channels are possible, the pattern should not be used. More modern technologies work much more effectively, and no proprietary implementation is required.

Consequences

Clients that can only communicate with the server via limited HTTP connections can be supplied with data almost in real time. Short and resource-consuming polling of data is no longer necessary, which reduces the load on the server and client components.

6.4 Patterns for Interface Versioning

The communication between distributed systems only works if both systems agree on how information is exchanged and, above all, agree on the significance of this information.

The meaning of the individual transferred data is often defined as a data schema, for example, using XML schemas, Protobuf (Protocol Buffers) descriptions, or JSON schemas.

Listing 6.27 shows the contents of a Protobuf file in which the structure for a person is defined.

```
message Person {
  string name = 1;
  int32 id = 2;  // Unique ID for the person.
  string email = 3;

  message PhoneNumber {
    string number = 1;
    PhoneType type = 2;
  }

  repeated PhoneNumber phones = 4;

  google.protobuf.Timestamp last_updated = 5;
}

enum PhoneType {
  PHONE_TYPE_UNSPECIFIED = 0;
  PHONE_TYPE_MOBILE = 1;
  PHONE_TYPE_HOME = 2;
  PHONE_TYPE_WORK = 3;
}

// Our address book file message AddressBook {
  repeated Person people = 1;
}
```

Listing 6.27 Protobuf Definition of a Person and an Address Book

However, formats such as *OpenAPI* or *AsyncAPI*, which are used to describe an interface, can also contain parts to describe the data.

Listing 6.28 shows how the returned user data of an HTTP GET call is specified as a list of string values.

```
openapi: 3.0.0
info:
  title: Sample API
  version: 0.1.9
paths:
  /users:
    get:
      summary: Returns a list of users.
      responses:
        "200": # status code
          description: A JSON array of user names
          content:
            application/json:
              schema:
                type: array
                items:
                  type: string
```

Listing 6.28 OpenAPI.yml Containing a Data Schema

However, the requirements for the transferred data usually change during the term of a software project. New fields may need to be included in the data to be transferred, or their format or structure may need to be fundamentally revised.

Without versioning the interface, serious problems can quickly arise in the communication between different parties. Some possible errors include, for example, the following:

- **Lack of backwards compatibility**
  Existing clients do not support the new format after the change and no longer function correctly. The interface contains incompatible changes, such as a restructuring of the data, so that an adjustment to the client software is necessary to continue working with the data.

- **Unclear communication**
  The communication between several teams is made more difficult because it is not clear or comprehensible what the status of the changes being discussed is.

- **Lack of control over changes**
  Without a log of changes, it quickly becomes unclear when a particular change was introduced or which clients could potentially have problems with an upcoming change.

- **Parallel development impossible**
  If only one version of an interface is published, no new functions or adaptations can be implemented without jeopardizing existing clients.

▶ **Frustrated and dissatisfied API users**
If changes to APIs are made without notice and without further information, these APIs can quickly become unreliable and replaced by alternatives.

For these reasons, the versioning of interfaces is unavoidable, regardless of whether the interface is synchronous or asynchronous. In messaging, message formats must be described as well as, in the case of a synchronous transmission, the entire transmission.

In this discussion, we'll only refer to "the interface," meaning both communication styles. The transmitted "messages" therefore represent the exchanged information. In the case of an RPC, these messages are the transmitted parameters; in the case of messaging, these are the actual transmitted messages.

> **Semantic Versioning**
>
> With *semantic versioning (SemVer)*, Tom Preston-Werner, one of the cofounders of GitHub, has developed and described a versioning schema that is now used by many software projects.
>
> Essentially, the specification defines the following three components of a version number:
>
> ▶ MAJOR: This digit within a version number should always be incremented if incompatible changes have been made to the last version.
> ▶ MINOR: Incrementing this digit represents a new version with extensions, but no incompatible changes.
> ▶ PATCH: This digit identifies new versions that contain compatible bug fixes.
>
> Consider the following example: Version 1.3.5—MAJOR 1, MINOR 3, PATCH 5.
>
> You can access the entire SemVer specification at *https://semver.org/*.

Every time you make a change to an interface, you must consider whether the change represents a *breaking change* (i.e., communication breaks down so much that at least one system no longer works correctly), or whether the impact of the change is not so great and all systems continue to work correctly.

Recognizing *breaking changes* is sometimes not as easy as it seems. Caution is advised because you should not only consider technical aspects. Changes can also be semantic. In this context, changes can result in at least one system no longer working correctly despite error-free transmission. For example, these values might be mandatory in a new version of the interface but had been marked as optional in an earlier version. Additional, unnecessary fields that are transmitted, but not ignored by the receiver, can also result in problems.

From a purely technical point of view, five patterns are well suited for providing interfaces with version information and for ensuring that changes to the format do not lead to *breaking changes*, namely, the following patterns:

- Endpoint for version
- Referencing message
- Self-contained message
- Message with referencing metadata
- Message with self-describing metadata

I will briefly describe these patterns next. Figure 6.23 shows where these patterns can be used.

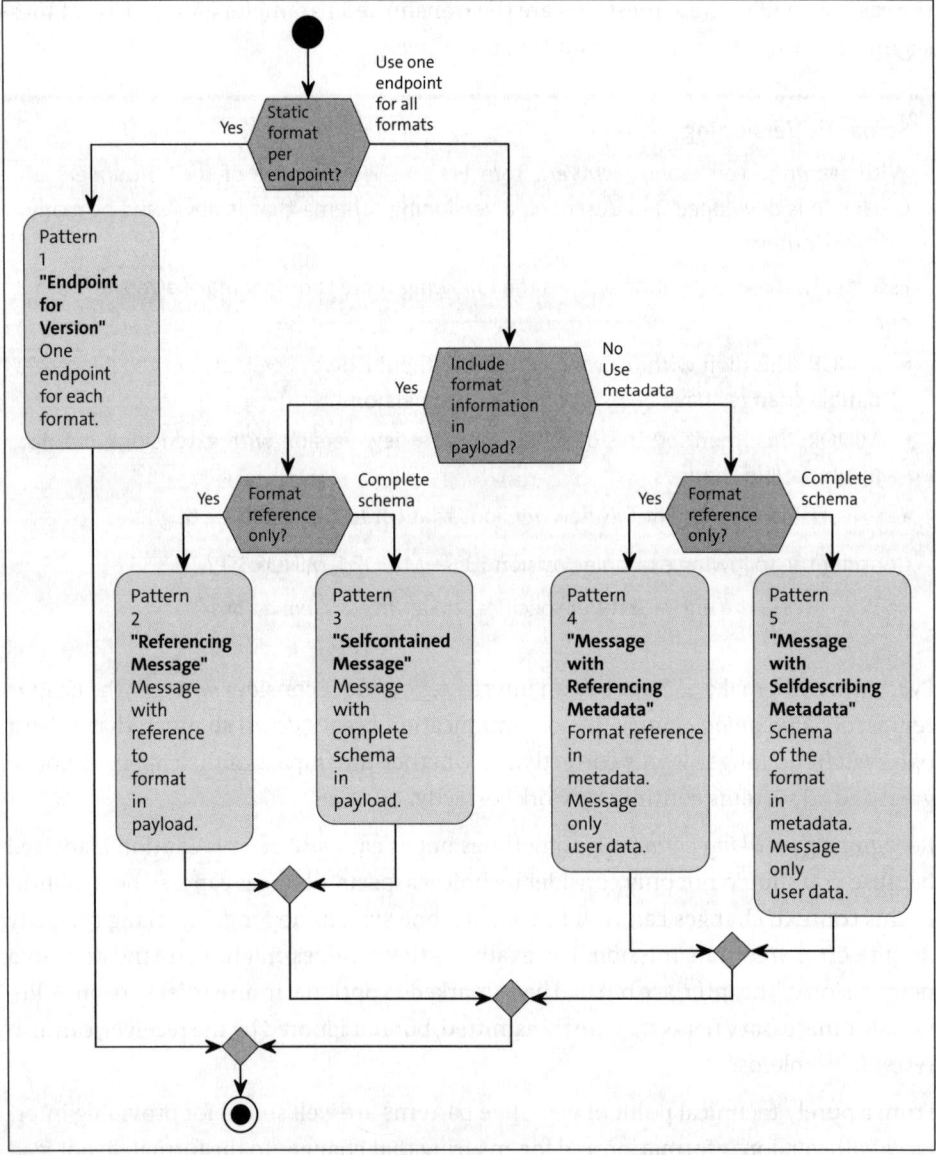

Figure 6.23 Overview of Patterns for Versioning

6.4.1 Endpoint for Version Pattern

When using the *endpoint for version pattern*, a separate endpoint is used for each version of the interface. For HTTP-based REST calls, therefore, a separate URL is defined for each version. In the case of asynchronous communication, a separate channel is used for each version, for example, a topic. In Section 6.3 on messaging patterns, I have already mentioned the *datatype channel pattern*, which corresponds to this approach.

Let's look at some examples of the endpoint for version pattern.

Compatible Version	Corresponding URL
Version 1	*http://host:port/api/v1/article*
Version 2	*http://host:port/api/v2/article*

Table 6.3 Examples of HTTP-Based Communication with REST

Compatible Version	Corresponding Channel or Topic Name
Version 1	`finance.transactions.v1`
Version 2	`finance.transactions.v2`

Table 6.4 Asynchronous Communication with Messaging

Some benefits include the following:

- The implementation is simple, and parallel operations are possible.
- Version information can be seen directly from the endpoint address and therefore provides direct information on the version used.

Some disadvantages include the following:

- Basically, many endpoints must be managed and operated in parallel if older clients are not migrated to newer versions.
- Maintaining and managing the endpoints can be quite time-consuming, for example, in terms of security or load distribution.

6.4.2 Referencing Message Pattern

With the *referencing message pattern*, a message is transmitted that contains a reference to the version of the schema used in addition to the payload. The receiver of the message can use this information to validate the message or use the corresponding code to evaluate the data. The reference can either be a *dereferenceable URL* via which the schema can be retrieved, or a specification that can be linked to a schema by the receiver, for instance, via a schema registry.

> **Dereferencable URL**
>
> A *dereferenceable URL* refers to a URL that, when accessed, leads to a retrievable resource. In simpler terms, if you put a dereferenceable URL into your web browser, it should take you to an actual webpage, file, image, or some other form of digital content.

The CloudEvents format, described earlier in Section 6.3.2, follows this approach but without prescribing a precise definition of the attributes used or their format. The specification only provides a recommendation for certain applications.

If the recommendation of the CloudEvents specification is followed, a reference to the schema or a version specification is transferred within the message, as shown in Listing 6.29. In our example, the `type` attribute contains the `com.example.someevent` specification for the definition of the schema used. In the event of an incompatible change to the data format of the messages, the information is supplemented by a version specification, such as `com.example.someevent.v2`.

```
{
    "specversion" : "1.0",
    "type" : "com.example.someevent",
    "source" : "/mycontext",
    "id" : "A234-1234-1234",
    "time" : "2018-04-05T17:31",
    "comexampleextension1" : "value",
    "comexampleothervalue" : 5,
    "datacontenttype" : "application/vnd.apache.thrift.binary",
    "data_base64" : "... base64 encoded string ..."
}
```

Listing 6.29 Referencing Message Pattern: CloudEvents Example

Alternatively, the defined `dataschema` or `datacontenttype` attributes can also be used to reference the schema since this is possible according to the specification.

The advantages include the following:

- Different messages or versions can be sent to one endpoint. The client can process the messages based on the information they contain.
- For each message, it is clear how it can and should be processed.
- Messages can be stored temporarily and processed later since the schema specification defines how the message is to be processed.

Some disadvantages include the following:

- The receiver may need to filter out or ignore unknown formats.

- The receiver must know the schema or download it from an external source.
- Parallel management of multiple endpoints is not necessary.

6.4.3 Self-Contained Message Pattern

The *self-contained message pattern* describes messages that contain the complete schema within the messages in addition to the payload and can therefore be processed without further queries or additional information from external sources. As part of the Apache Software Foundation's *AVRO* serialization format, for example, a *container format*, the *AVRO object container*, is defined, with which the schema information for the data contained in each message is supplied in addition to the payload.

The example shown in Listing 6.30 shows an AVRO schema description for data records consisting of two fields: `time` and `customer`.

```
{
  "type": "record",
  "name": "test_schema",
  "fields": [
    {
      "name": "time",
      "type": "long"
    },
    {
      "name": "customer",
      "type": "string"
    }
  ]
}
```

Listing 6.30 Sample AVRO Schema

If a new corresponding message with two data records is created in *object container format (OBF)* file and saved, a binary file is created. Figure 6.24 shows a screenshot with the hex dump of this binary file. The first part of the file contains the schema information, followed by encoding information and the actual payload.

Some benefits include the following:

- Messages can be processed independently; querying external sources is not necessary.
- Messages can be stored temporarily and processed later since the schema specification defines how the message should be processed.
- Different formats can be transferred to one endpoint, which can be useful when updating the data format.

- High flexibility and independence of the receivers: You can react dynamically to changes.

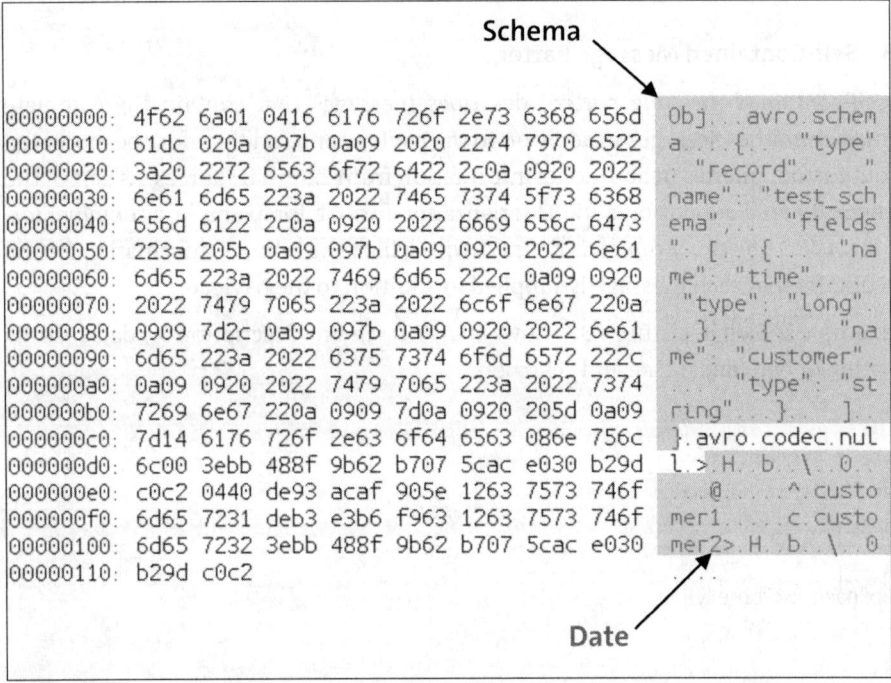

Figure 6.24 AVRO Data Format

Some disadvantages include the following:

- The amount of message data to be transmitted is larger.
- Data transmission becomes redundant if there are many messages of the same type.

6.4.4 Message with Referencing Metadata

With the *message with referencing metadata pattern*, the data and the format or version are separated from each other, but still transferred together.

In this case, the transmitted message consists of two components: a payload area that contains only this data and a metadata area for additional information about the message. In the case of transmission via HTTP, for example, this metadata consist of the HTTP headers that are transmitted with every request. In messaging-based communication, APIs also offer the option of transferring metadata for each message.

The following example shows the approach taken by the HTTP protocol. Listing 6.31 shows a request that contains the Content-Type key in the header with a corresponding MIME type (see information box). As a result, the receiver knows what data can be expected in the body of the message.

```
PUT http://www.source-fellows.com/api/audi
Content-Type: application/vnd.mydatatypeV1+json

{
  "License plate":"RT-X-123"
}
```

Listing 6.31 HTTP Request with an HTTP Header Field for the Format

When the server responds, it can again provide information on the format in the `Content-Type` header.

The MIME type information can also be extended using parameters, such as a version number:

`Content-Type: application/vnd.api+json;` **`version=2.0`**

MIME Types

MIME (Multipurpose Internet Mail Extensions) types are widely used, standardized type or format designations for documents and messages exchanged on the internet. The format designations are subdivided into a media type (e.g., `text`, `image`, or `application`) and a subtype (e.g., `plain`, `jpeg`, or `json`).

Some examples of MIME types include the following:

- `text/plain` for unformatted text
- `application/pdf` for PDF documents
- `application/json` for JSON documents
- `text/html` for HTML documents
- `application/vnd.ms-excel` for Microsoft Excel files

In addition to predefined types, which ensure good interoperability and comprehensibility of various kinds of documents, you can also define your own formats using the prefix `application/vnd`. The `vnd` abbreviation stands for "Vendor" and indicates that this MIME type is a proprietary format.

Some benefits include the following:

- Utilization of the basic idea of the HTTP protocol and thus the use of an established standard.
- Payload is free of format or version information.
- You can use different endpoints for different formats.
- The data transfer overhead is minimal.

Some disadvantages include the following:

- When saving the pure payload, format information may be lost, and the data may be difficult to process later.
- The way in which the information is provided depends on the protocol or API used.

6.4.5 Message with Self-Describing Metadata

The *message with self-describing metadata pattern* describes the entire data format within the metadata for a message using a schema.

Some benefits include the following:

- Receivers of the messages can interpret the data regardless of their prior knowledge of the data format provided, which increases flexibility and can also decouple systems.
- The payload is free of format or version information.
- The ability to transfer different data formats to one endpoint makes it easier to update the data format.

Some disadvantages include the following:

- The volume of data to be transferred is larger since a complete schema is transferred with each request.
- In most cases, schema information cannot be provided directly, so the data must be encoded in Base64 format, for example.

Chapter 7
Patterns and Concepts for Distributed Applications

In modern distributed systems, scalability, consistency, fault tolerance, and transparency are essential for successful deployment. This chapter provides you with basic concepts and suitable design patterns to help you optimize modern distributed architectures with regard to these challenges.

Modern applications are usually part of complex systems that themselves are connected to various other systems and are often operate in cloud environments. Key factors to a successful deployment include scalability, consistency, fault tolerance, and transparency.

Although these requirements are primarily relevant for the development of distributed applications, these demands also apply to the development of standalone applications because these usually access central services as well. Effective communication and collaboration with these services requires an understanding of the server concepts on the client side, too. Some of these concepts and design patterns can also be transferred to client applications.

The patterns presented in this chapter can be divided into the following categories, as shown in Table 7.1.

Pattern	Category	Focus
Stateless architecture	State management	Scalability and simplification
Database per service	Database	Decoupling and autonomy
Optimistic locking	Consistency	Concurrency control
Saga (distributed transactions)	Consistency	Error and rollback handling
Transactional outbox	Consistency	Consistency between events/data and secure delivery
Event sourcing	State management	Historization and traceability

Table 7.1 Assignment of Patterns to Categories

Pattern	Category	Focus
Command query responsibility segregation	Performance	Efficient read transactions
Distributed tracing	Monitoring	Debugging and performance analysis

Table 7.1 Assignment of Patterns to Categories (Cont.)

First, however, we'll look at the topic of consistency and some related theories.

7.1 Consistency

In systems, the term *consistency* refers to a state in which all copies or replicas of a data record are identical and comply with defined rules or restrictions.

Consistency is important in a system to ensure that all processes in the system have a uniform and, above all, correct view of the data. Decisions based on data can only be made correctly if the data is correct. In banking applications, for example, the correctness of the data is crucial, and consistency must always be ensured.

In consistent systems, every read operation always returns the last written and therefore current value, regardless of the copy of a data record used. Write operations are only considered successful if all copies of the existing data records have the latest value after the write operation.

For example, if a client updates a value in an application that is operated on multiple nodes in a cluster, this state must be replicated on all nodes before another client can read the data. As shown in Figure 7.1, this process follows these steps:

❶ A client updates a value on one of the application's nodes. At the same time, another client attempts to read the data from a second node. Before the two operations of client 1 and client 2 return, the data update must be distributed to all nodes.

❷ The values are distributed to all nodes.

❸ The write operation of client 1 and the read operation of client 2 provide the current value.

In this example, the cluster behaves as a closed system with strong consistency, and no client receives outdated data.

The data affected by the need for consistency can also extend over multiple data records. To stick with our banking example, a debit entry in one account must be matched by a credit entry in another account, and only when both actions have been completed will the data be consistent: The data has been transferred from one consistent state to another consistent state.

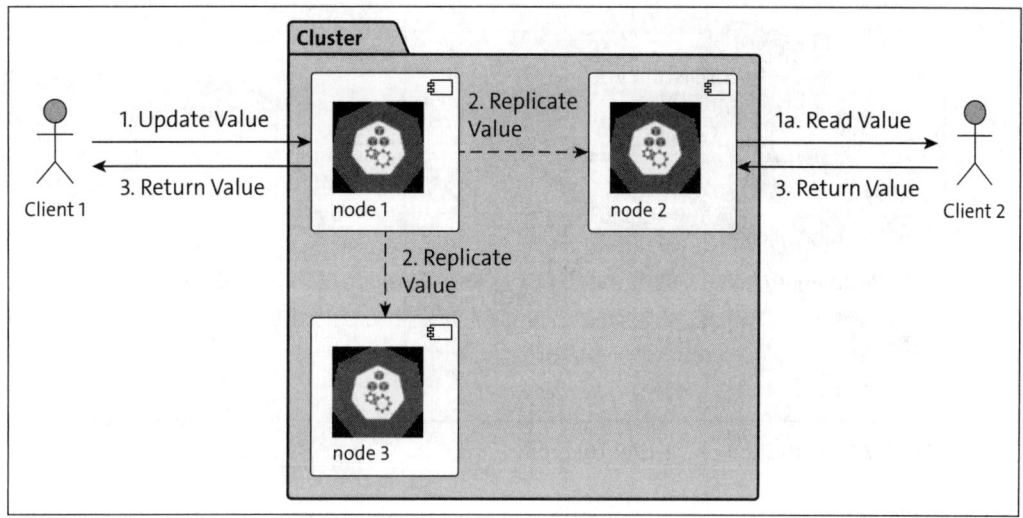

Figure 7.1 Update in a Consistent System

7.2 The CAP Theorem

The *CAP theorem*, presented by Eric Brewer, originates from database theory and states that it is impossible to guarantee the following three properties simultaneously in distributed systems:

- Consistency: Every read access that is executed at the same time receives either the current data or an error. After a write access to data, every copy of this data is also updated.
- Availability: Every request to a system that is not down receives a response.
- Partition tolerance (or failure tolerance): Systems continue to work even if they are divided into multiple partitions and messages are lost between them. This problem may arise, for example, if network components fail.

According to the CAP theorem, a maximum of two out of the three properties can be guaranteed in distributed systems.

This results in three possible combinations you can choose from, as shown in Figure 7.2:

- CA: Consistency and availability
- CP: Consistency and partition tolerance
- AP: Availability and partition tolerance

In distributed systems (described in more detail in Chapter 6), it must be assumed that networks are unreliable and that each system must be equipped to deal with the resulting problems and therefore with partition tolerance.

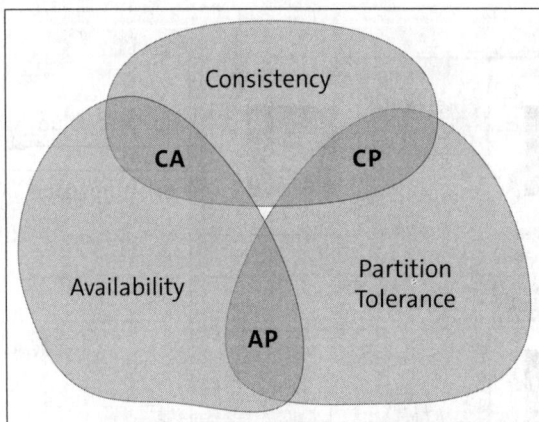

Figure 7.2 Illustration of the CAP Theorem

In relation to the CAP theorem, only two properties left to choose between—in addition to partition tolerance—in distributed systems:

- Consistency leads to CP (consistency/partition tolerance)
- Availability leads to AP (availability/partition tolerance)

If you select consistency as the primary objective, the system must respond to requests with an error message in the event of partitioning caused by a network failure, for example, since the synchronization of data between the single nodes is no longer guaranteed for write access.

In the other case, if you select high availability as the goal, the system will continue to deliver data even if it is partitioned, even if this data is no longer up to date, because the data was updated on another node and could not be replicated due to a network error.

7.3 The PACELC Theorem

The *PACELC theorem* is an extension of the CAP theorem. In addition to consistency and availability, PACELC also takes latency into consideration and therefore also includes a performance analysis.

The theorem states that, even if the system is running without errors and is not partitioned by a network error, a decision must always be made between consistency and latency.

If a system guarantees the consistency of the data, latencies inevitably increase since the data must always be distributed, which can take a long time under certain circumstances.

An adapted decision diagram with the extension for PACELC is shown in Figure 7.3. (The E in the theorem stands for *else*.)

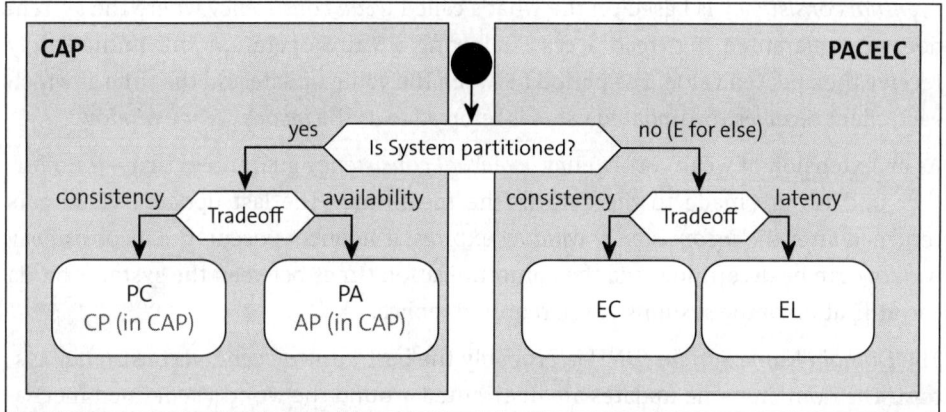

Figure 7.3 Decision Tree for the PACELC Theorem

Multiple systems (e.g., NoSQL databases) decide to proceed differently at runtime and thus accept different compromises.

Thus, if partitioning exists, they decide between consistency or availability, and in situations where there is no partitioning, they decide between consistency or latency in terms of the system's performance.

Accordingly, a different partitioning of the system results in the following possible combinations from which applications can choose:

- PA/EL: If the system is partitioned (P), availability (A) is prioritized; otherwise (E), latency (L) is prioritized. Consistency plays a subordinate role in this approach.
- PA/EC: If the system is partitioned (P), availability (A) is prioritized; otherwise (E), consistency (C) is prioritized.
- PC/EL: If the system is partitioned (P), consistency (C) is prioritized; otherwise (E), latency (L) is prioritized.
- PC/EC: If the system is partitioned (P), consistency (C) is prioritized; otherwise (E), consistency (C) is prioritized.

7.4 Eventual Consistency

I mentioned briefly that achieving strong consistency, as described in the preceding paragraphs, is difficult or nearly impossible in distributed systems. In addition, the CAP theorem states that consistency within a distributed system can only be achieved if either availability or partition tolerance plays a subordinate role.

For this reason, the principle of *eventual consistency* describes a consistency model that can be understood as a compromise solution for all the challenges of the CAP theorem.

Eventual consistency is based on the what's called *weak consistency* where the system does not guarantee that read access following a write operation will immediately receive the updated value. The period between the value update and the time at which each client receives the updated value is referred to as the *inconsistency window*.

As an extension of weak consistency, *eventual consistency* guarantees that—if no further updates are made to the data in the meantime—the last updated value gets returned after the *inconsistency window* expires. If no errors occur, the *inconsistency window* can be determined via the communication times between the systems or via the utilization of the systems or the required replicas.

The *Domain Name System (DNS)* is probably the best-known *eventual consistency* system: The domain name updates are distributed around the world via defined mechanisms and time-controlled caches. However, the data is not always consistent across all systems.

The process of an update in an eventual consistency-based system can proceed, as shown in Figure 7.4, through these steps:

- Client 1 initiates an update (❶ on the left), which is received by a node in the cluster.
- If client 2 requests the same data record at the same time, it receives the value that has not yet been updated (❶ and ❷ on the right).
- After the write operation of client 2, it immediately receives a response (❷ on the left), and the replication of the data is triggered.
- The data gets replicated to all nodes ❸.
- If client 3 makes a request after the replication, it receives the updated value accordingly.

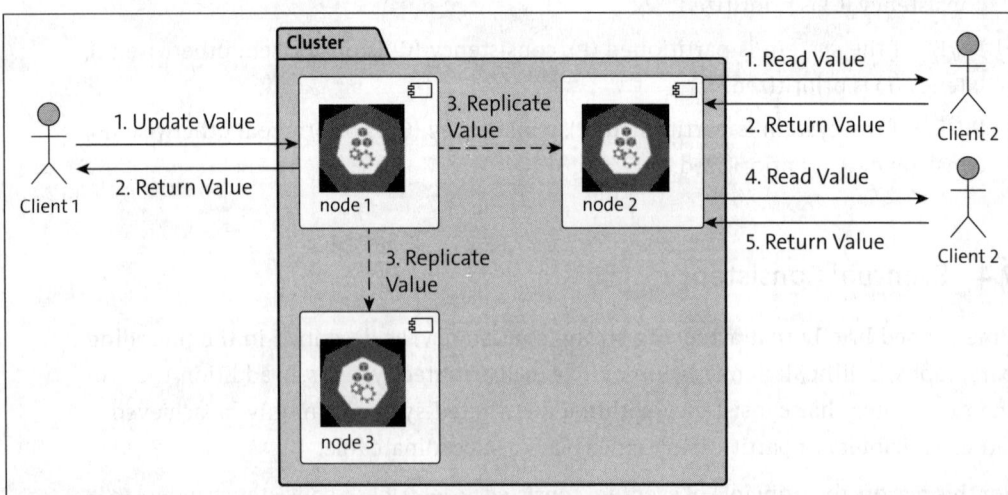

Figure 7.4 Eventual Consistency Process

Eventual consistency is a compromise that can also be used in applications, not just in distributed microservices, to improve performance and scalability. Let's consider the following example of an online store:

- Customer A searches a web store and finds the last available copy of a patterns book. They place it in the shopping cart and complete the order.
- At the same time, customer B also finds the patterns book and adds it to their shopping cart. They also sees that it is the last one available.
- After placing an order, customer A immediately receives an order confirmation. However, the stock has not yet been updated since this update is performed by an asynchronous process in the background. As long as customer B continues to see the book as available, there is no consistency in the overall system.
- After a short time, the stock levels are updated, and the consistency of the data is restored. Customer B now also sees that the last book is already out of stock and cannot complete his order.

Further examples of use cases include updating the order status, displaying product ratings, or synchronizing data between individual distributed services.

The principle of *eventual consistency* is used to bridge short-term failures of system components, but you can also use it to improve the latency of a system since not all states have to be distributed directly and requests can be answered immediately.

The advantages of *eventual consistency* can therefore be summarized as follows:

- **Ensuring high availability**
  Despite network component failures and system partitioning, requests can still be processed without enforcing strong consistency.
- **Good scalability of a system**
  If new nodes are added or existing ones removed from a system, they do not have to be available and immediately synchronously supplied with data during every write operation.
- **Improved, low latency**
  Write access does not have to be replicated across all system components. Read access also does not require any locks in the data sources and can be carried out with local data.
- **Improved partition tolerance**
  The system remains functional even if individual nodes are unavailable.

Some of the following patterns use the principle of *eventual consistency*:

- *Optimistic locking pattern* (Section 7.7)
- *Event sourcing pattern* (Section 7.10)
- *Command query responsibility segregation pattern* (Section 7.11)

▶ *Materialized views*, in which precalculated results of database queries are stored in a physical table (see also Section 7.11 on the command query responsibility segregation pattern).

7.5 Stateless Architecture Pattern

The *stateless architecture pattern* describes a fundamental (and possibly the most important) design principle for cloud-based applications, in which an application does not save any state between individual requests from a client. Each request is processed completely independently of the requests already made, and all required values are provided as parameters.

7.5.1 Problem and Motivation

In modern, distributed applications, for example, in *microservice architectures*, scalability, partition tolerance, and maintainability are at the heart of software architecture. Stateful applications pose a particular challenge in these respects. They store a client-dependent state between the individual requests of a client, which must be managed. In an e-commerce application, for example, this state contains the items a user has placed in their shopping cart or the progress in a multipage form dialog.

If you only have one instance of the application, managing states can be performed in the application's main memory. However, if multiple instances are used, as is common in modern cloud environments, the effort required to manage states increases and makes horizontal scaling of the application more difficult. The state must either be synchronized between the instances, or individual clients are bound to dedicated servers in what are called *sticky sessions* so that their states don't need to be replicated. Such a configuration can be implemented using a load balancer, as shown in Figure 7.5.

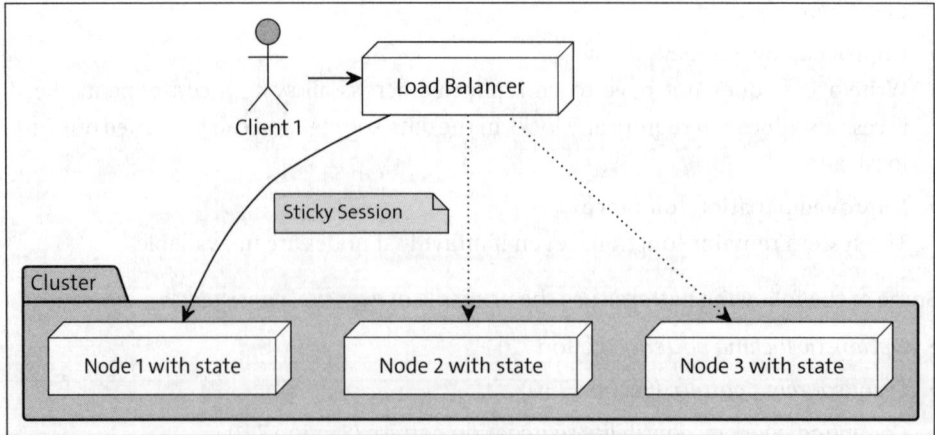

Figure 7.5 Load Balancer and Sticky Session

As shown in Figure 7.6, databases are also often used to store a state, such as a session state for clients. Therefore, a status no longer requires explicitly synchronization between the individual nodes, but instead, a new dependency on a database is created.

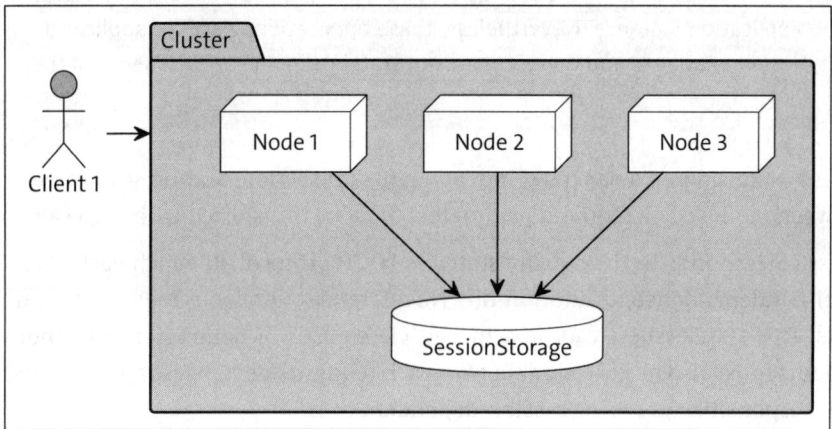

Figure 7.6 SessionStorage as Central Data Storage

These types of dependencies increase the complexity of the application and can cause problems in the event of system failure or during peak loads. In this example, the load on the application and the data volume in the database automatically increase as the number of simultaneous clients increases. The possibilities for scaling are restricted.

7.5.2 Solution

Stateless applications do not have a user-dependent status and therefore do not require status management. The term *stateless* does not refer to the complete absence of data storage, but rather to the fact that no user-dependent data is stored.

The following distinctions are important for differentiation:

- **Application state**
  The application state is the status that is held on the server side *for a client* and contains information about the current context of that client. These are, for example, the shopping carts of customers.
- **Resource state**
  The resource state is the status of a resource managed by the server, which is independent of any interaction by a client. This state could be item data or stock levels, for example. Of course, a client can place an order, for example, and the stock level changes, but this data is *not linked to a client*.

Stateless applications do not use the application state.

> **Stateless Despite a Status in the Database?**
>
> Applications that save a user-dependent status in a database (e.g., in a session database) are often referred to as "stateless" as well because they do not save any status within the application instance. Nevertheless, these applications have an application state that must be managed and therefore are not stateless in the strictest sense of the word.

Each request is considered a separate, independent transaction, and no user state is stored between requests. All required parameters must be transferred with each call.

This process corresponds to the equally stateless HTTP protocol, in which each HTTP request is also independent and autonomous. No reference to previous requests or contexts exists. This statelessness can be achieved via cookies or parameters, but then, these elements must also be passed again with each request. For stateless applications, the client is responsible for saving session-dependent data.

For stateless applications that require authentication, this means, for example, that the relevant authentication information must be transmitted again with each request,

Let's look at two HTTP requests to illustrate this point. Listing 7.1 shows a stateful communication with the server. The first request performs a search and returns a result set and a cookie in the corresponding HTTP header. If the client wants to jump to the second page, the client sends a request to a generic URL that contains the returned cookie in the `Cookie` HTTP header. This session cookie identifies the client to the server so that the corresponding data can be sent back.

```
# Search
GET https://www.google.com/search?q=Golang


# Query the next page
GET https://www.google.com/nextPage
Cookie: ...


# Query the next page
GET https://www.google.com/nextPage
Cookie: ...
```

Listing 7.1 Stateful HTTP Requests (Without Responses)

The communication shown in Listing 7.2, on the other hand, is stateless. The client provides all information with each request, and the server can respond to the request accordingly.

```
# Stateless requests
GET https://www.google.com/search?q=Golang&page=1
GET https://www.google.com/search?q=Golang&page=2
GET https://www.google.com/search?q=Golang&page=3
```

Listing 7.2 Stateless HTTP Requests (Without Responses)

A stateless architecture benefits from the following advantages:

- The application can quite easily be scaled horizontally.
- The caching of data is easier since all information is included with the individual requests, and these requests can therefore be differentiated.
- The application architecture is simpler compared to stateful solutions and therefore easier to maintain.
- No replication between individual application nodes is required.
- Fault tolerance is high for failures of individual nodes, as each instance can seamlessly take over the tasks of a failed node.
- Deployment can be carried out flexibly because no status data requires synchronization.

7.5.3 Sample Solution

As an example, let's take a look at an application for to-do lists that access a database in the background. A user-dependent status or session is not used. The client must re-authenticate with a security token for each query, and the data is saved as resources for all clients.

> **Stateless Despite Authentication?**
>
> In our example, authorization credentials in the form of a *token* are provided with every request within the HTTP header. The token typically contains all the information required by the server to authenticate the user. In the case of *JSON web tokens (JWT)*, a basic validation of the token can also be carried out using a contained signature without access to an external service.
>
> Although the state of an authorization is used within the header, the application itself is not stateful in the classic sense. As long as the server does not store any *session* information about the client and all relevant information is contained in the requests (including the token), the application is still considered stateless.
>
> However, if the server saves the state or information about the token in a database (e.g., to manage a revocation list), then the application becomes stateful.

Figure 7.7 shows a class diagram for a Spring-based application.

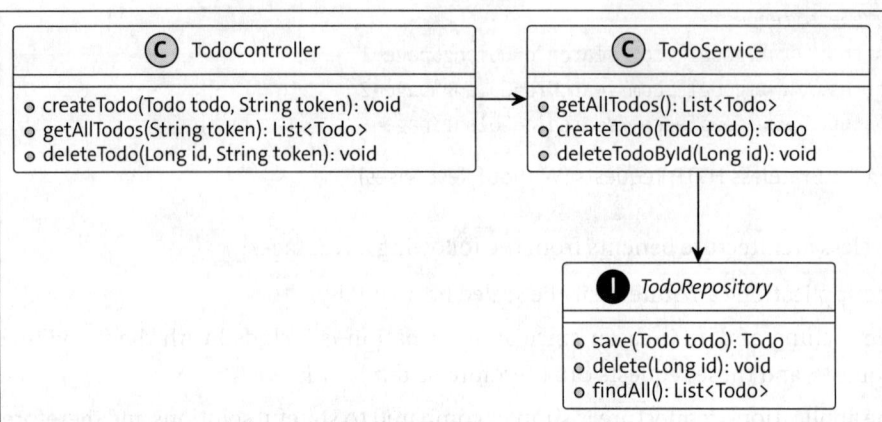

Figure 7.7 Class Diagram of the Application

The application has a stateless HTTP interface that can be used to create, display, and delete Todo objects. A new entry can be created with the HTTP request shown in Listing 7.3. In this context, an important step is that the request is given a token for authentication since no HTTP session is established.

```
# Add a new entry
POST http://localhost:8082/api/todos
Authorization: Bearer xxx
Content-Type: application/json

{
    "title": "Cleanup",
    "completed": false
}
```

Listing 7.3 HTTP Request to Create a New Entry

The code on the server side must check the authentication again for each call. In the example shown in Listing 7.4, this check is implemented separately in each method. Of course, an interceptor or filter could and should be used in this case.

```
@PostMapping
public ResponseEntity<Todo> createTodo(@RequestBody Todo todo,
                        @RequestHeader("Authorization") String token) {

    validateToken(token); // JWT validation
    Todo createdTodo = todoService.createTodo(todo);
    return ResponseEntity.status(HttpStatus.CREATED).body(createdTodo);
}
```

Listing 7.4 Spring Controller Implementation for Accepting the Data

The remaining HTTP requests and controller implementations are similar, as shown in Listing 7.5.

```
# Query all existing todo entries
GET http://localhost:8082/api/todos
Authorization: Bearer xxx

###
# Delete an existing entry
DELETE http://localhost:8082/api/todos/1
Authorization: Bearer xxx
```

Listing 7.5 Stateless Requests for Servers

7.5.4 When To Use the Pattern

Some examples of when to use the pattern include the following:
- The server does not have or should not have a user-dependent state.
- Good horizontal scalability of the application should be given.
- The environment must react to changing load requirements by dynamically starting additional instances.
- Reliability should be achieved across multiple instances, and the failure of individual instances should be insignificant.
- Simple structures in the application should increase maintainability.

7.5.5 Consequences

The use of the pattern has the following consequences:
- The application is easily scalable.
- Fault tolerance is greater since individual instances can be replaced without any problems.
- Maintenance is easier because the application is less complex.
- Applications designed with this pattern can be easily deployed in cloud technologies, for example, in container environments (with Kubernetes) and in serverless environments.
- Overhead is greater for individual requests since all parameters must always be transferred.
- No client-dependent status management is necessary. Workarounds may be required for session management, for instance, if session resources are to be created.
- No session support is available.

7.6 Database per Service Pattern

With the *database per service pattern*, each service is assigned its own independent storage location. The service can only use the data from this data storage (e.g., a database).

7.6.1 Problem and Motivation

One of the main objectives when developing microservices or services for the cloud is to loosely couple them with each other. It should be possible to develop, expand, and operate the individual services as independently as possible.

However, if multiple components use the same data from a shared database, dependencies arise that inevitably link the individual components with each other. Changes to a table or its structure can lead to all dependent components being affected. This interconnectedness also influences the independence of the originally isolated services, which inevitably will require adaptation as well.

If references are defined between individual tables in the database, these dependencies can scaling the database as planned difficult or can lead to performance losses, for example, because the database *locks* or because transactions in the database must be checked and adhered to across multiple services.

If a central database fails, then due to its nature as a *single point of failure*, the functionality of many components may be affected and their functionalities impaired. The same applies to data issues, such as data inconsistencies, which might be caused by one service but can affect multiple services.

The loose coupling and independence of the individual microservices should also make it possible for each service to use the best possible technology for its purpose. This approach applies not only to the programming language used to implement the service, but also, for example, to the persistence technologies used to manage the data. For some services, storing the data in an SQL-based database is advantageous, while other services may benefit from storing the data in a NoSQL-based database. If a common, central database is used, this goal cannot be achieved, and other solutions must be found.

7.6.2 Solution

One option is to implement the *database per service pattern*, where each service is assigned its own independent storage location.

Different strategies can be used to store data in a SQL database, such as the following:

- **Private tables per service**
  With this storage strategy, the data is stored in a shared database but in separate tables. The tables of the individual services are not linked to each other by references (such as foreign keys).

- **Schema per service**
  With this strategy, a central database is still used, but the data of the individual services is stored in different database schemas. This enables explicit security settings and prevents references between the data.

- **Database server per instance**
  If each service uses its own instance of a database, the services are maximally separated from each other. The question as to how the instances are managed is an infrastructure task that will not be discussed here.

These strategies can of course be applied similarly to alternative persistence technologies.

Some advantages of using the pattern include the following:

- The result is a loose coupling since the database does not build up any dependencies between the applications.
- Each service can choose a suitable database technology.
- Changes to the data structure can be made for each service and without dependencies on other services.
- Access to the data of a service takes place exclusively via the application programming interface (API) published by the service.
- Access to the data can be better controlled.
- Resilience increases because no single database can become a single point of failure.

7.6.3 Sample Solution

When implementing the pattern, you must pay particular attention to the separation of the data and the references between them.

In the example shown in Figure 7.8, two services originally used a shared database with two tables. The two tables were linked by the foreign key, in this case, CUSTOMER_ID.

In the solution, the two tables have been split between their respective services, and the interface of the OrderService now contains the CUSTOMER_ID as an additional parameter. Both services are thus separated from each other.

7   Patterns and Concepts for Distributed Applications

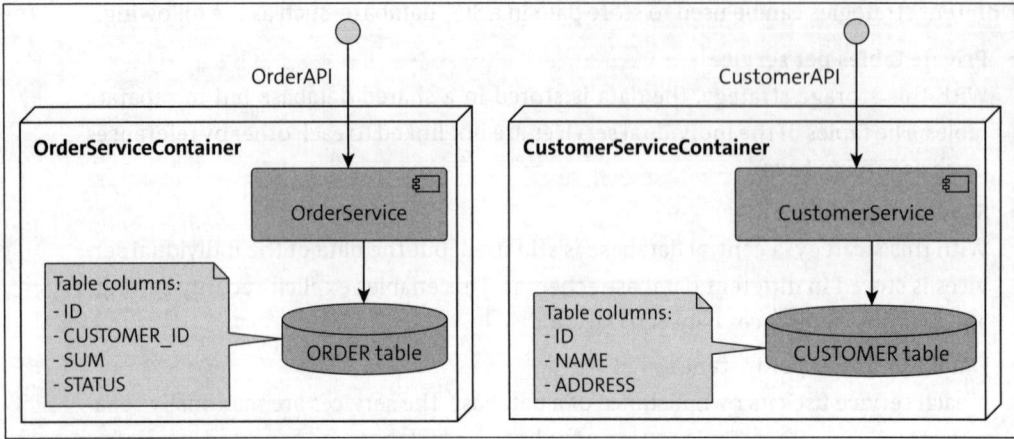

Figure 7.8 Sample Solution Using the Database per Service Pattern

7.6.4   When To Use the Pattern

Some examples of when to use the pattern include the following:

- Services should be coupled as loosely as possible, but this loose coupling is prevented by a common data structure.
- A centrally defined and implemented persistence strategy cannot efficiently fulfill the requirements of all services.
- Access to different datasets should be better separated.
- Scaling problems occur due to too many references in the database.
- Incorrect data generated by one service can have a direct impact on many services.

7.6.5   Consequences

The use of the pattern has the following consequences:

- Services are no longer linked via persistence technology.
- Each service can use the most suitable persistence technology.
- Transactions may be more difficult to implement as they extend across multiple services (see also the optimistic locking pattern for this).
- Data queries may become more difficult since the data is no longer stored centrally in a database.
- If multiple persistence technologies are used, the corresponding products must also be managed.
- Data may need to be replicated between services.
- The pattern can also be used in modularized applications. It does not necessarily have to involve different services.

7.7 Optimistic Locking Pattern

The *optimistic locking pattern* helps avoid data inconsistencies with clients working in parallel. It is based on checking whether the original data has changed since it was last accessed before the data is updated.

7.7.1 Problem and Motivation

In distributed systems or in environments with multiple simultaneous users, situations can arise in which different users try to read and change the same data in parallel. Without a suitable synchronization mechanism between the various write processes, simultaneous updates could result in data inconsistencies. A change that arrives later could then inadvertently overwrite values that have already been adjusted, leading to the *lost update problem*.

The flow of an item management application is shown in Listing 7.15 to illustrate this problem. Two clients read the availability of an item from a shared service and update it. Let's look at the steps of this flow:

- Client 1 reads the item data record with ID 1. In this query, the client receives the value 3 as the current availability of the item.
- In the meantime, client 2 also reads the data in order to update it. This client also receives the current value of 3 for the availability of the item.
- Client 2 updates the local copy of the data and reduces the availability by 1 to the new value 2.
- With another call to the shared service, client 2 transfers the updated item and thus sets the new availability to the value 2, which is now stored in the database.
- Once client 2 has completed its work, client 1 also reduces the availability value by 1. However, it does this operation on the basis of 3, which is what it knows (it read this value at the beginning). Client 1 therefore also sets the new availability to the value 2.
- Client 2 also uses the interface of the shared service to transfer the updated item data and then updates the database.

The result is inconsistent data since both clients have reduced the availability of the item by 1, but the overall database status has only been changed by 1 and not by the value 2.

Basically, using locks in the database is an easy way to solve the problem. Each time a data record is accessed, it would be locked by the client. Other clients can only use the record again after the first client releases the lock.

In distributed multiuser systems, this locking mechanism would lead to considerable effort and performance losses: Clients would have to wait for each other, and in the event of an error—for example, if a client fails to release the lock it set due to an error—such situations must be handled in a specific manner.

7 Patterns and Concepts for Distributed Applications

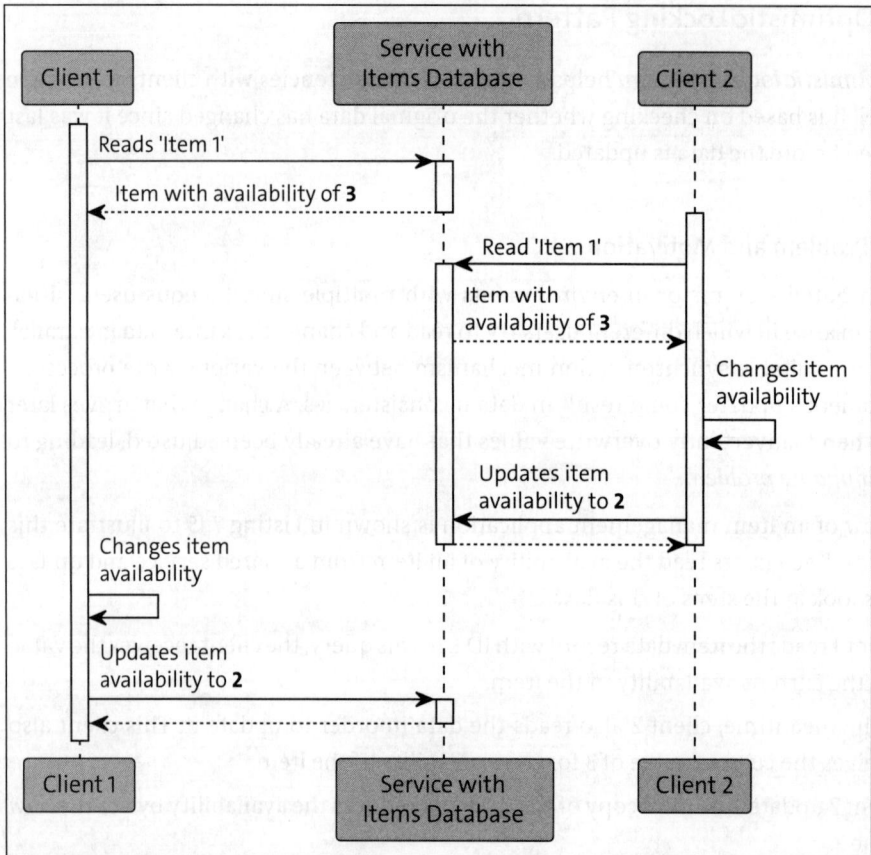

Figure 7.9 Problems with Missing Locks

7.7.2 Solution

With the *optimistic locking pattern*, you can set up an effective conflict detection mechanism whenever a data record is changed simultaneously by concurrent users.

With the *optimistic locking pattern*, each data record receives an additional field containing either a version number or a change timestamp. The corresponding value is also updated each time the data record is updated. Before saving, the system checks whether the status of the data record has remained unchanged since the data was read. If another write operation has already updated the data and the values do not match, the change will be rejected.

Figure 7.10 shows how the process in our example changes by using the *optimistic locking pattern*:

▶ Client 1 reads the item data record with ID 1. In the context of this query, this client receives the current availability of the item of 3 as well as the version number 1 of the item data record.

438

- In the meantime, client 2 also reads the data to update it. This client also receives the current value 3 for the availability of the item as well as the version number 1 for the data record.
- Client 2 updates the local copy of the data and reduces the availability by 1 to the new value 2.
- With another call of the shared service, client 2 transfers the updated item with the version number 1 included.
- The service compares the version number of the transferred item data record with that from the database. In this case, both have the value 1, and the update can be carried out. After the update, the data record in the database is given the new version number 2.
- After client 2 has completed its work, client 1 also reduces the availability value by 1 to the new value 2.
- Client 1 also uses the interface of the shared service to transfer the updated item data and finally enter it into the database. The transferred data record still has the version number 1.
- Prior to the update, the service checks the version information in the database again with the information from the transferred data record. These values are different because the database already has version 2, and the transferred data record is based on version 1; therefore, the update is rejected.

In addition to a version number, a useful practice might be storing information about who last updated the data record. This data can then be used to inform that user if their update is rejected.

The *optimistic locking pattern* can be applied to SQL-based databases either directly via SQL commands or alternatively via extensions to a large number of *object-relational mapping (ORM) libraries*.

For an implementation using SQL, you can add a version number to the UPDATE statements, for example. The current or known version is used within the WHERE condition, and the update includes the version number that has already been increased by 1. If the statement does not return a changed database row, the data has already been updated in the meantime, and the update must be rejected, as shown in Listing 7.6.

```
UPDATE seminars.trainer
SET email="dan.develop@source-fellows.com" and version = 4214
WHERE id = 4711
   AND version = 4213
```

Listing 7.6 SQL Statement for Optimistic Locking

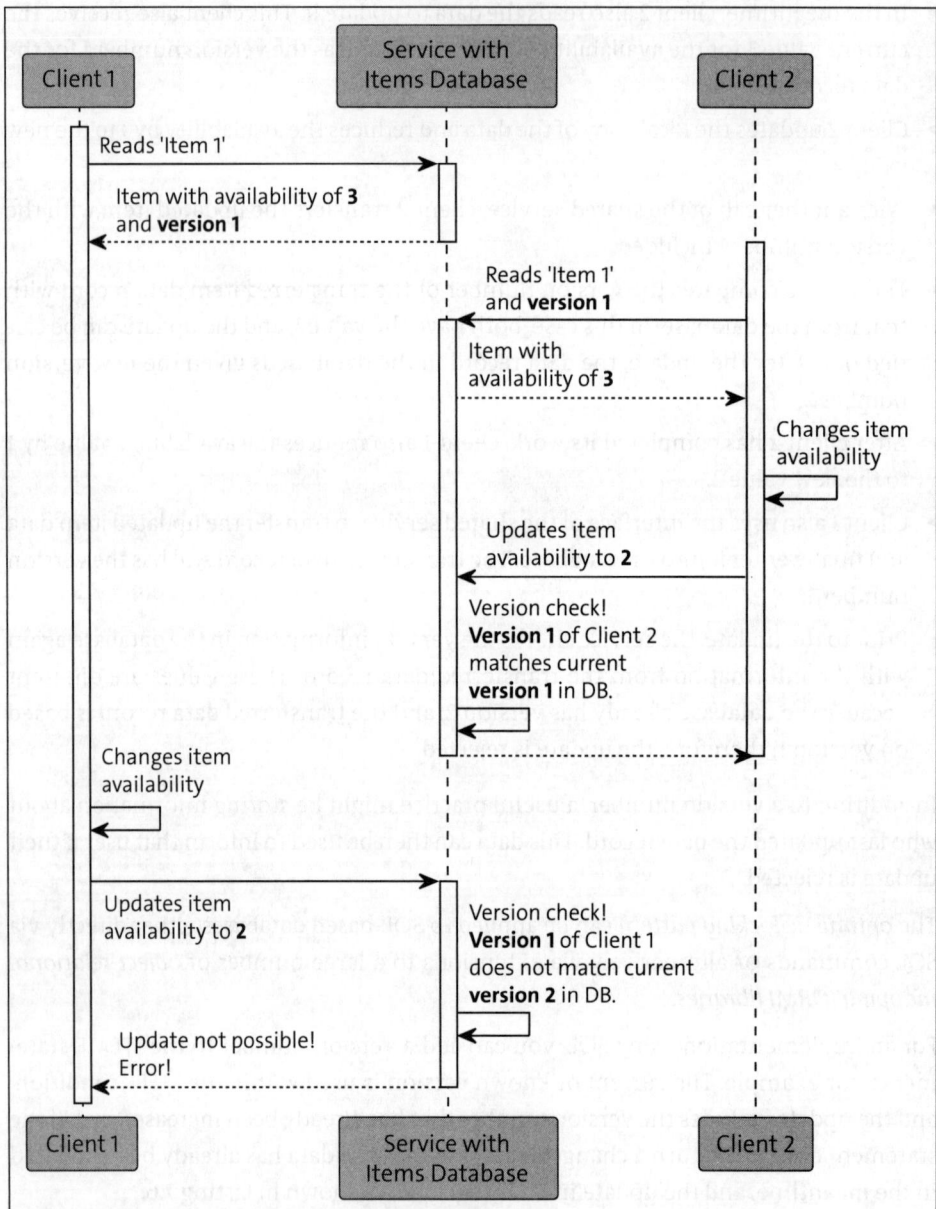

Figure 7.10 Optimistic Locking with Version Numbers

By using an *object-relational mapper (OR mapper)* such as *Hibernate* or the *Java Persistence API (JPA)*, a version specification can often be integrated into entities via declarations. This version information is evaluated during updates and leads to the corresponding messages if necessary.

```
@Entity
public class Student {

    @Id
    private Long id;

    private String firstName;

    private String lastName;

    @Version
    private Integer version;
    …
}
```
Listing 7.7 JPA Entity Class with a Version Specification for Optimistic Locking

7.7.3 Sample Solution

Based on a Go-based example with the *Object Relational Mapper (ORM)* library *GORM* and the additional *optimisticlock* plugin, the following sections describe how you can implement the *optimistic locking pattern*. This plugin automatically manages the version information of the entities and checks these values during updates.

In this application, item data with stock availability is stored in a database. Listing 7.8 shows the corresponding Item struct, which has a gorm.Model as its *EmbeddedType* and thus "contains" its fields. Further, the Item struct has a Version field of the optimisticlock.Version type of the *optimisticlock* plugin.

```
import (
   "gorm.io/gorm"
   "gorm.io/plugin/optimisticlock"
)

type Item struct {
   gorm.Model
   Version      optimisticlock.Version
   Availability int
}
```
Listing 7.8 Item Data Record

Accordingly, the items are stored in the database with the structure shown in Figure 7.11.

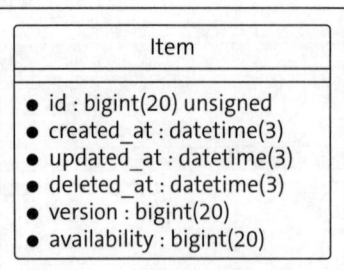

Figure 7.11 Database Table for an Items Database

If data is updated via the GORM library, the plugin automatically adds the version number to the SQL statements and thus ensures adherence to *optimistic locking*. For this purpose, the Version struct from the *optimisticlock* library implements the CreateClausesInterface and UpdateClausesInterface interfaces of the GORM library, whose implementations can be called at the appropriate times. A corresponding class diagram is shown in Figure 7.12.

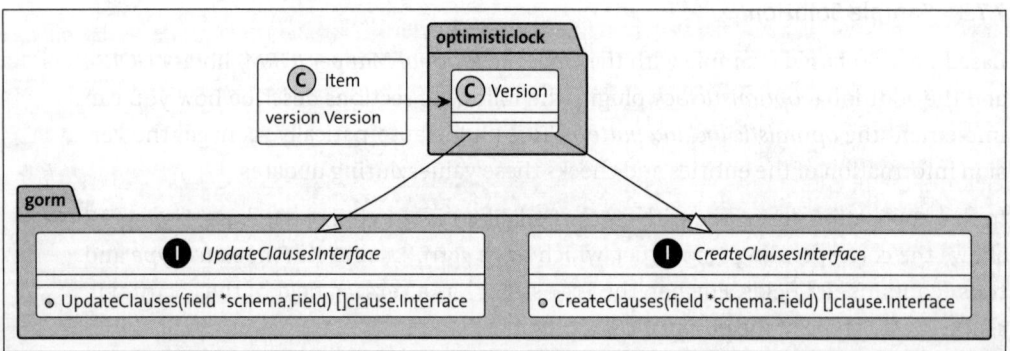

Figure 7.12 Class Diagram of the optimisticlock Plugin

Within the application, the availability or stock level of the item is updated, as shown in Listing 7.9.

```
var item model.Item
err := db.First(&article).Error
if err != nil {
    return err
}
firstItem.Availability = 3
update := db.Updates(&article)
```

Listing 7.9 Code for Updating an Item

This update leads to the SQL statement, shown in Listing 7.10, to update the database.

```
UPDATE `item`
SET `availability`=?,
    `created_at`=?,
    `id`=?,
    `version`=`version` + 1,
    `updated_at`=?
WHERE `item`.`deleted_at` IS NULL
    AND `item`.`version` = ?
    AND `id` = ?
```

Listing 7.10 Sample SQL for Updating an Item

Once the database has been updated, a still necessary step is to check whether data has been changed by the statement since the version information is automatically included in the WHERE condition. This step ensures that the version number has not changed in the meantime.

```
update := db.Updates(&firstItem)
if update.Error != nil {
    return update.Error
}
if update.RowsAffected != 1 {
    return errors.New("no update - version mismatch")
}
```

Listing 7.11 Check for Changes to Database Rows

7.7.4 When To Use the Pattern

Some examples of when to use the pattern include the following:

- Competing access to data records is possible in a distributed multiuser system.
- The *lost update problem* for a distributed multiuser application should be avoided.
- The number of read access operations in a system significantly exceeds the number of write operations, and database locks are to be avoided.
- Applications should remain scalable despite consistency checks.

7.7.5 Consequences

The use of the pattern has the following consequences:

- Data consistency can be ensured in a distributed multiuser application with concurrent access.
- All entities that are checked must have at least one additional field for versioning.
- Version information must be transferred to the client.

- ▶ An audit can be completed with little effort.
- ▶ Every update in the database must include a version check.
- ▶ The possibility of deadlocks occurring in the database is minimized since clients do not have to block access to data.
- ▶ Conflicts can often occur with high write loads or many concurrent accesses.
- ▶ Conflicts during read operations continue to be an issue. If multiple clients retrieve data and this data is updated later, the clients continue to work with the outdated data.
- ▶ The responsibility for resolving data conflicts lies with the client who must decide how to deal with and resolve the conflict in this situation.

7.7.6 Pessimistic Locking

In contrast to *optimistic locking*, *pessimistic locking* mechanisms lock data exclusively for the access of one user. In this case, other clients either have limited or no access to the data.

To avoid data inconsistencies, a client sets a lock when reading the data, which it releases again once it has completed its work. This setup allows the client to work with the data while editing it without any conflicts arising in the overall system.

Pessimistic locking can be achieved explicitly in SQL-based databases using statements such as `SELECT FOR UPDATE`. This statement locks all data that is returned during reading for further access. The lock is removed at the end of the associated transaction.

Since most SQL database drivers use an *isolation level* of `SERIALIZABLE` by default, transactions are already executed in isolation and therefore use implicit pessimistic locks. For this reason, when configuring your database connection, you should pay attention to how this affects your application system.

Pessimistic locking is generally used in applications with a high conflict rate, for example, in booking systems where data integrity is absolutely critical. This pattern is common because locking procedures can lead to performance problems and deadlocks with many parallel data access.

> **Isolation Levels**
>
> In relational database systems, *isolation levels* define the degree of isolation between simultaneously executed transactions. In a way, they can be a mitigation of the *ACID* properties (described in detail in Section 7.8), which require transactions to be executed independently of each other.
>
> The isolation levels determine the extent to which concurrent access is isolated from each other in order to avoid conflicts and inconsistencies. You can choose from among four standard isolation levels:

- Read uncommitted: Transactions can read *uncommitted* changes to other transactions.
- Read committed: Transactions can read *committed* changes to other transactions.
- Repeatable read: Repeated reads of the same data within a transaction provide consistent results.
- Serializable: Transactions are treated as if they were executed one after the other, that is, *serialized*. This isolation level represents the highest level of isolation and is often the default setting for databases or database drivers.

The use of isolation levels can weaken the isolation between transactions, which can speed up the flow of data because less synchronization is necessary, but can also contribute to data inconsistencies.

Figure 7.13 shows a comparison of the *read uncommitted* and *serializable* isolation levels:

- A transaction (Tx1) starts and reads a data record from the database.
- Shortly thereafter, the two transactions Tx2 and Tx3 start.
- Tx1 updates the data record in the database to a new value. The state of the data record is now described as *dirty* in the database.
- Tx2 and Tx3 also attempt to read the data record from the database.
- Tx2 has the *read uncommitted* isolation level, which leads to what's called a *dirty read*. Tx2 receives the value Y, which is not yet committed.
- Tx3 is executed with the *serializable* isolation level and—depending on the database implementation—must either wait until transaction Tx1 has been completed, or Tx3 receives the value before the transaction, as in our example.

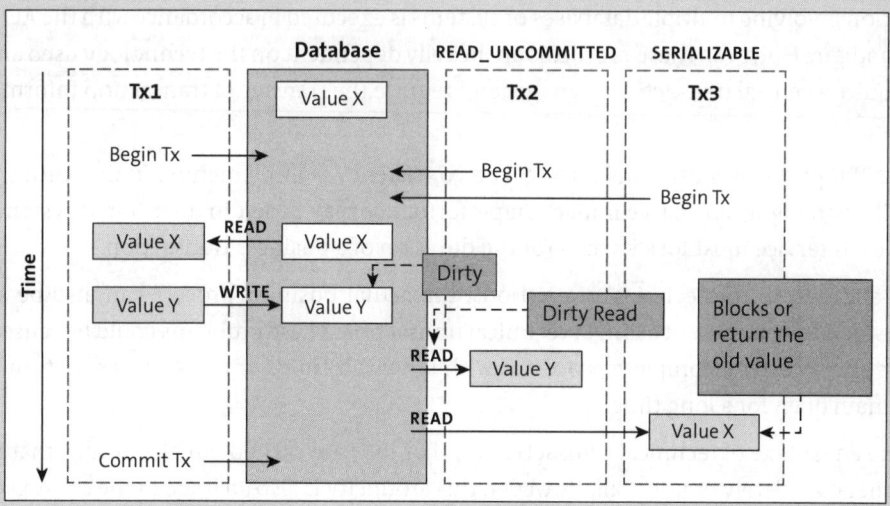

Figure 7.13 Isolation Level Process of Three Transactions

7.8 Saga Pattern: The Distributed Transactions Pattern

The *saga pattern* enables you to achieve data consistency across multiple transactionally protected services in distributed environments.

7.8.1 Problem and Motivation

In modern architectures, systems are often divided into many small, loosely coupled, and independent services or modules. If the *database per service pattern* has been applied, each service also has its own persistence layer to store its data.

These circumstances make it almost impossible to carry out a classic technical transaction that includes data access for several services and thus to consistently implement complex business processes that extend across multiple services or modules in the various databases.

This limitation is partly due to the *ACID paradigm* that governs databases. This paradigm states that a transaction must be characterized by the following features:

- Atomic
- Consistent
- Isolated
- Durable

The ACID paradigm is designed for individual databases and is quite difficult to implement in distributed environments.

One solution is the use of *two-phase commit (2PC) protocols*, which ensure that a transaction involving multiple databases or systems is executed in accordance with the ACID paradigm. However, these protocols are heavily dependent on the technology used and require a central transaction manager and require the transfer of transaction information.

If a 2PC protocol is used and strong data consistency is implemented in line with the ACID paradigm, a negative impact on performance may be felt in distributed systems, as each service must lock its data for the duration of a business transaction.

In addition to the technical restrictions, the actual business process can also be an obstacle to carrying out a longer technical transaction. These problems could be caused, for example, by incomplete processes with long wait times or transactions that must remain open for a long time.

One advantage of technical transactions is that they are carried out atomically, that is, either completely or not at all. However, this atomicity is also an issue in the context of distributed systems: If the process extends across multiple calls, each of which can fail, a business process or transaction can only be performed successfully if all called systems are available and usable at the same time.

7.8.2 Solution

The *saga pattern* makes it possible to ensure data consistency across multiple services in a distributed environment and to manage the corresponding transactions.

> **Saga: Its Origin Story**
>
> You might assume that the name *saga* is an acronym. However, it's inspired by Norse mythology and literature, the epic stories and chronicles that often describe complex and lengthy events spanning long periods of time with many different characters. In most cases, several storylines are interwoven.
>
> In software development, the term *saga* is used to describe the idea that complex, long-running processes are divided into many small, independent steps. The individual steps must be coordinated and, similar to an epic, form an overall picture or the overall process.

Every business process that spans multiple services is implemented as a saga, that is, as a sequence of local, completed transactions.

The *saga pattern* can be implemented in two ways:

- Event-based or choreography: The different services communicate via messages.
- Orchestrated: A central instance coordinates the overall transaction and controls the individual services through calls. Shorter local transactions take place within the individual services.

Event-based communication is implemented via *messaging*, whereby messages or events with information about the process are exchanged between the services.

If a message is processed within a service, a local technical transaction is executed, and the service data is updated. However, if errors occur during local processing, *compensation actions* are defined to undo the previously performed local transactions in other systems. Corresponding messages are also sent for this purpose. Technical errors are repeated within the services and do not directly lead to the overall transaction being aborted.

In an orchestrated implementation, the business process is implemented within a central instance that controls the individual services (and thus the local transactions) and reacts to their results.

7.8.3 Sample Solution

Let's now implement the *saga pattern* through our training registration example: An overall registration process was divided into individual steps using multiple independent services that communicate with each other by exchanging messages.

7 Patterns and Concepts for Distributed Applications

The following four services are involved:

- RegistrationService: This service is responsible for receiving and storing training registrations.
- AdministrationService: All training courses and training dates are managed in this service.
- CustomerService: All participant data is stored and managed in this service.
- NotificationService: This service sends messages to customers, trainers, and administrators.

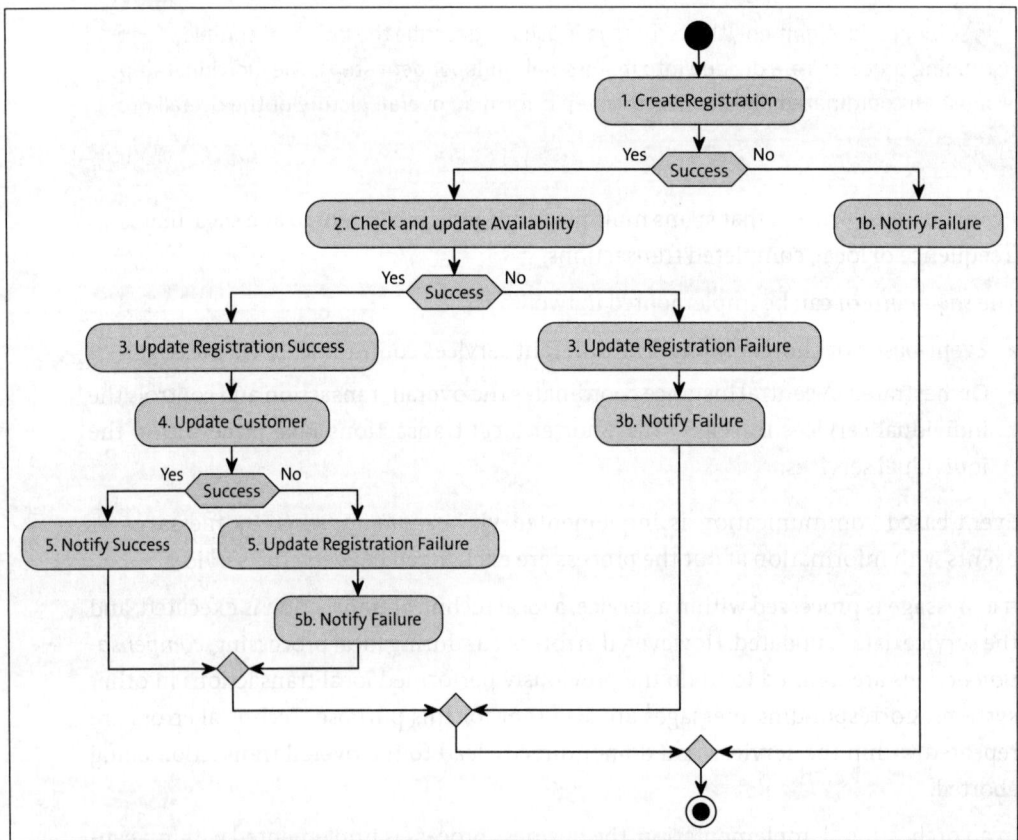

Figure 7.14 Procedure for Booking a Training Course: Registration

The overall process, shown in Figure 7.14, follows these steps:

❶ The customer registers for a training course. This step takes place within the RegistrationService, which sends a message about this event once the customer has been successfully created. If an error occurs, the customer is informed using the NotificationService, which is also connected via messaging.

❷ The `AdministrationService` processes the message about the new registration and checks whether seats are still available for the selected training courses. This service updates the availability in its database accordingly and sends a corresponding message via messaging.

❸ The registration will then be adjusted according to the availability of the training. If no place is available, the registration is canceled, and a suitable message is sent to the customer via the `NotificationService`. In this case, the cancelation represents the *compensation logic* for the error case within an overall transaction.

❹ If seats are still available for the training, the customer data record is updated in `CustomerService` so that an invoice can be generated later.

❺ If errors occur when updating the customer, the registration will be canceled (compensation). Regardless of the outcome, the customer is notified of the result of the registration via the `NotificationService`.

By subdividing the transaction into several steps, sections can be repeated, and retry mechanisms can be used for technical errors within a service. Only in the event of technical errors must the process be terminated.

7.8.4 When To Use the Pattern

Some examples of when to use the pattern include the following:

▶ Distributed applications should remain loosely coupled.
▶ Long-running transactions that span multiple services must be implemented in a distributed application.
▶ Data consistency must be ensured across all services involved.
▶ The possibility of reversal should be provided if a process fails.

7.8.5 Consequences

The use of this pattern has the following consequences:

▶ Long-term transactions can be carried out across multiple services.
▶ Data consistency between different services can be ensured.
▶ The additional compensation actions that must be implemented in the event of errors mean that more work is required. You cannot automatically roll back a transaction.
▶ The long-term split transaction is not isolated because the ACID paradigm would be violated.
▶ Asynchronous communication must take place between the individual services.
▶ Message delivery must be transactionally secured.

7.9 Transactional Outbox Pattern

The *transactional outbox pattern* enables reliable communication with external systems that is transactionally linked to the application logic.

7.9.1 Problem and Motivation

In distributed applications, messages are often exchanged between the individual applications via a messaging system. The messages inform the other systems about changes or the status of a business process.

If the messages are sent within a database transaction, an error that occurs later can cause a rollback. Thus, the data can be rolled back within a database, but messages already sent cannot be revoked or retrieved.

As shown in Figure 7.15, RegistrationService can saves a registration in the database at the start of a local process and then transmit a message to the messaging system. The database actions run within a local transaction.

However, an error might occur after the message is sent, and so, the enclosing database transaction is rolled back. Since the messaging system is not part of the local transaction, however, the message cannot be retrieved, and the information on the newly created registration has already been sent. As a result, the risk of data inconsistency is rather high.

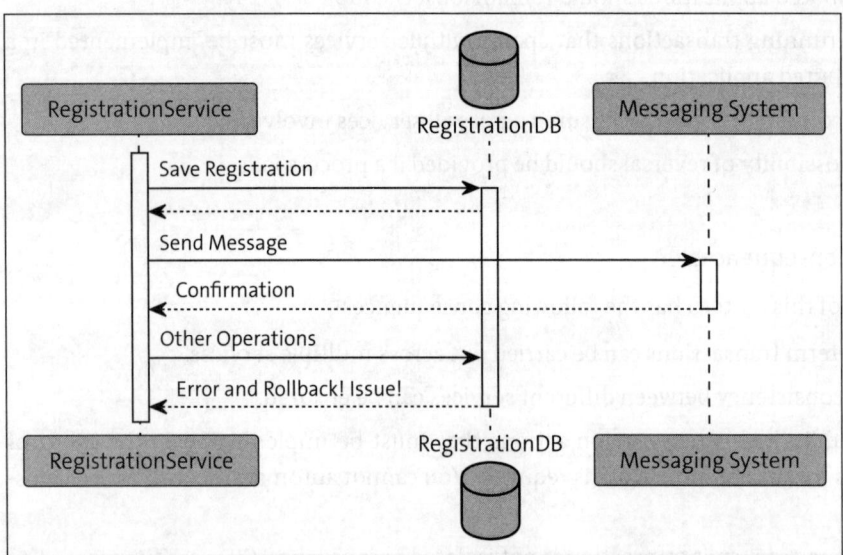

Figure 7.15 Problem Without Transactional Outbox Pattern

A solution to this problem is to introduce a 2PC protocol, which we discussed in Section 7.8 in the context of the *saga pattern*. The database and the messaging system must be addressed within a common distributed transaction, which both of them use for transaction control. First, this configuration is complex since a separate transaction manager is required, and second, few messaging systems support this kind of integration with distributed transactions.

7.9.2 Solution

Related operations that exchange both data in a database and messages via a messaging system can be performed reliably and transactionally protected using the *transactional outbox pattern*.

This pattern divides the overall process, which includes updating the data and sending messages, into two parts, each of which is transactionally protected by the database. In the first step, the data is changed in the database, as shown in Figure 7.16. However, the corresponding message is not sent directly via the messaging system as part of the process as before but instead is stored in a separate database, called the *outbox*.

Both database operations run within a common transaction, and the message (or information that a message should be sent) can be undone together with the data change in the event of an error. If successful, the database contains the data changes and the information on the message dispatch.

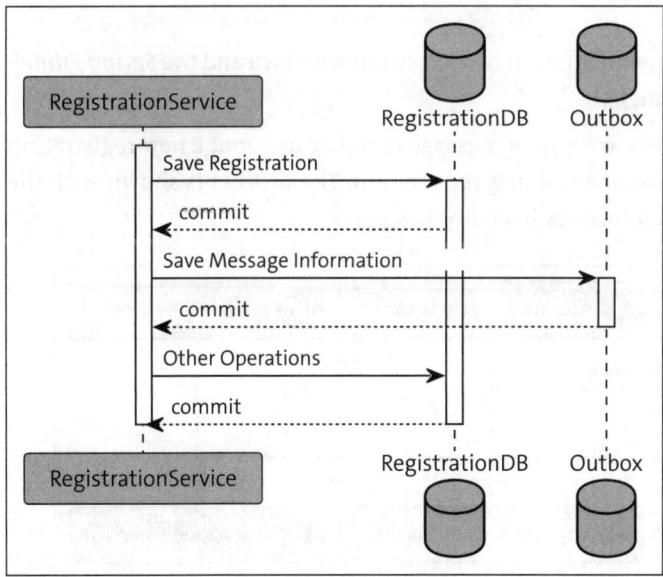

Figure 7.16 Solution Using the Transactional Outbox Pattern: Data Creation

In the second step, as shown in Figure 7.17, a separate component, the *outbox processing* component, takes over the dispatch of the messages. It checks whether there are messages in the outbox database that still need to be transferred to the messaging system. If so, these messages are transferred to the messaging system. The message is only deleted from the outbox or marked as sent once it has been successfully sent via the messaging system. It is therefore also possible to try to send it again.

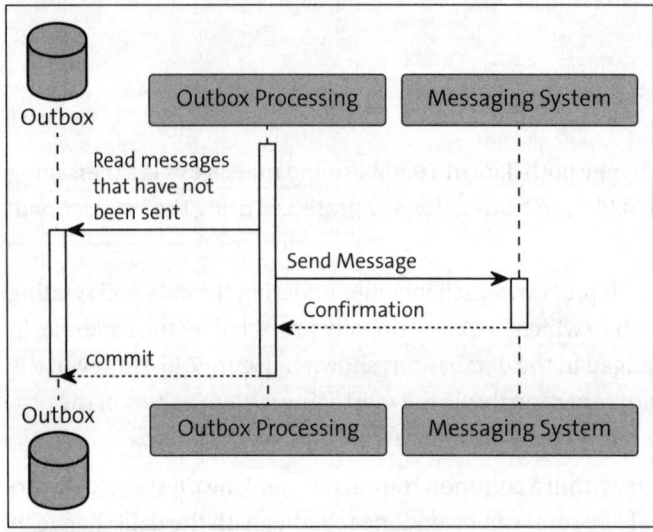

Figure 7.17 Solution Using the Transactional Outbox Pattern: Message Dispatch

7.9.3 Sample Solution

This example shows an implementation of the pattern with Java and the *Spring Framework* or with the *Spring Data JPA*.

During a registration process, we want a message to notify us about a new registration after the successful creation of a training registration. The process is shown with the corresponding classes and interfaces in Figure 7.18.

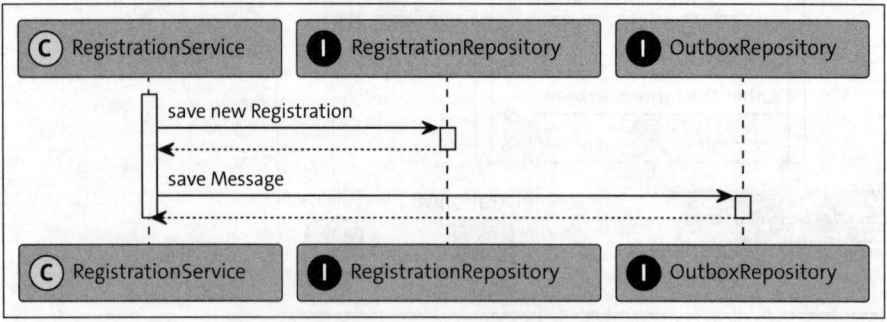

Figure 7.18 Procedure of the Registration Process

As shown in Figure 7.19, this implementation contains the two repository interfaces: RegistrationRepository and OutboxRepository.

Both interfaces extend the CrudRepository interface of *Spring Data* and therefore automatically have functionality for storing and reading data records. In our example, we'll use only the save method.

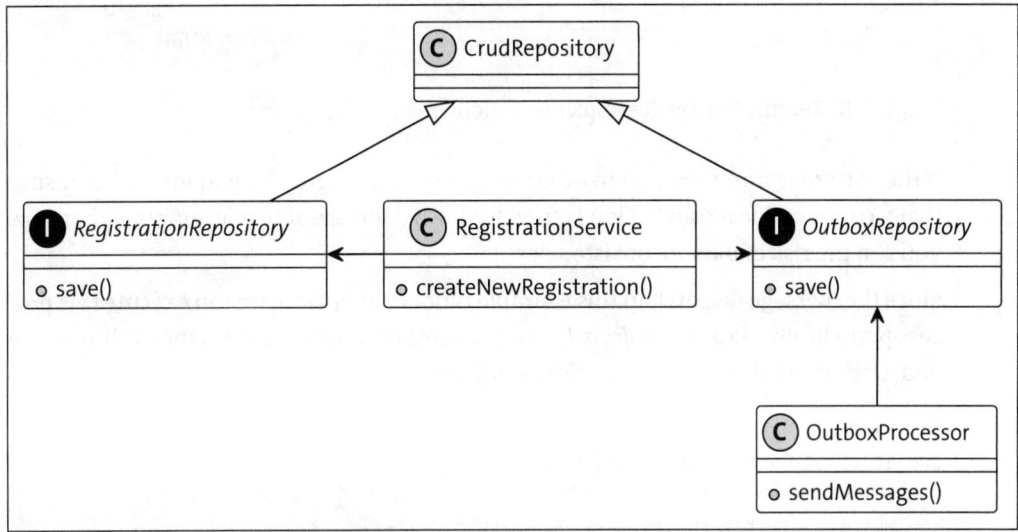

Figure 7.19 Class Diagram for the Transactional Outbox Pattern

The code of the corresponding service implementation is shown in Listing 7.12. The @Transactional annotation is mandatory so that the two database access attempts are actually executed within a joint transaction. Only with this specification are both calls of the corresponding save methods managed together in one transaction and the basic requirement of the pattern fulfilled.

```
import org.springframework.transaction.annotation.Transactional;

@Service
public class RegistrationService {

    private final RegistrationRepository registrationRepository;
    private final OutboxRepository outboxRepository;

    @Autowired
    RegistrationService(OutboxRepository outboxRepository,
                        RegistrationRepository registrationRepository) {
        this.outboxRepository = outboxRepository;
        this.registrationRepository = registrationRepository;
    }

    @Transactional
    public void registerNewAttendee(Registration registration) {
```

```
            Registration saved = this.registrationRepository.save(registration);
            Message message = new Message(...);
            this.outboxRepository.save(message);
    }
    ...
}
```

Listing 7.12 RegistrationService Implementation

If the pattern is implemented in other programming languages, you must also ensure that a common local transaction is used for both database access attempts. Otherwise, you run the risk of data inconsistencies.

Since the message dispatch in this example is not time critical, we can execute this process periodically via a *ScheduleTask*, and the message is deleted from the outbox at the end. Listing 7.13 shows a corresponding implementation.

```
@Service
public class OutboxProcessor {

    private final OutboxRepository outboxRepository;

    @Autowired
    public OutboxProcessor(OutboxRepository outboxRepository) {
        this.outboxRepository = outboxRepository;
    }

    @Scheduled(fixedRate = ...)
    public void sendMessages() {
        log.info("sendMessages");
        this.outboxRepository.findAll().forEach(outbox -> {
            log.info("sendMessages ");
            ...//real transmission logic
            this.outboxRepository.delete(outbox);
        });
    }
    ...
}
```

Listing 7.13 OutboxProcessor Implementation

7.9.4 When To Use the Pattern

Some examples of when to use the pattern include the following:

- Database updates within a messaging-based distributed application lead to the sending of messages, and data integrity must be maintained.
- The commit of a database transaction must lead to a message dispatch, and rolling back the transaction should not initiate any dispatch.
- The use of a 2PC protocol is not possible because either the database or the messaging system does not support integration.

7.9.5 Consequences

The use of the pattern has the following consequences:

- Operations that combine database updates and message dispatch can be performed within a transaction.
- No 2PC protocol is required, which is not supported by many messaging solutions anyway.
- The application is made more complex by the new outbox processing component.
- The order in which messages are sent may need to be checked and ensured. Due to the asynchronous processing of the message dispatch, the sequence of the database changes may not match the sequence of the message dispatch.
- Depending on the implementation of outbox processing, messages can be sent twice and contain a unique ID for idempotent processing so that they are not processed twice.

7.10 Event Sourcing Pattern

The *event sourcing pattern* is an architectural approach in which the complete change history of data objects is stored in the form of events instead of their current status. The current status is reconstructed from the totality of the events.

7.10.1 Problem and Motivation

Applications often use databases to store the current status of a data object. In the case of fleet management software, for example, this status could be the current position of a particular ship. If new position information is received, the status of the ship (i.e., its position) is updated accordingly in the database.

However, this approach also has disadvantages in terms of traceability and auditability: The old status is overwritten with each update, and the history of the objects is lost. In our fleet management example, the paths taken by your ships cannot be fully traced later.

If only the last reported status is saved, subsequent adjustments to the processing algorithm can no longer be applied to past changes. If, for example, not only the current position of a ship is to be saved, but also the distance it has traveled up to that point, you would need more recent data.

In our example of ships' positions, changes to a ship's position are unlikely to occur at the exact same time as changes to any other ship. However, when a bank customer's account balance is updated, concurrent access cannot be ruled out, and all status changes must be synchronized. In databases, this synchronization can only be achieved with a lot of effort, many transactions, and the use of transaction locks. As already described, many database locks can negatively affect the execution speed of an application.

7.10.2 Solution

When using the *event sourcing pattern*, the application data is stored in the database as a list of changes and not as a single state.

For our example of tracking ships, we don't just store the last position; we retain the entire list of position changes. This approach preserves the history of events that has led to the current position, and the status can be reconstructed from all events. However, the current status no longer needs to be saved separately in the application.

The idea behind event sourcing is that storage space and computing power are available indefinitely (at least to a certain extent), and therefore, the current state can be calculated from the history without any problems. In addition, access to individual statuses no longer needs to be synchronized, as only new data records are saved and are no longer modified.

To optimize these state calculations, intermediate states can also be calculated and saved, as shown in Figure 7.20. In this way, you won't need to process the entire history to determine the current state.

This pattern is particularly suitable for use in distributed applications in which state changes are exchanged between the components via messages. A central messaging platform stores all events that have occurred for entities, and each connected consumer can process the complete messages to arrive at the current state.

However, using this pattern also means you can decouple read and write operations within an application. If, for example, you expect a large number of read operations, but only a few write operations, will take place in an application, the current state can be kept in memory, and the updates can be processed asynchronously by messages.

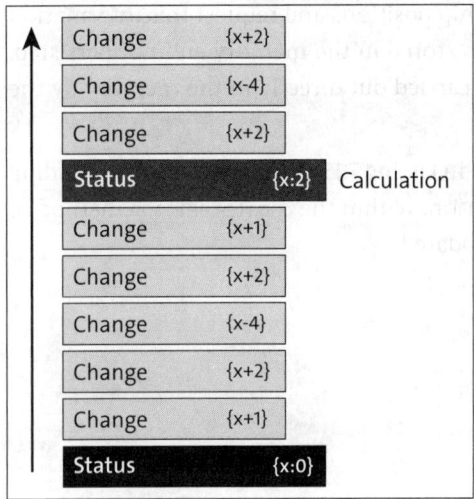

Figure 7.20 The State Results from the Total of Changes

7.10.3 Sample Solution

Our next example shows an implementation of the *event sourcing pattern* in Go to manage our fleet. This example uses *NATS.io*, a messaging system.

> **What Is NATS.io?**
>
> *NATS.io* is a high-performance, lightweight, and scalable open-source messaging system developed for distributed systems and cloud-native applications.
>
> With the *JetStream* extension, messages can be persisted, and thus *event sourcing* can be implemented effectively.
>
> For more information, see *https://www.nats.io*.

The sample application receives updated position information from an external system for the respective ship at regular intervals in the form of ship movements.

Each ship movement is received as a *CloudEvents* event and contains the name, a timestamp, and the current position of the ship. This data is converted into objects of the Movement struct in Go in the application, as shown in Listing 7.14.

```
type Movement struct {
    ShipName   string
    TimeStamp  time.Time
    ToPosition Position
}
```

Listing 7.14 Movement as an Event

7 Patterns and Concepts for Distributed Applications

Many clients are interested in the current ship positions and request this information quite frequently. For this reason, the data is stored in the memory and not persisted. Updates are triggered asynchronously and carried out directly in the memory by the events.

The two `Ship` and `Position` Go structs shown in Listing 7.15 represent the corresponding section of the domain model in the application. Within the `UpdatePosition` method of the `Ship` type, only the current position is updated.

```
type Position struct {
    Lat float64
    Lon float64
}

type Ship struct {
    name        string
    position    Position
}
func (s *Ship) UpdatePosition(pos Position) {
    s.position = pos
}
```

Listing 7.15 Domain Model of the Application

Figure 7.21 shows the corresponding class diagram.

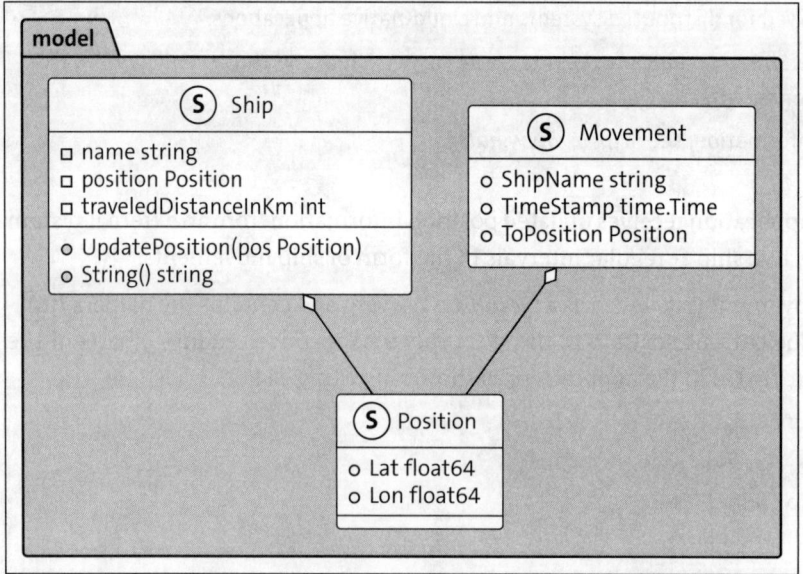

Figure 7.21 Class Diagram of the Application

As I have already mentioned, the Movement objects are exchanged via messaging. The corresponding Go code for receiving messages is shown in Listing 7.16. All errors in the printed source code are consistently ignored for the sake of readability.

```go
nc, _ := nats.Connect(...)
js, _ := jetstream.New(nc)
stream, _ := js.Stream(ctx, "positionStream")
consumer, _ := stream.CreateConsumer(ctx, jetstream.ConsumerConfig{
    Durable:       "cons1",
    AckPolicy:     jetstream.AckExplicitPolicy,
    FilterSubject: "position.new",
})

iter, _ := consumer.Messages()
for {
    msg, _ := iter.Next()
    event := cloudevents.Event{}
    event.UnmarshalJSON(msg.Data())
    handleEvent(event)
    msg.Ack()
}
```

Listing 7.16 Receiving Messages via NATS.io

The objects are processed in the handleEvent function, as shown in Listing 7.17, which uses the data to update the Ship objects in the memory. Any errors that occur are also ignored here.

```go
func handleEvent(event cloudevents.Event) {

    movement := model.Movement{}
    json.Unmarshal(event.Data(), &movement)

    ship, ok := ships[movement.ShipName]
    if !ok {
        ship = model.NewShip(movement.ShipName, movement.ToPosition)
        ships[movement.ShipName] = ship
    } else {
        ship.UpdatePosition(movement.ToPosition)
    }
}
```

Listing 7.17 Message Processing

In the nats.io instance, the *JetStream* extension is used which persists all Movement messages in the sequence in which they occur so that they can be retrieved again at a later point in time.

If the clients are not only interested in the current position of the ships, but also in the total distance they have traveled, then you must adapt the logic and data structure of the application.

The Ship type then receives an additional traveledDistanceInKm field, which can be filled accordingly in a customized UpdatePosition method. Listing 7.18 shows the implementation:

```
type Ship struct {
    name                 string
    position             Position
    traveledDistanceInKm int
}

func (s *Ship) UpdatePosition(pos Position) {
    if s.position.Lat == 0 {
        s.position = pos
        return
    }
    s.traveledDistanceInKm += calculateDistance(s.position, pos)
    s.position = pos
}
```

Listing 7.18 Customized Business Logic in the Application

Old events remain stored in the nats.io messaging system and can be consumed again by the application. This makes it possible to synchronize the database again and thus update it locally. In this way, each client can query not only the position of a ship but also the total distance traveled.

7.10.4 When To Use the Pattern

Some examples of when to use the pattern include the following:

- Reconstructing the current statuses of entities should be possible from their histories.
- Adjustments to the processing of events should be possible retrospectively, for testing purposes or for simulations.
- Traceability and the ability to audit should be provided.
- No persistence of the current state is necessary, which can be determined again if required.

- A decoupling between update operations and read operations is desired. Updates can be entered asynchronously, while the read operations are not affected.
- You expect high load requirements for read operations, but only a few write operations are expected.
- Different representations of the same data and their histories should be available.

7.10.5 Consequences

The use of the pattern has the following consequences:

- The application does not need to save a state. Only unchangeable state changes are saved.
- Traceability and the ability to audit the changes are ensured by a complete history of the changes made.
- Read and write operations can be separated from each other. Write operations can be carried out asynchronously.
- The memory and computing requirements for storing the changes and calculating the current state are constantly increasing. You must ensure sufficient capacity for both.
- Restoring old events can take an enormous amount of time.
- Problems may occur due to changing message formats. All formats must be supported.

7.11 Command Query Responsibility Segregation Pattern

The *command and query responsibility segregation (CORS) pattern* separates read and write operations for a data store to increase the performance, scalability, and security of an application.

7.11.1 Problem and Motivation

The core of every application should consist of a business data model that is represented by classes or data types. The business rules and validations for the individual data objects should be implemented within this model.

Many applications use these models, for example, via *ORM libraries*, to define the database structure and in interfaces for saving or reading the data.

> **Object-Relational Mapping**
>
> *Object-relational mapping (ORM)* is a technique that enables you to manage data from a relational database using object orientation. Tables are represented by classes, and the individual rows are stored in instances of the respective class.
>
> The mapping process is usually performed by libraries. Some well-known examples of libraries include the following:
>
> ▶ Java Persistence API (Java)
> ▶ Hibernate (Java)
> ▶ Django ORM (Python)
> ▶ Entity Framework (.NET)
> ▶ NHibernate (.NET)
> ▶ Doctrine (PHP)
> ▶ GORM (Go)

This uncomplicated path to persistence is suitable for numerous applications, especially those primarily based on basic CRUD operations.

However, as soon as the application performs various database queries and the calling applications receive different data objects in return, for example, data transfer objects (DTO), converting the data can become increasingly complex. You might need to load data via the ORM framework that is otherwise not required for transmission to the client, or you might need to perform validations that are otherwise irrelevant for the current use case. This structure is shown in Figure 7.22.

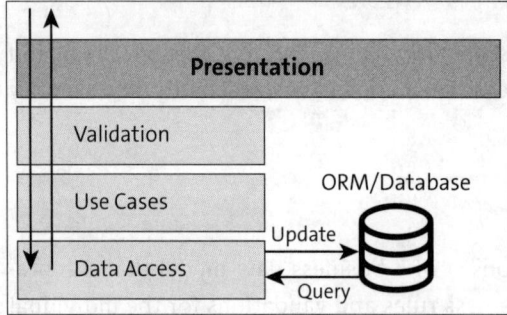

Figure 7.22 All Access to Data via ORM

When larger, interrelated data models are used, database references (such as foreign keys) are usually also required within the database to ensure consistency. Concurrent read and write processes of the data can lead to noticeable speed losses, as each access must be serialized for transactional backup. This serialization forces a sequential execution of the processes, which means that the competing processes must wait for each other until one of the processes is completed.

In addition to the technical restrictions, compliance with access rights can also pose a challenge, as these rights must be checked in the data model or when mapping the data.

7.11.2 Solution

The *command and query responsibility segregation (CQRS) pattern* separates the read and write processes and uses two independent data models. In this pattern, *commands* update the data, and *queries* carry out the read operations.

The separation of the two models facilitates their design and implementation since the data models that are returned to the client as the result of a data query, for example, can be optimized for the respective use case. Because *queries* cannot update any additional data, validating the data in the corresponding models is not necessary.

The individual *commands* that update the data are implemented on a task-related basis, not a data-centric basis. Thus, each action implements a use case and doesn't individual data fields. *Commands* should therefore implement the task "Register user XY for a training course" and not "Set availability of training course to value X."

In addition to the synchronous execution of the write operation, *commands* can also be executed asynchronously by using a messaging solution. This approach additionally decouples the read and write processes and can further optimize write access since access attempts can be performed sequentially, as shown in Figure 7.23.

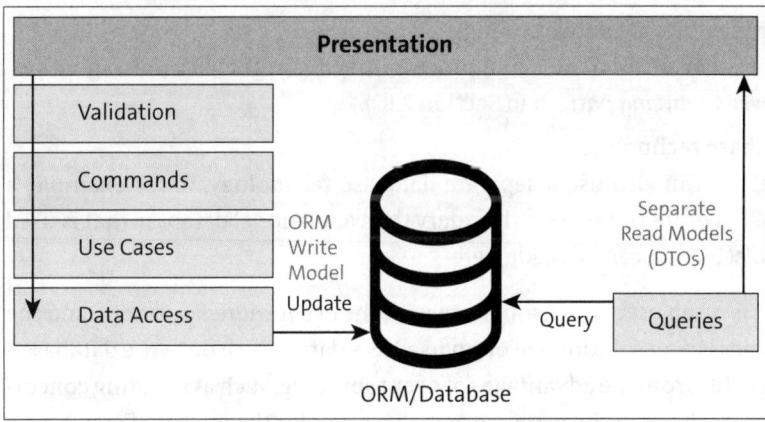

Figure 7.23 Different Models for Reading and Writing

To achieve even better separation of read and write access, the databases can also be separated into two instances: a read database and a write database. You can achieve this separation, for example, by replicating the data from the database for write access using a periodically running job in a read-only database that is only responsible for read access. Figure 7.24 shows a structure of this type.

7   Patterns and Concepts for Distributed Applications

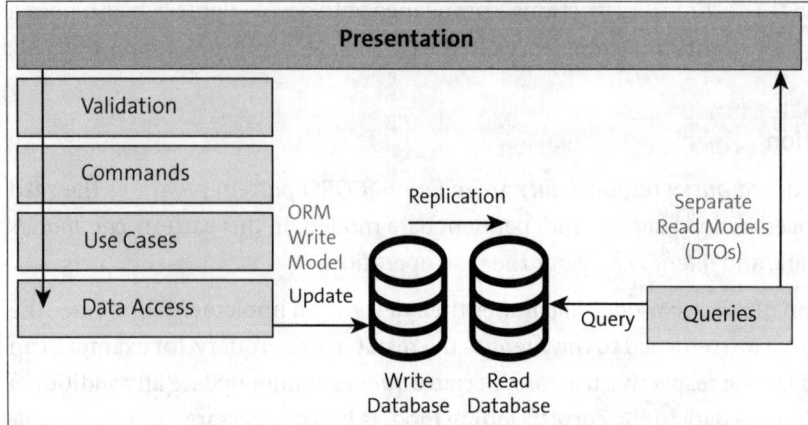

Figure 7.24 Two Separate Databases with CQRS

Even if a 1-to-1 replication is usually simple to implement, various options are available for separating these databases, for instance:

- **Different schemas**
  The structure of the read database may differ from the structure of the write database. *Materialized views*, for example, can save the data directly in the form in which it is to be queried and read later. This saves database joins when querying or unnecessarily complex object mappings when reading.

- **Messaging with events for data exchange**
  The updates can also be distributed and processed as messages via a messaging system (see the *event sourcing pattern* in Section 7.10).

- **Different database technologies**
  The read database can also use a separate database technology. If, for example, a NoSQL database is more suitable for the query than a relational database that is used for storage, a distinction can be made here.

The *CQRS pattern* is often used in combination with the *event sourcing pattern*, and the messages in the messaging solution are often used as a data source or "write database." This approach benefits from the advantages of event sourcing, such as avoiding concurrent access and inconsistent write accesses, but also provides the option of generating *materialized views* anew at any time, which hold the data specifically for individual use cases.

7.11.3   Sample Solution

Let's say a training portal was developed for a training service provider in which the training attendees can register and carry out various actions for their training courses. For example, they can download the training material or their attendee certificates. Figure 7.25 shows the corresponding section of the application's class diagram.

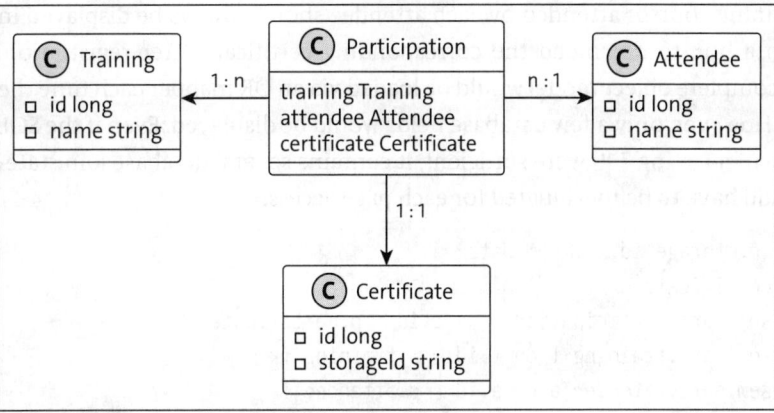

Figure 7.25 Class Diagram for the Sample Application

The data is stored in a relational database, according to the entity relationship diagram shown in Figure 7.26. Mapping is performed by an OR mapper, and a corresponding validation logic is also implemented within the implemented types or classes.

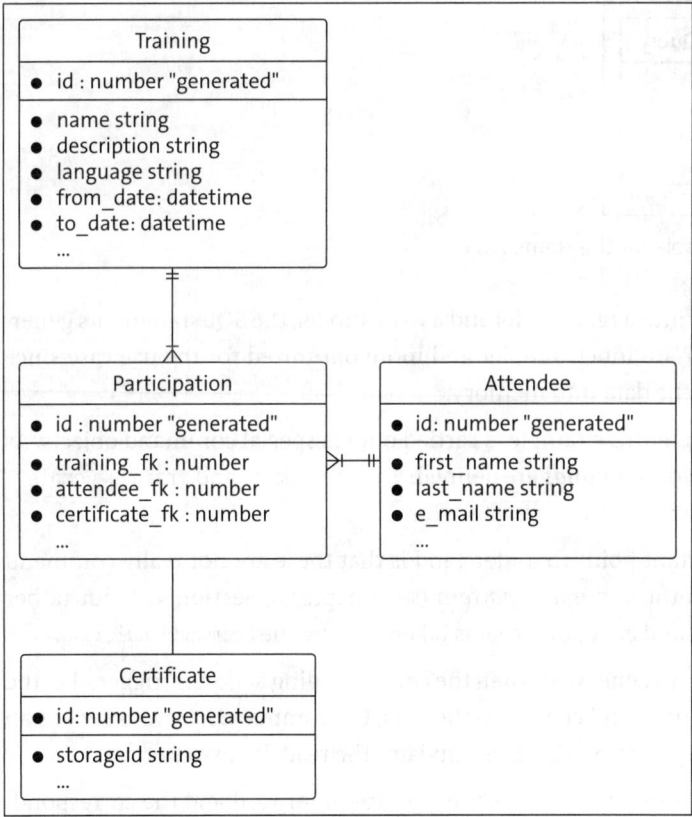

Figure 7.26 Data Model for the Sample Application

A list of the training courses attended by each attendee should always be displayed in the portal with a link to download the corresponding certificate after registration. Therefore, the complete object model would be loaded via an OR mapper each time the view is loaded. However, only a few database fields would be displayed. Even if the SQL statement shown in Listing 7.19 were sufficient, it contains several database join statements that would have to be reevaluated for each page access.

```
select t.name, c.storage_id, t.from_date
from seminars.certificate c
        join seminars.participation p on c.id = p.certificate_fk
        join seminars.training t on t.id = p.training_fk
        join seminars.attendee a on a.id = p.attendee_fk
where a.e_mail = 'John.Doe@source-fellows.com'
```

Listing 7.19 SQL Query for Displaying the Download Link

To optimize access, a second, optimized table (a *materialized view*) is created for this query. This table only contains the required data, as shown in Figure 7.27.

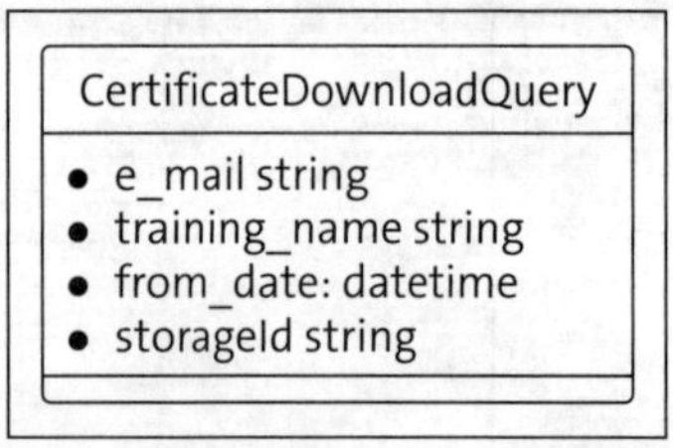

```
┌─────────────────────────────────┐
│   CertificateDownloadQuery      │
├─────────────────────────────────┤
│  • e_mail string                │
│  • training_name string         │
│  • from_date: datetime          │
│  • storageId string             │
└─────────────────────────────────┘
```

Figure 7.27 Optimized Table for the Home Page

By splitting the models into a read model and a write model, the SQL statements generated by the OR mapper are much simpler and more optimized for the use case since they no longer load all the data into memory.

Each update of the data in this example is carried out via special command objects. As shown in Figure 7.28, two *commands* are defined: `BookingCommand` and `TrainingCompletedCommand`. Both are executed by a `CommandHandler`.

At this point, an important point to understand is that these are not really command objects as described in the *command pattern* (see Chapter 4, Section 4.5), but rather parameters for a command execution that is taken over by the `CommandHandler`.

The commands are always generated when the corresponding action is triggered in the user interface. After a successful change to the data, the commands are also sent as an event via the messaging system, which also updates the read database.

The messages are processed by a `DownloadQueryUpdater` afterward and the corresponding information is entered into the table if required.

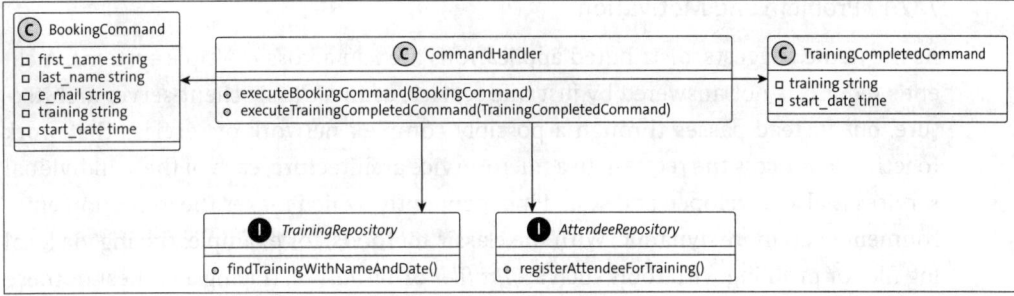

Figure 7.28 Class Diagram of the Commands

7.11.4 When To Use the Pattern

Some examples of when to use the pattern include the following:

- Read and write access in an application should be separated for performance, security, or scalability reasons.
- Only a few write operations are available in the application, and the read speed of the data is extremely important.
- Task-based updates can be used in the business logic, and the domain models must be laboriously checked for consistency and validity.
- You desire a clear separation between the developers who are responsible for the write access (and the associated complex logic) and the developers who are responsible for the optimized display.
- Other systems will be connected via event sourcing.

7.11.5 Consequences

The use of the pattern has the following consequences:

- The complexity of the application increases because separate read and write models must be maintained independently.
- If multiple databases are used, they must be synchronized and maintained.
- If the data is processed asynchronously or synchronized between multiple databases, data consistency issues may arise because it is possible that not all changes have arrived or been processed everywhere.

7.12 Distributed Tracing Pattern

The *distributed tracing pattern* enables you to trace requests in distributed systems across different services. This pattern creates transparency, enables performance to be analyzed, and improves fault diagnosis.

7.12.1 Problem and Motivation

As the name suggests, distributed applications are run across multiple systems. A client's request is not answered by just one service, as in a classic client-server architecture, but instead passes through a possibly complex network of services that work together to process the request. In a microservice architecture, each of these individual services is also developed and scaled independently, which makes the execution environment even more dynamic. With the classic methods, for example, tracing via local log files or profiling with tools such as *JProfiler* or *VisualVM*, tracing a request in these systems is almost impossible. These methods are all designed for local evaluation only.

For example, if the individual services each write their log statements to a local log file, all files from all services must first be collected and then aggregated before they can finally be analyzed.

Even central logging solutions, in which the individual services transfer their logs to a central database, are only advantageous if the logs of the services can be meaningfully aggregated and are not viewed as independent data streams.

Without a common identifier for the individual calls or a standardized format for the log statements, aggregation with subsequent analysis is time-consuming and difficult to implement.

7.12.2 Solution

With the *distributed tracing pattern*, calls can be tracked across multiple services and systems, providing detailed insights into the process and performance.

Distributed tracing works by bringing together multiple individual sections of a complete call at a central point to form a common process. The following terms are important in this context:

- **Trace**
  A *trace* describes the entire process or life cycle of a request that is answered by a distributed system. An example is a search query sent by a client to a service but answered by multiple systems.

- **Span**
  A *span* is a single operation within a *trace*, which in turn can consist of multiple spans. Each span contains the start time and end time, the duration, and other metadata about the operation performed. An example is an HTTP call to an external system.

- **Trace ID**
  Each trace has a unique identifier by which it can be identified. All spans contained in the trace also receive this trace ID to enable subsequent correlation, often a generated UUID.

- **Span ID**
  Each span also has a unique ID by which it can be identified and referenced. You can use these IDs to define dependencies between individual spans. If a span is part of a larger sequence, for example, a span can also reference the span ID of the higher-level span, called a *parent span*.
- **Tag or attribute**
  Each span can contain additional data as key-value pairs. These context-dependent values are referred to as *tags* or *attributes*. For an HTTP request, for example, a tag can be the information about which HTTP method was used to call an action.

As shown in Figure 7.29, distributed tracing works in the following way:

- A client calls service A to start a process that spans multiple services.
- Within service A, a *trace ID* is generated for each request, which identifies this request.
- A *span* is also created for the called operation as part of the higher-level *trace* and sent to a central tracing system.

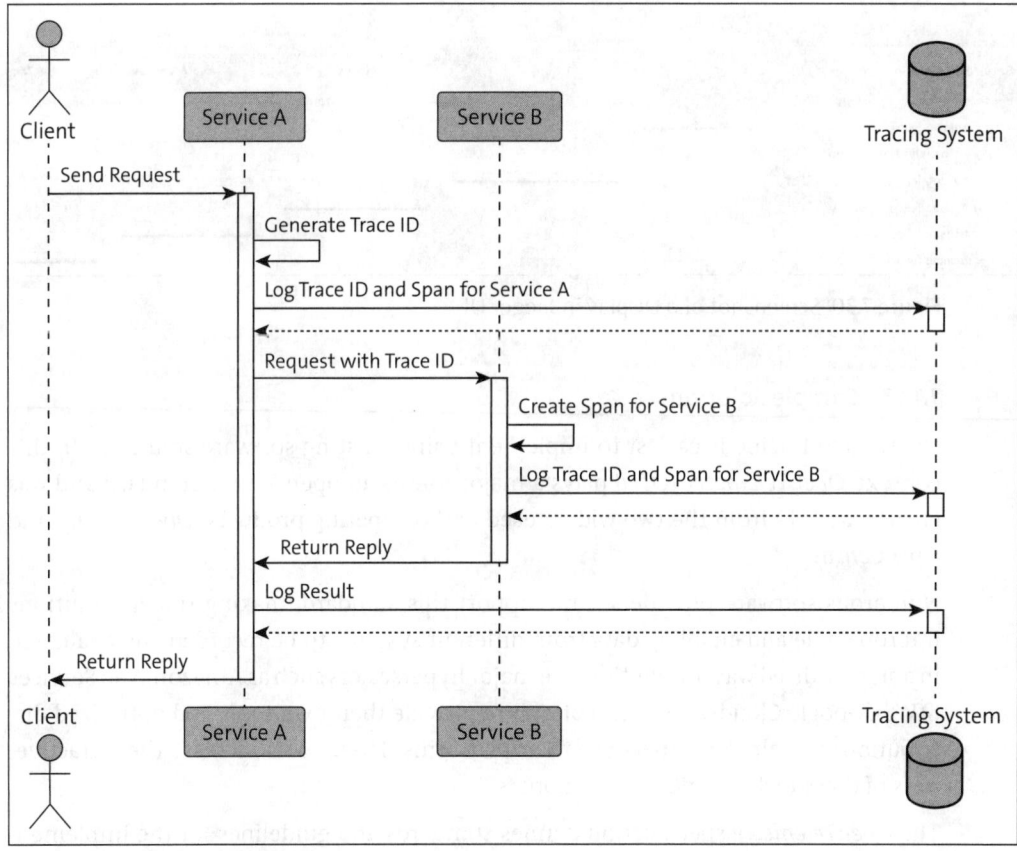

Figure 7.29 Sequence Diagram for Distributed Tracing

- Service A calls service B via a remote interface as part of its operation. During this call, service A transfers the relevant span and trace information to service B. As a result, the individual operations on the two services can be combined with each other in a subsequent evaluation.
- A new span is created within the second service, service B, which is linked to the calling span. This information is also transmitted to the central tracing system.
- The operations are completed, and the results of the operations are also transmitted to the tracing system.

Within the Tracing System, the information is collected in a database and can be analyzed. Figure 7.30 shows a screenshot of the *Jaeger UI*, an open-source tracing solution, in which a simple trace across two systems is shown.

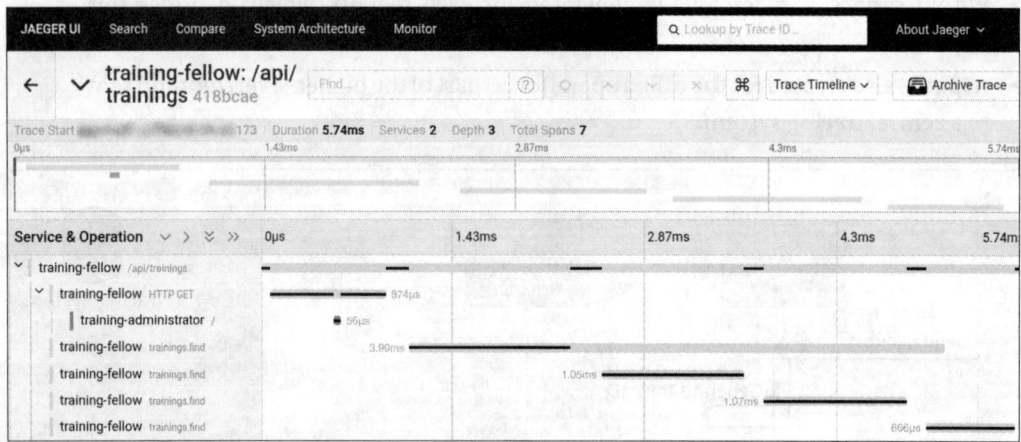

Figure 7.30 Screenshot of a Display in Jaeger UI

7.12.3 Sample Solution

Distributed tracing is easiest to implement using existing software solutions. In this context, *OpenTelemetry (OTel)* plays a major role as an open-source standard and was created in 2019 from the two widely used and competing products: *OpenTracing* and *OpenCensus*.

Numerous software providers now support this standard, making tracing solutions interoperable and enabling data from different systems to be recorded and evaluated in a standardized way. In addition, the major hyperscalers such as Amazon Web Services (AWS), Google Cloud, and Microsoft Azure provide their own tools and optimized distributions to help developers use their platforms. These tools increase the attractiveness of these technologies for developers.

The *OpenTelemetry* specification defines standards and guidelines for the implementation of distributed tracing—including the *OpenTelemetry Protocol (OLTP)*, which describes how tracing information should be transferred between different systems. In

7.12 Distributed Tracing Pattern

addition, an Application Programming Interface (API) and software development kits (SDK) are provided as well.

One possible software package that you can use for storing and analyzing tracing data is the *Jaeger* open-source product. This *distributed tracing observability platform* can process OpenTel data and offers multiple client libraries and instrumentation options.

The setup using Jaeger, shown in Figure 7.31, follows these steps:

- As soon as a part is reached within the application code that is to be traced, an integrated *OpenTelemetry* library generates a *span* and, if necessary, a new *trace*. This step is performed either through explicit method calls in custom code or via woven-in aspects (see Chapter 2, Section 2.6.1, on *aspect-oriented programming*).
- The information collected is sent to a *collector* via the *OpenTelemetry SDK*, which is also included in the application. In our example, a *Jaeger collector* is executed within a Jaeger installation.
- The *Jaeger collector* saves the data in a database.
- The data can be evaluated via a user interface; in our example, we're using the *Jaeger UI*.

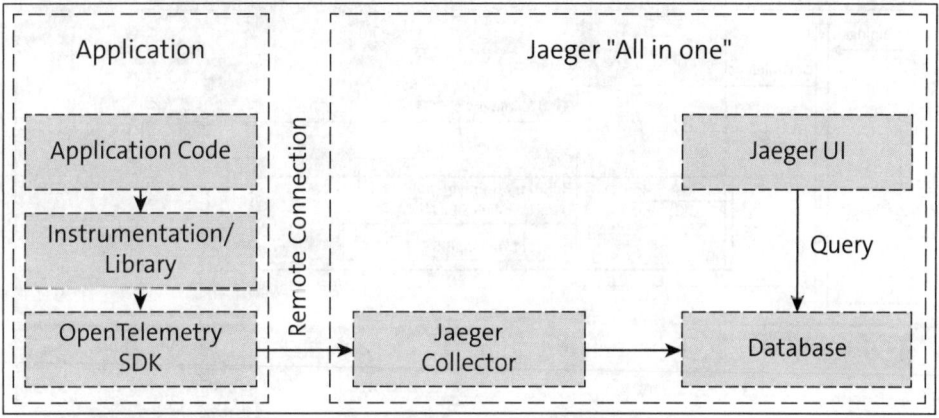

Figure 7.31 Setup for Tracing Using Jaeger

Local Jaeger Installations

A simple installation process via a Docker environment is ideal for local tests. The Jaeger project has published an all-in-one Docker image that is easy to use.

Start the image locally via the following command:

```
docker run -p 16686:16686 -p 4318:4318 jaegertracing/all-in-one:1.56
```

Traces can then be saved and retrieved or analyzed via a simple web UI at *http://localhost:16686/*.

7 Patterns and Concepts for Distributed Applications

Using an example microservice-based training management software, this section demonstrates the use of *OpenTelemetry* and distributed tracing in a Go application.

Let's say training courses are published on the training provider's website together with the dates on which they will be held. The data required for this website is loaded by the *Training Fellow* service via an HTTP GET request. The service has a database in which the respective training dates are stored.

However, as the training company has decoupled the technical training creation from the scheduling, the descriptions and materials associated with the training courses are managed in a separate service. For this reason, the Training Fellow service must access the *Training Administrator* service and retrieve the current descriptions there in order to achieve a complete representation of the training courses. Figure 7.32 shows a corresponding business-related sequence diagram. To ensure that training courses that are no longer managed will no longer be displayed, the list of all current training courses is retrieved from the *Training Administrator* at the start of the process.

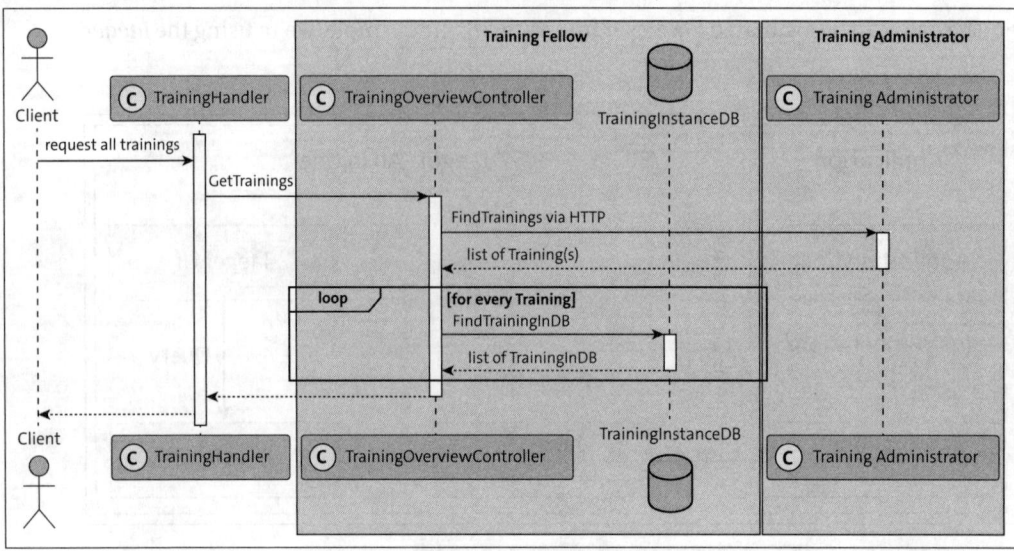

Figure 7.32 Sequence Diagram of the Training Application

To process the HTTP requests, the application uses the *Gin* framework and implements the TrainingHandler, which in turn calls the TrainingOverviewController. The corresponding code is shown in Listing 7.20.

```
func (t *TrainingHandler) handleGetTrainings(c *gin.Context) {
    ctx := c.Request.Context()
    trainings, err := t.controller.GetTrainings(ctx)
    if err != nil {...}
```

```
    internalTrainings := ...//Mapping logic
    c.JSON(http.StatusOK, internalTrainings)
}
```

Listing 7.20 TrainingHandler Implementation (Go)

To ensure that the business logic within the handler implementation remains decoupled from the tracing logic, the application relies on an aspect-oriented approach and uses the *OTel* middleware provided by the OpenTelemetry project, which is integrated into the processing chain. Figure 7.33 shows the corresponding sequence diagram for this excerpt.

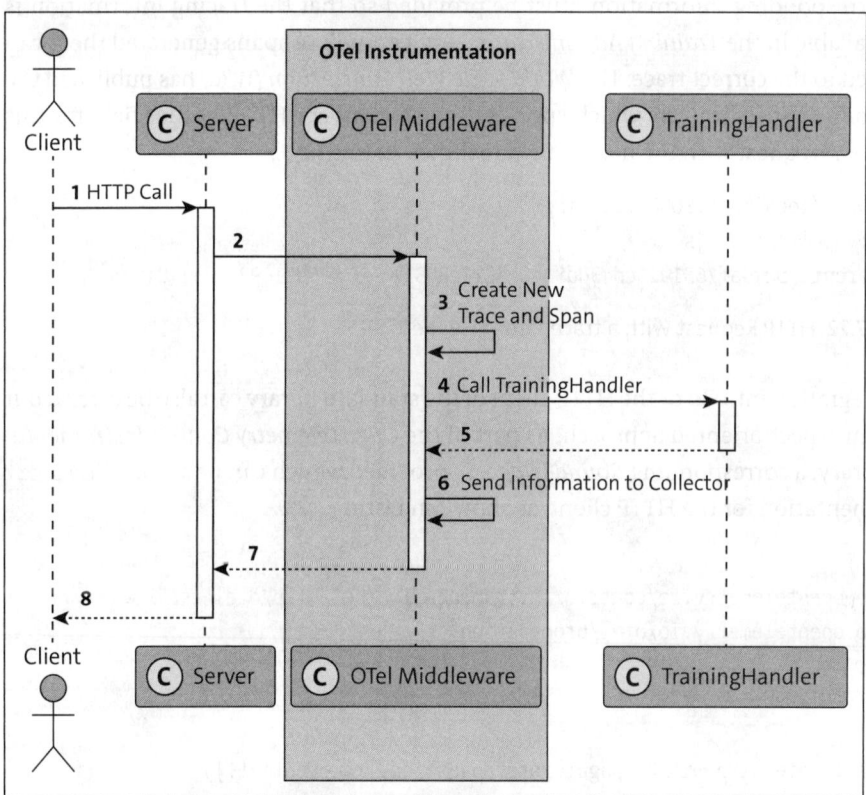

Figure 7.33 OTel Middleware in the Gin Framework

This integration takes place for all HTTP handlers contained in the application when the server is started. As a result, every incoming HTTP request is directly assigned a span. The middleware implementation also handles the redirect to the *collector*.

```
import (
    "go.opentelemetry.io/contrib/instrumentation/github.com/gin-gonic/gin/otel-gin"
    ...
```

```
)
...
trainingHandler := TrainingHandler{controller: controller}
r := gin.Default()
r.Use(otelgin.Middleware("api"))
r.GET("/api/trainings", trainingHandler.handleGetTrainings)
```

Listing 7.21 Registration of the Gin Handler and the OTel Middleware (Go)

As is typical for Go, the trace context is passed on or saved within the `context.Context` object that is passed on with every method call.

The corresponding information must be provided so that the tracing information is also available in the *Training Administrator* service, and the spans generated there can be linked to the correct trace. The *World Wide Web Consortium (W3C)* has published the *trace context* specification, which specifies the `traceparent` HTTP header field for this purpose. A request with the header field is shown in Listing 7.22.

```
GET http://localhost:8085/trainings
Accept: application/json
traceparent: 00-0af7651916cd43dd8448eb211c80319c-b7ad6b7169203331-01
```

Listing 7.22 HTTP Request with a traceparent Header

The integration into Go or the HTTP client of the standard library can also be carried out using an aspect-oriented approach. As part of the *OpenTelemetry Contrib Instrumentation* library, a corresponding *RoundTripper* is provided, which can be set as a `Transport` implementation for the HTTP client, as shown in Listing 7.23.

```
import (
    "go.opentelemetry.io/contrib/instrumentation/net/http/otelhttp"
    "go.opentelemetry.io/otel/propagation"
    ...
)
...
options := otelhttp.WithPropagators(propagation.TraceContext{})
httpClient := &http.Client{
    Transport: otelhttp.NewTransport(http.DefaultTransport, options),
}
```

Listing 7.23 RoundTripper for OTel Transmission of the Trace Context (Go)

Within the *Training Administrator* service, the header field can then be evaluated and used with the OpenTelemetry implementation. This step allows the spans to be merged into a trace in the Jaeger database.

The application uses a *MongoDB* database and the official *MongoDB* database driver for Go to store the data. These database accesses can also be created as separate spans.

Also implemented as part of the *OTel Contrib* package is the *MongoDB monitor,* which can take over the creation of the spans. This tool can be set up during the initialization of the MongoDB client. At this point, no more work is necessary. Listing 7.24 shows the implementation.

```go
import (
    "go.opentelemetry.io/contrib/instrumentation/go.mongodb.org/mongo-driver/mongo/otelmongo"
)
...
opts := options.Client().ApplyURI("mongodb://...")
opts.Monitor = otelmongo.NewMonitor()

client, err := mongo.Connect(ctx, opts)
if err != nil {
    return nil, err
}
```

Listing 7.24 Integration of the OTel Monitor for MongoDB (Go)

The OTel environment still needs to be configured within the Go application to ensure that all parts work together correctly, as shown in Listing 7.25.

```go
import (
   "go.opentelemetry.io/otel"
   "go.opentelemetry.io/otel/exporters/otlp/otlptrace"
   "go.opentelemetry.io/otel/exporters/otlp/otlptrace/otlptracehttp"
   "go.opentelemetry.io/otel/sdk/resource"
   "go.opentelemetry.io/otel/sdk/trace"
   semconv "go.opentelemetry.io/otel/semconv/v1.27.0"
    ...
)
...
headers := map[string]string{
    "content-type": "application/json",
}

exporter, err := otlptrace.New(
    ctx,
    otlptracehttp.NewClient(
        otlptracehttp.WithEndpoint("jaegerserver:4318"),
        otlptracehttp.WithHeaders(headers),
        otlptracehttp.WithInsecure(),
```

```go
    ),
)
if err != nil {...}

tracerprovider := trace.NewTracerProvider(
    trace.WithBatcher(
        exporter,
        trace.WithMaxExportBatchSize(trace.DefaultMaxExportBatchSize),
        trace.WithBatchTimeout(trace.DefaultScheduleDelay*time.Millisecond),
        trace.WithMaxExportBatchSize(trace.DefaultMaxExportBatchSize),
    ),
    trace.WithResource(
        resource.NewWithAttributes(
            semconv.SchemaURL,
            semconv.ServiceNameKey.String("training-fellow"),
        ),
    ),
)
otel.SetTracerProvider(tracerprovider)
```

Listing 7.25 Configuration of the OTel Library (Go)

Finally, the entire call of individual clients can be traced. The result is shown in Figure 7.30.

7.12.4 When To Use the Pattern

Some examples of when to use the pattern include the following:

- Distributed systems must be monitored in a uniform way.
- Better identification of performance bottlenecks is required.
- The possibility of tracking calls that extend across several systems should be implemented.
- Debugging and troubleshooting in systems needs to be facilitated.
- Multi-cloud environments are used, and monitoring needs to be simplified.
- A standardized solution is required for tracing through the use of OpenTelemetry.

7.12.5 Consequences

The use of the pattern has the following consequences:

- Processes in distributed systems become transparent and monitorable.
- Troubleshooting in distributed systems is simplified and may be possible in the first place.

- Additional infrastructure (such as the central tracing solution) is required, and the application's resource requirements increase due to the creation and transmission of tracing data.
- Tracing is standardized and interoperable.
- The complexity of the application is increasing, and developers must use tracing consistently and correctly in order to create any added value at all.

The Author

 **Kristian Köhler** is a software architect and developer with a passion for solving problems using efficient, well-structured software. He is the managing director of Source Fellows GmbH.

Index

A

Abstract class	68, 70
Abstraction	64, 131
Abstraction boundary	107
Abstraction layer	126
ACID paradigm	446, 449
Acronyms	159
Actor	109, 170
Adapter pattern	259
Adapters (hexagonal architecture)	317
Advice	135
Aggregation	174
Agile methods	16
Amazon Simple Notification Service (SNS)/Simple Queue Service (SQS)	303, 307
Ambiguities	160
Anti-patterns	283
Apache Camel	302
Apache HttpComponents	214
Apache Kafka	303
Apache Software Foundation	47
Application business rules	322
Application layers	115, 128
Application programming interface (API)	47
Application server	50
Application state	429
arc42	162
template	161
Architectural concepts	35
Architectural patterns	36, 165, 168
Architectural styles	35, 290, 292
monolithic	307
Architecture	31
Architecture decisions	168
Architecture documentation	157, 162
Architecture styles	168
AsciiDoc	189, 191
AspectJ	135
Aspect-oriented programming (AOP)	85, 135
framework	135
AssertJ	99
AsyncAPI	411
Asynchronous communication style	388
Atlassian Confluence	189
Autorecovery	360
Autoscaling	360
Availability	423
AVRO object container	393, 417

B

Backoff	359
Backus-Naur form	15
Base class	72
Bean	259
Behavioral patterns	40
chain of responsibility	223
command pattern	232
iterator pattern	268
observer pattern	243
strategy pattern	216
Behavioral state	176
Big ball of mud	283
Big blue book	311
Big picture	184
Big red book	311
Black box	165
Blender plugins	305
Böhm-Jacopini theorem	18
Booch method	42
Bounded context	308, 311, 313–314
Box-and-line diagram	160, 180
Boy scout rule	107
Breaking change	117, 129, 413
Broadcasting	399
Broadcast mechanism	391
Broken windows theory	107
Broker topology	300
Builder pattern	206
Building block view	165
Build pipeline	87, 138
Build process	189
Bulk accessor methods	341
Business logic layer	297
Business transaction	357

C

C#	64
C++	64
C4 model	160, 180
code	188
component	187
container	186
system context	184
Caching	431
CAP theorem	423, 425

Index

Cargo cult programming 287
Chain of responsibility 223
Channels 300, 388
Class 109
   abstract 68, 70
   base class 72
   derived 72
   diagram 169, 171
   superclass 72
Class-responsibility-collaboration (CRC) 37–38
   cards 37–38
Clean architecture 320
Clean code 32, 43, 75, 144, 149
Client-server architecture 36, 296
Client-server model 293
Closed layer 298
Closure 118
Cloud 55, 60
Cloud application 177
CloudEvents 392–393, 401, 457
Cloud Native Computing Foundation (CNCF) 392–393
Code comments 143
Code review 108
Code smell 90, 96, 100, 123
Cohesion 109
Command query separation princple 96
Commands 232, 396, 463, 466
Comment templates 156
Common Object Request Broker Architecture (CORBA) 45
Compensation logic 447, 449
Competing consumer 390
Compile-time weaving 135
Component 177–178, 180, 187
Component class 331
Component diagram 169, 177
Component Object Model (COM) 45
Component test 138
Composition 174
Consistency 422–423
Constructor 65, 117
Consumer 388
Container 46, 132, 180, 187
Container diagram 186
Content-Type (HTTP header) 418
Context 170
Context boundary 164
Context-free grammar (CFG) 15
Continuous documentation 189
Continuous improvement 108
Contract 120
Correlation identifier 401
Coupling 110, 123
CQRS 464
Cracks 358
Crackstoppers 358
Create, read, update, and delete (CRUD) 386, 462
Creational patterns 40
   factory method pattern 198
   singleton 251
Cross-cutting concerns 134–135, 167, 188

D

Database locks 434
Database server per instance 435
Data context and interaction 321
Data storage layer 297
Dave LeBlanc law 63
Decisions
   documenting 158, 161
   unconscious 161
Decoupling the business logic 320
Definition of done 162
Dependency
   layers 116
   management 132
   injection 132, 139
   inversion 108, 128
   rule 322
Deployment 46
Derived class 72
Design patterns 16, 29, 84, 165, 168, 197
Design principles 34, 43, 108
Diagrams-as-code 194
Disconnectors 358
Distributed tracing 402, 467
Distribution view 167
Django ORM 462
Doc-as-code 188, 194
Doctrine 462
Documentation 145
   of a software 157
   of software architectures 157, 159
   target group-oriented 159
   template 156
Document template 162
Domain 312
Domain-driven design (DDD) 78, 311
Don't repeat yourself (DRY) 104
Dynamic binding 74

E

EasyMock	140
Eclipse plugins	305
Efficient communication	157
Embedded Ruby	328
Encapsulation	64, 66, 113, 131
Endpoint	389
Enterprise business rules	322
Enterprise integration patterns	54
Enterprise Java Beans (EJB)	53
Enterprise service bus (ESB)	309
Entity Beans	53
Entity-control-boundary	321
Entity Framework	462
Error avoidance -> see Resilience patterns	356
Event broker	300
Event bubbling	229
Event channel	245
Event-driven architecture	36, 299
Events	299
Event sourcing pattern	456
Eventual consistency	425
Example tests	146
Exponential backoff	368
Expressive code	149
Extension points	304
Extract method	91

F

Factory methods	117
Fat client	293
Filters	85
Flow of control	132
Fluent API	213
Format indicator	392
Format indicator and datatype channel	392
Framework	72
Framework conditions	164
Frameworks and drivers	322
Functional programming	22
Function type	118

G

Gang of Four (GoF)	29, 39, 197
Generics information	172
Glossary	159, 169
God object	284
Google Cloud Pub/Sub	303
Google Java Style Guide	87

GORM	441, 462
Go RoundTripper	474
goto statement	18

H

Hamcrest	99
Health checks	359
Heartbeats	359
Hexagonal architecture	315
Hibernate	440, 462
Hillside Group	29
History comments	156
Hollywood principle	132
HtmlSanityCheck	163
HTTP polling	403–404

I

IBM App Connect Enterprise	302
IBM MQ dead-letter queues	395
Imperative programming	18
Implementation rules	168
Inconsistency window	426
Information hiding	131, 335, 339
Inheritance	64, 72
Inline method	93
Integration	54
Integration point	356
Integration test	138
Interface	68, 178
Interface adapter	322
Interface segregation principle (ISP)	106, 108, 134, 335, 339
Inversion of control (IoC)	132, 249, 318
ISO/IEC-2382-15 (standard)	21
Isolation level	444
Ivory tower architects	290

J

Jaeger UI	470
Jakarta EE	50, 135, 296, 299
Jakarta EE Technology Compatibility Kit (TCK)	46
Java	16, 50, 64, 296
Java 2 Platform, Enterprise Edition (J2EE)	50
javadoc	145
Java Messaging System (JMS)	402
Java Persistence API (JPA)	440, 462
entity classes	322
Java records	337

JavaScript Object Notation (JSON)
  schemas 411
JetStream (NATS.io) 457
Join point model 135
JProfiler 468
JSON web token (JWT) 431
JUnit 98–99, 137, 140

K

Knowledge transfer 157
Kubernetes 381

L

Lambda calculus 22
Lambda functions 24
Lapsed listener problem 249
Latency 424
LaTeX 190
Layer
  closed 298
  open 298
Layered architecture 36, 294
Layers of isolation 294
Library 72
License information 150
Linear backoff 368
Linter 87
Liskov substitution principle (LSP) 108, 120
Load balancing 361
Load shedding 360
Load-time weaving 135
Location (HTTP header) 405
Log files 468
Lombok 336
Long parameter list 99
Lost update problem 437

M

Maintainability of software 158
Markdown 189
Materialized view 466
Mediator topology 300
Merge 111
Mermaid 194
Message 389
Message broker 300, 302
Message channels 388–389
Message construction patterns 395
Message header 401

Message-oriented middleware (MOM) 351
Message processor 300
Message router 389
Message transformers 389
Messaging 388
  channels 300
  concepts 388
  query 400
  remote procedure call (RPC) 400
  system 351, 388
Method signature 74
Microkernel architecture 304
Microservices 307, 347
  database per service 434
  distributed tracing 468
  stateless architecture 428
Middleware 85
MIME (Multipurpose Internet Mail Extensions)
  type 418
Mind map 191
MinIO 123
Model view controller pattern (MVC) 324
Model View ViewModel (MVVM) 329
Modularization 165
Module 109, 177
MongoDB 475
Monolithic architectural style 307
Move Method refactoring 103

N

NATS.io 457, 460
NHibernate 462
NoSQL databases 425
Notify/acknowledgement 400
N-tier architecture 294

O

Object Management Group (OMG) 39
Object message sequence charts 37
Object Modeling Technique (OMT) 37, 42
Object-oriented programming 16, 20, 64
Objectory 42
Object-relational mapping (ORM) . 441, 462, 465
Object store 123
One-way communication 399
Onion architecture 321
OpenAPI 411
OpenCensus 470
Open-closed principle 108, 113, 204, 297
Open layer 298

Open source messaging system	457
Open source tracing solution	470
Open Systems Interconnection Model (OSI)	295
OpenTelemetry (OTel)	472, 474
middleware	473
OTel Contrib package	475
OpenTelemetry Protocol (OTLP)	470
OpenTracing	470
Outbox	451
Outbox processing	451, 455
Output formats	190
EPUB	190
HTML	190
PDF	190
Word	190
Over-engineering	53
Overloading methods	74
Overwriting methods	74

P

PACELC theorem	424
Package	178
Paging mechanism	385
Parameter object	100
Partitioning	361
Partition tolerance	423
Patterns	
adapter	259
anti-patterns	283
asynchronous request-reply	402
Big Ball of Mud	283
builder	206
bulkhead	378
cargo cult programming	287
chain of responsibility	223
circuit breaker	359, 362, 372, 383
command	232
command and query responsibility segregation (CQRS)	461
command message	395–396, 400
composite	41, 276
database per service	434, 446
datatype channel	415
document message	395, 397–398, 400
endpoint for version	415
error propagation	357
event message	396, 398, 400
event sourcing	455
facade	112
factory method	198
failover	360
fast lane reader	299
fixed backoff	368
for error avoidance	358
for interface versioning	411
glue	41
God object	284
graceful degradation	360
hub-and-spoke	301
iterator	268
long polling	407
message with referencing metadata	418
message with self-describing metadata	420
observer	243, 391, 398
optimistic locking	427, 437–438, 444
options	117
plugin architecture	304
presentation model	329
referencing message	415
Reinventing the Wheel	286
remote facade	339–340
request reply	399
retry	359
saga	446–447, 450
self-contained message	414, 417
singleton	251
spaghetti code	285
stateless architecture	428–429
steady state	383
strategy	216
structural patterns	40
timeout	361, 383
transactional outbox	450–451
wrapper	41, 259
People	
Adrian Cockroft	308
Alan Kay	20
Alexander Wolf	35
Alistair Cockburn	315, 321
Alonzo Church	22
Barbara Liskov	119
Bertrand Meyer	113
Bjarne Stroustrup	16
Bobby Woolf	54
Christopher Alexander	27
Conrad F. D'Cruz	54
Corrado Böhm	18
Dan Malks	50
Dave LeBlanc	63
David Garlan	157
Deepak Alur	50
Dewayne Perry	35
Doug Lea	239

Edsger Wybe Dijkstra 18
Eric Brewer 423
Eric Evans 311
Erich Gamma 29, 39
Felix Bachmann 157
Florian Deissenboeck 77
Frank Buschmann 36–37
Fred Brooks 76
George L. Kelling 107
Gernot Starke 162
Giuseppe Jacopini 18
Grady Booch 29, 169
Gregor Hohpe 54
Gregor Kiczales 135
Hal Hildebrand 30
Ivar Jacobsen 169
Ivar Jacobson 169, 321
James Coplien 43, 76, 321
James Ivers 157
James Lewis 307
James Q. Wilson 107
James Rumbaugh 169
Jeffrey Palermo 321
Jim Coplien 30
John Backus 16
John Crupi 50
John Ousterhout 77
John Vlissides 29
Jonathan Simon 54
Joshua Bloch 114
Judith Stafford 157
Jürgen Holler 53
Ken Auer 30
Kent Beck 28, 38, 75, 89, 134
Konrad Ernst Otto Zuse 15
Kristen Nygaard 20
Kyle Brown 54
Len Bass 157
Mark Richards 297
Markus Pizka 77
Martin Fowler 54, 289–290, 307
Mary Shaw 35
Michael Feathers 108
Michael J. Rettig 54
Michael T. Nygard 356
Mike Cohn 138
Ole-Johan Dahl 20
Paul Clements 157
Paulo Merson 157
Peter Hruschka 162
Peter Naur 16
Phil Karlton 77
Ralph Johnson 29
Reed Little 157
Richard Helm 29
Robert Baden Powell 107
Robert C. Martin 32, 43, 63, 75, 94, 104, 108, 116, 123, 128, 144, 320
Robert Nord 157
Rob Pike 105, 126
Rod Johnson 53
Sean Neville 54
Simon Brown 290
Thom Holwerda 76
Tom Benner 77
Tom Preston-Werner 413
Trygve Reenskaug 321
Vaughn Veron 311
Ward Cunningham 28, 38
William Croft 28
PEP 8 87
Pessimistic locking 444
Pillars (AWS) 61
Pipes and filters 389
Plain text 189
PlantUML 170, 191
   diagram types 192
   sequence diagram 192
Plugins 304
Pointcut 135
Point-to-point channel 390, 397, 399
Polly 382
Polymorphism 73, 114, 124
Port 178
Ports and adapters 315, 317
Presentation layer 297
Private tables per service 435
Producer 388
Profiling 468
Project Lombok 336
Protobuf 411
Protocol state 176
Publish-subscribe channel 391, 397–398
Pull model 399
Push model 399
Python 64

Q

Quality assurance 108
Quality features 158, 165
Quality requirement 168
Quality tree 168
Queries 463

R

RabbitMQ	303, 402
Dead Letter Exchange	395
queues	391
Reactive application	300
Reactive programming	25
Read committed	445
Read uncommitted	445
Receiver	388
Recovery	360
Redux, middleware	302
Redux Thunk	302
Refactoring	90, 99
extract method	91
Introduce Parameter Object	100
long parameter list	99
Move Method	103
Preserve Whole Object	101
Replace Parameter with Method	100
techniques	149
Reflection	67
Regression test	138
Reinventing the wheel	286
Remote procedure call (RPC)	325, 390, 413
Repeatable read	445
Replier	399
Requestor	399
Requests for comments (RFC)	160
Requirement levels	160
may	160
must	160
must not	160
should	160
should not	160
Resilience4J	382
Resilience patterns	356
Resource state	429
Retry-After (HTTP header)	405
Retry mechanisms	362, 449
Review process	189
Rich clients	296
Routing	389
Runtime environment	46, 132
Runtime view	167

S

Schema per service	435
Screaming architecture	321
Scrum	162
Self-explanatory code	144
Semantic versioning	413
Sender	388
Separation of concerns (SoC)	134, 295, 341
Separation of layers	316
Sequence diagram	170, 174
Serializable	445
Service	293
Service component	188
Service layer	298
Service Provider Interface (SPI)	47
Servlet	47
Side effects	96, 153
Single page application (SPA)	330
Single point of failure	302, 434
Single responsibility principle	97, 108–109, 134, 138, 158, 214, 252, 286
Singleton beans	259
SLF4J	323
Smart endpoint and dumb pipes	308
Snapshot	360
SOAP-RPC	396
Software architect	33, 157, 290
Software architecture	31, 34–35, 114, 162, 289
documentation	157
Software components	158
Software crisis	16, 64
Software design	63, 75, 131
Software documentation	143
Software engineering	16
SOLID	108, 134
Spaghetti code	285
Span	468
Span ID	469
Spring Framework	53, 67, 135, 385, 452
Spring Data	385
Spring Data JPA	385, 452
Spring Integration Error Channel	395
Spring framework	
Spring Integration	302
Standard header	150
State diagram	170, 175
Stepdown rule	94
Stereotype	171, 178
Sticky session	428
Stored procedures	296
Strategic design	315
Structural patterns	
adapter pattern	259
composite pattern	276
wrapper pattern	259
Structured programming	18
Structuring template	161

Index

Structurizr 183, 194
   diagram 195
Style guide 87
Subclass 72
Substitution principle 119
Sun Microsystems 16
Superclass 72
System context 170, 184
System landscape 184

T

Tactical design 315
Target group-oriented documentation 159
Technical jargon 159
Technology decisions 165, 186
Telescope constructor 214
Template 161
Testing
   mock framework 139
   mock object 139, 319
   testing pyramid 138
Thick client 293
Thin client 293
Three-tier architecture 51, 296, 316
Throttling strategy 366
Time-to-live (TTL) information 363
TODO comments 153
Toxiproxy 365
Trace context 474
Trace ID 468
Tracing 468
Transformers 389
Trigger message 301
Truth in code 157
Turing Award 119
Two-phase commit (2PC) protocol 446, 450, 455
Two-pizza team size 309
Two-way binding 335
Two-way communication 399
TypeScript 64, 330, 337

U

Ubiquitous language 312–313
Uncle Bob -> see People, Robert C. Martin 43
Unified Modeling Language (UML) ... 39, 68, 115, 164, 169
   class diagram 169, 171
   component diagram 169, 177
   diagram 170
   sequence diagram 170, 174
   state diagram 170, 175
   use case diagram 169–170
Unit of work 357
Unit test 137, 146
Unix philosophy 308
Use case diagram 169–170

V

Value binding 330
Variadic parameter 118
Version control system (VCS) check-in 154
Vertical slices 305
Virtual machine 50
Visualization of software systems 169
Visual Studio Code (VS Code) extensions 305
VisualVM 468

W

Waterfall model 16
Weak consistency 426
Weaving 135
Web Application Archive (WAR) 47
Web container specification 46
WebSockets 406, 410
Well-architected cloud 60
Whiteboard 160
White box 165
Wiki systems 188
Wireframe 191
WordPress plugins 305

X

X-Request-ID 402